AF428256

Hydraulic Components Volume C

Hydraulic Transmission Lines

Dr. Medhat Kamel Bahr Khalil, Ph.D, CFPHS, CFPAI.
Director of Professional Education and Research Development,
Applied Technology Center, Milwaukee School of Engineering,
Milwaukee, WI, USA.

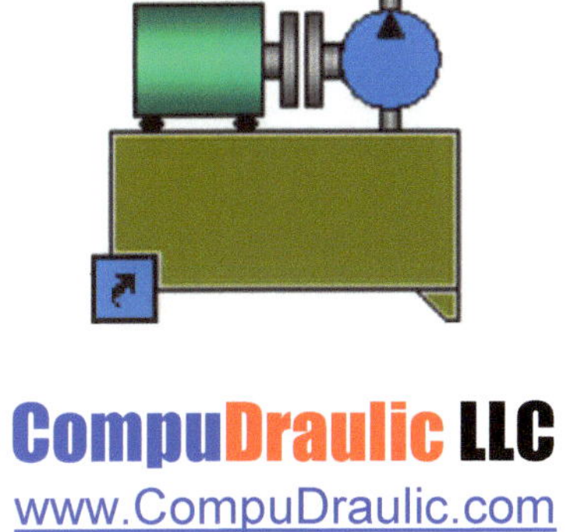

CompuDraulic LLC
www.CompuDraulic.com

Hydraulic Components Volume C

Hydraulic Transmission Lines

ISBN: 979-8-9870268-0-9

Printed in the United States of America
First Published by March 2023
Revised by Sept 2023

Disclaimer

It is always advisable to review the relevant standards and the recommendations from the system manufacturer. However, the content of this book provides guidelines based on the author's experience.

Any portion of information presented in this book might not be suitable for some applications due to various reasons. Since errors can occur in circuits, tables, and text, the author/publisher assumes no liability for the safe and/or satisfactory operation of any system designed based on the information in this book.

The author/publisher does not endorse or recommend any brand name products by including such brand name products in this book. Conversely the author/publisher does not disapprove any brand name product not included in this book. The publisher obtained data from catalogs, literatures, and material from hydraulic components and systems manufacturers based on their permissions. The author/publisher welcomes additional data from other sources for future editions. This disclaimer is applicable for the workbook (if available) for this textbook.

Hydraulic Components Volume C
Hydraulic Transmission Lines

PREFACE, 3

ACKNOWLEDGEMENT, 4

ABOUT THE BOOK, 5

ABOUT THE AUTHOR, 7

Chapter 1-Basic Types of Hydraulic Transmission Lines, 8
1.1. Basic Types and Contribution of Hydraulic Transmission Lines
1.2. Sizing of Hydraulic Transmission Lines
1.3- Rated Pressures for Hydraulic Lines
1.4- Hydraulic Pipes
1.5- Hydraulic Tubes
1.6- Hydraulic Hoses
1.7- Flanges for Transmission Line Connections
1.8- Rubber Expansion Fittings
1.9- Test Points
1.10-Pressure Measurement Hoses
1.11 Manifolds

Chapter 2-Contamination Control in Hydraulic Transmission Lines, 113
2.1- Contamination in Hydraulic Transmission Lines
2.2- Projectile Cleaning
2.3- Pickling of Hydraulic Transmission Lines
2.4- Flushing of Hydraulic Transmission Lines

Chapter 3: Safety and Maintenance of Transmission Lines, 132
3.1-BP-Transmission Lines-01-Selection and Replacement
3.2-BP-Transmission Lines-02-Maintenance Scheduling
3.3-BP-Transmission Lines-03-Installation and Maintenance
3.4-BP-Transmission Lines-04-Standard Tests and Calibration
3.5-BP-Transmission Lines-05-Transportation and Storage
3.6- Oil Injection Avoidance and Treatment

Chapter 4: Troubleshooting and Failure Analysis, 160
4.1- Hydraulic Transmission Lines Inspection
4.2- Hydraulic Transmission Lines Troubleshooting
4.3- Hydraulic Transmission Lines Failure Analysis

Chapter 5: Energy Losses in Transmission Lines, 171
5.1- Flow Patterns in Transmission Lines
5.2- Conservation of Mass in Fluid Flow
5.3- Conservation of Energy in Fluid Flow
5.4- Pressure Drop in Transmission Lines

Chapter 6: Modeling of Hydraulic Transmission Lines, 196
6.1- Modeling of Seamless Hydraulic Transmission Lines
6.2- Modeling of Hydraulic Fittings
6.3- Modeling of Hydraulic Orifices
6.4- Modeling Hydraulic Transmission Line Assembly

Chapter 7: Water Hammer and Noise Control, 217
7.1- Water Hammer
7.2- Noise Control in Hydraulic Transmission Lines

APPENDIXES, 225

APPENDIX A: LIST OF FIGURES, 225

APPENDIX B: LIST OF TABLES, 231

APPENDIX C: LIST OF REFERENCES, 232

INDEX, 242

PREFACE

Hydraulic transmission lines have crucial effect on the performance of hydraulic systems. They are basically used for power transmission, but improper design, selection, and routing can drastically affect the system performance. This book overviews the basic types, contamination control, safety, maintenance, troubleshooting, failure analysis, noise control, and modeling of hydraulic transmission lines.

Dr. Medhat Kamel Bahr Khalil

ACKNOWLEDGEMENT

All praise is to Allah who granted me the knowledge, resources, and health to finish this work.
To the soul of my parents who taught me the values of ISLAM
To my family: wife, sons, daughters in law, and grandchildren
To my best teachers and supervisors

The author thanks the following companies (listed alphabetically) for permitting him to use portions of their copyrighted literatures in this book.
- American Technical Publishers
- ArgoHyots
- Assofluid
- Bosch Rexroth
- Brennan Industries
- CEJN
- Fluid Power Training Institute
- Gates
- Hydraulic and Pneumatic Magazine
- International Fluid Power Society
- International Hydraulic Safety Authority
- Lightening Reference Handbook (IFPS)
- Parker Hannifin
- Spectroline
- Ultra Clean Technologies

Lastly, the author extends his thanks to the following sources of public information used to enrich the contents of the book.

www.ifps.com
www.redl.com
www.stauffusa.com
www.sapphirehydraulics.com
www.epha.com
www.sealsaver.com
www.hydrotechnik.com
https://flanges-pipe.com
www.grainger.com
www.daman.com
www.bondfluidaire.com
www.mac-hyd.com
www.new-line.com
www.Flaretite.com
www.aircraftsystemstech.com

ABOUT THE BOOK

Book Description:

The book is targeting students and professionals who are looking to advance their fluid power careers. The book is colored and has the size of standard A4. This book is the third in a series that the author plans to publish to offer separate book for every hydraulic component. This book introduces knowledge foundation about hydraulic transmission lines. This book overviews the basic types, contamination control, safety, maintenance, troubleshooting, failure analysis, noise control, and modeling of hydraulic transmission lines.

Book Objectives:

Chapter 1: Basic Types of Hydraulic Transmission Lines

This chapter focuses on browsing the construction and the features of the three main hydraulic transmission lines, Pipes, Tubes, and Hoses. For each transmission line, the following topics are presented: sizing, material, construction and pressure rating. This chapter also presents information about fittings and manifolds.

Chapter 2: Contamination Control in Hydraulic Transmission Lines

This chapter discusses best practices for controlling contamination in hydraulic transmission lines including projectile cleaning and hydraulic system flushing.

Chapter 3: Safety and Maintenance of Transmission Lines

This chapter provides guidelines for transmission lines selection, replacement, maintenance scheduling, installation, testing, storage and transportation. This chapter is supported by examples and figures granted by leading fluid power manufacturers.

Chapter 4: Troubleshooting and Failure Analysis

This chapter discusses hydraulic transmission lines inspection, troubleshooting, and failure analysis. In this chapter, a troubleshooting chart for transmission line faults is presented. The chapter also presents examples of defective transmission lines.

Chapter 5: Energy Losses in Transmission Lines

This chapter discusses the two common flow patterns in hydraulic transmission lines, laminar flow, and turbulent flow. The chapter presents reviews of fundamental concepts conservation of mass and conversation of energy as applied to fluid flow in transmission lines. of power losses in hydraulic transmission line. The chapter presents both mathematical and graphical methods of quantifying the power losses in transmission lines, fittings, and orifices.

Chapter 6- Modeling of Hydraulic Transmission Lines

This chapter presents concepts of modeling transmission lines, fittings and orifices. Model for a transmission line considers compressible fluid so that effect of line capacitance can be investigated. Developed models were validated based on other software.

Chapter 7- Water Hammer and Noise Control

This chapter presents a summarized idea about water hammer, its effect, and some techniques to minimize its damaging effects as applicable to hydraulic control systems. This chapter also presents the construction and operating principle of using shock suppressors to reduce the noise that are transmitted by hydraulic transmission lines.

Book Statistics:

The table shown below contains interesting statistical date about the textbook:

Chapter #	Pages	Figures	Tables	Equations	Words	Editing Time (Hours)
Chapter 1	105	122	115	5	13152	235
Chapter2	19	20	0	1	3873	198
Chapter 3	28	27	4	23	4859	124
Chapter 4	11	16	2	0	1402	96
Chapter 5	25	23	1	0	5505	174
Chapter 6	21	20	0	15	2673	110
Chapter 7	8	7	0	7	1298	178
Total	**217**	**235**	**122**	**51**	**32762**	**1,115 Hour = 46 Days**

ABOUT THE AUTHOR

Medhat Khalil, Ph.D. is Director of Professional Education & Research Development at the Applied Technology Center, Milwaukee School of Engineering, Milwaukee, WI, USA. Medhat has consistently been working on his academic development through the years, starting from bachelor's and master's Degrees in Mechanical Engineering in Cairo Egypt and proceeding with his Ph.D. in Mechanical Engineering and Post-Doctoral Industrial Research Fellowship at Concordia University in Montreal, Quebec, Canada. He has been certified and is a member of many institutions such as: Certified Fluid Power Hydraulic Specialist (CFPHS) by the International Fluid Power Society (IFPS); Certified Fluid Power Accredited Instructor (CFPAI) by the International Fluid Power Society (IFPS); Member of Center for Compact and Efficient Fluid Power Engineering Research Center (CCEFP); Listed Fluid Power Consultant by the National Fluid Power Association (NFPA); and Listed Professional Instructor by the American Society of Mechanical Engineers (ASME). Medhat has balanced academic and industrial experience. Medhat has vast working experience in Fluid Power teaching courses for industry professionals. Being quite aware of the technological developments in the field of fluid power,

Medhat had worked for several world-wide recognized industrial organizations such as Rexroth in Egypt and CAE in Canada. Medhat had designed several hydraulic systems and developed several analytical and educational software. Medhat also has considerable experience in modeling and simulation of dynamic systems using Matlab-Simulink. Medhat has been selected among the inductees for Pioneers in fluid Power by NFPA (2012) and Hall of Fame in fluid Power by IFPS (2021).

Chapter 1

Basic Types of Hydraulic Transmission Lines

Objectives

This chapter focuses on browsing the construction and the features of the three main hydraulic transmission lines, Pipes, Tubes, and Hoses. For each transmission line, the following topics are presented: sizing, material, construction and pressure rating. This chapter also presents information about fittings and manifolds.

Brief Contents

1.1. Basic Types and Contribution of Hydraulic Transmission Lines
1.2. Sizing of Hydraulic Transmission Lines
1.3- Rated Pressures for Hydraulic Lines
1.4- Hydraulic Pipes
1.5- Hydraulic Tubes
1.6- Hydraulic Hoses
1.7- Flanges for Transmission Line Connections
1.8- Rubber Expansion Fittings
1.9- Test Points
1.10-Pressure Measurement Hoses
1.11 Manifolds

Chapter 1: Basic Types of Hydraulic Transmission Lines

1.1. Basic Types and Contribution of Hydraulic Transmission Lines

Hydraulic transmission lines are used to transmit the energy between system components. For energy saving and external leakage avoidance, new designs of hydraulic systems minimize use of transmission lines and rather rely much on manifolds. However, transmission lines are still needed at least to connect control valves to actuators. As shown in Fig. 1.1, there are three types of transmission lines, pipes, tubes, and hoses.

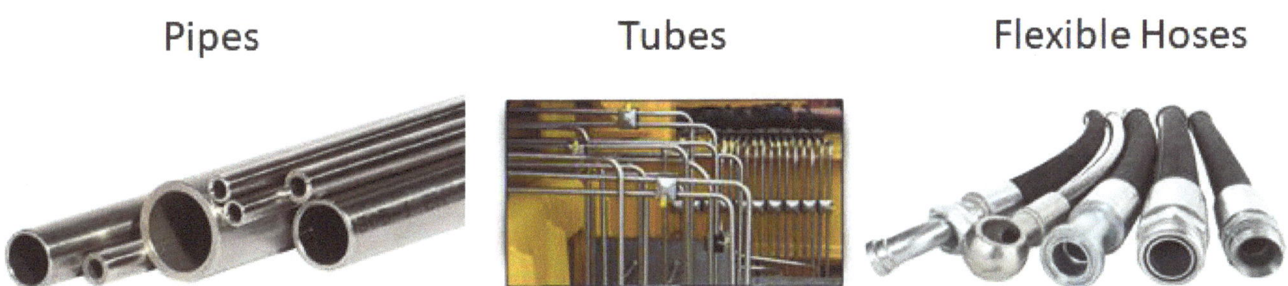

Fig. 1.1 - Types of Hydraulic Transmission Lines

May be the first question is, which type of transmission line is better to use? The following set of bullets answers that question:

- **Interchangeability:** Hydraulic hoses, tubes and pipes are limitedly interchangeable. This means that pipes, tubes, and hoses can't replace each other in every case. Each type has its own advantages that make it more adequate for specific applications.
- **Flexibility:** Pipes are not intended to bend. Tubes can be bend but offers no flexibility. Hoses are intended to flex. However, hoses are expanding and stretching under pressure and have limited working temperature range.
- **Stiffness:** Pipes and tubes are best because they have less wall elasticity. Because of a hose wall elasticity, the overall bulk modulus of the system is reduced; and so, the system stiffness, response, and bandwidth.
- **Service Life:** Pipes have the longest service life and hoses have the shortest service life and require routine replacement because they can fail without warning.
- **Cost:** Pipes are the most cost effective and hoses are the most expensive.
- **Applications:** Hoses are more recommended for mobile applications because of limited spaces requirements and relative motion between machine components.
- **Noise and Vibration:** Hoses are better for absorbing vibration and suppressing noise.
- **Heat Transfer:** Pipes and tubes dissipate more heat than hoses can dissipate.

1.2. Sizing of Hydraulic Transmission Lines

Improper sizing, selection, installation, and maintaining of *Hydraulic Transmission Lines* can lead to one or more of the following problems:
- Pump cavitation.
- Turbulent flow.
- System heating-up.
- External leakage.
- Line accidental breakage, system damage and loss of life risks.
- Increased maintenance costs both in replacement parts and labor.
- Increased machine downtime.
- Increased environmental cleanup costs.

Therefore, a hydraulic transmission line must be precisely sized. Sizing a transmission line means calculation of the *Inner Diameter* (ID). Size of transmission lines can be determined mathematically or by charts, tables, or software.

1.2.1 Sizing Hydraulic Transmission Lines Mathematically

- **Step 1:** <u>Comply with the recommended flow speed:</u>
    - Equation 1.1A (Metric Units) or 1.1B (English Units) is used to solve for a line effective area "**A**". Line inner diameter can then be easily calculated when the effective area is found. While using Equation 1.1, consider the following recommended flow speeds doe each line in the system:
        - Pressure Line (industrial machines) = 7 - 15 fps. = 2.1- 4.6 m/s.
        - Drain Line = 4 - 7 fps. = 1.2 - 2.1 m/s.
        - Suction Line = 2 - 4 fps. = 0.6 - 1.2 m/s.

$$Q\left[\frac{\text{lit}}{\text{min}}\right] = \frac{v\,[\text{cm/s}] \times A[\text{cm}^2] \times 60}{1000} \qquad \textbf{1.1A}$$

$$Q[\text{gpm}] = \frac{v\,[\text{fps}] \times A[\text{in}^2]}{0.321} \qquad \textbf{1.1B}$$

- **Step 2:** <u>Securing laminar flow:</u>
    - Equation 1.2A (Metric Units) or 1.2B (English Units) is used to solve for a line inner diameter "**D**". While using Equation 1.2 is used, consider a value of Reynolds Number R_e < 2000.

$$D[mm] = \frac{21231\, Q\left[\frac{l}{min}\right]}{v\,[Cst] \times R_e} \qquad\qquad 1.\,2A$$

$$D[mm] = \frac{3164\, Q[gpm]}{v\,[Cst] \times R_e} \qquad\qquad 1.\,2B$$

- **Step 3: Final Solution:**
  - The larger size out of applying the two equations 1.1 and 1.2 is considered.

1.2.2 Sizing Hydraulic Transmission Lines by Charts

As shown in Fig. 1.2, using the chart, the following steps are required to calculate the inner diameter of the transmission line:

- **Step 1:** Define the flow rate on the upper scale.
- **Step 2:** Define the recommended flow speed based on the transmission line type.
- **Step 3:** Draw a dashed line between the flow rate and the flow speed.
- **Step 4:** Intersection of the dashed line with the middle scale defines the inside area that can be used to calculate the inner diameter.

Example 1: As shown in the figure, assume flow rate of 30 liter/min is flowing in a pressure line with a maximum allowable flow speed is 5.3 m/s. the chart shows inside area = 0.95 cm² and an inner diameter ID = 11 mm.

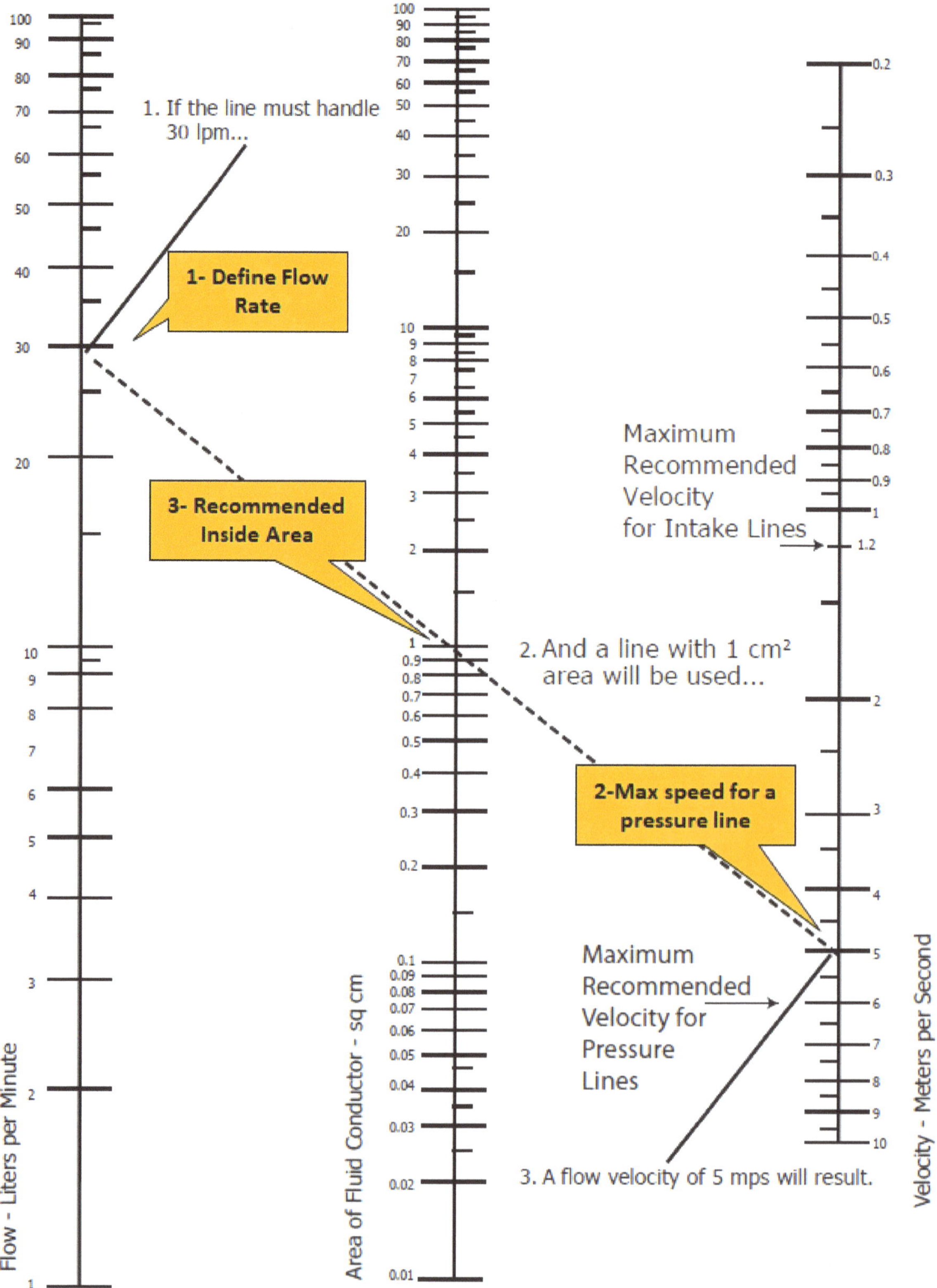

Fig. 1.2 - Example 1 of using Charts for Sizing a Transmission Lines

Example 2: Figure 1.3 shows an example where the flow requirements are increased from 20 to 25 GPM and a decision must be made whether to keep the same line or replace it by another larger size line to cope with an increased flow requirement.

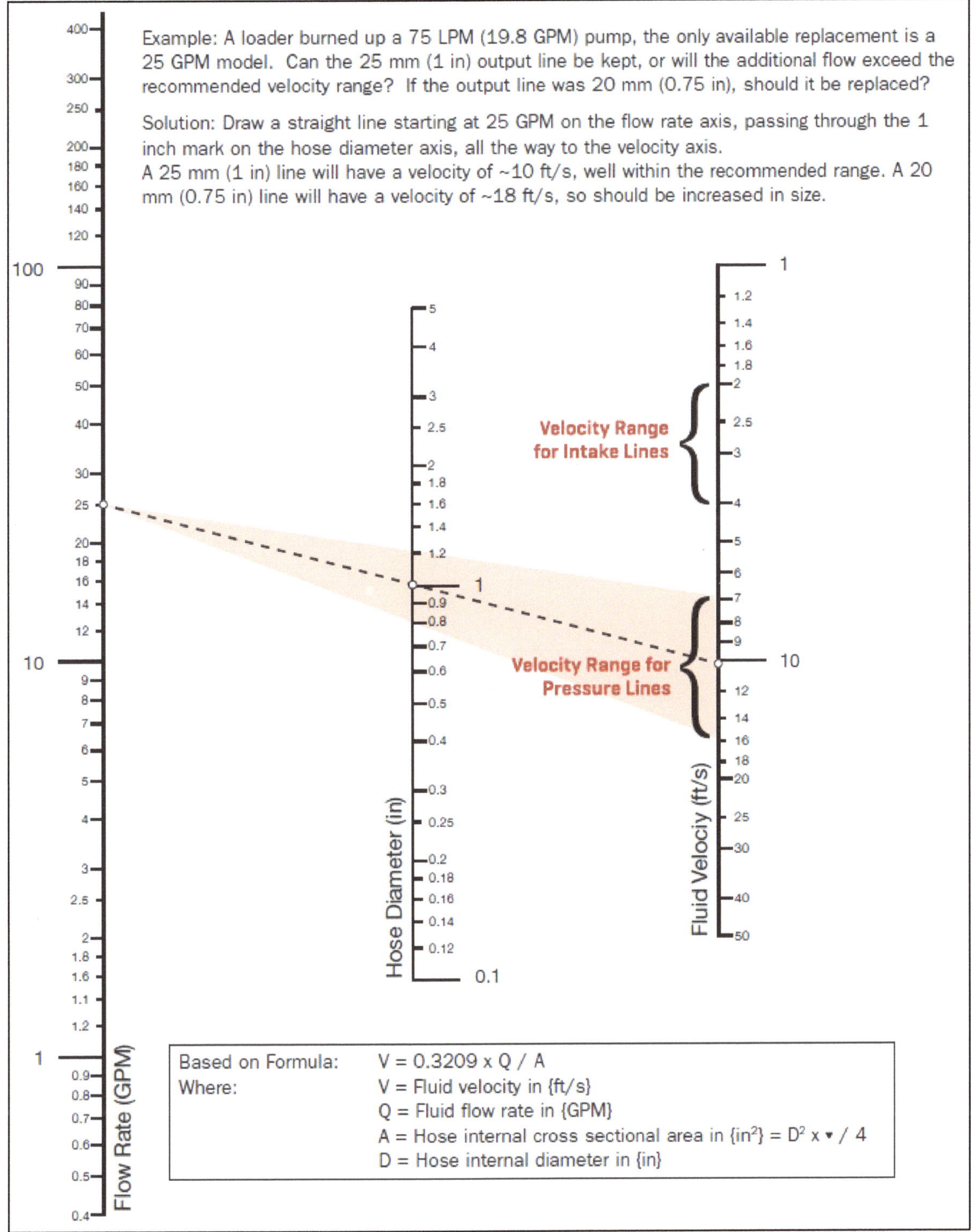

Fig. 1.3 - Example 2 of using Charts for Sizing a Transmission Lines (Courtesy of Gates)

1.2.3 Sizing Hydraulic Transmission Lines using Table

Sizing Hydraulic Lines Using Tables: Using the table shown in Table 1.1, the following steps are required to calculate the inner diameter of the transmission line:
1. Locate the vertical column that shows the recommended flow speed at the top.
2. Go down in this column until the flow rate through the line is met.
3. Go left to find the corresponding inner diameter.

Example 3: The figure shows that, for flow speed 15 fps (4.5 m/s) and flow rate of 8.9 gpm (40 lit/min), inner diameter = 3/8 inches (12.5 mm). Interpolating the results with the previously given example validates the results from the chart and the software for example 1.

STANDARD PIPE – SCHEDULE 40

PIPE SIZE	OD (INCHES)	WALL (INCHES)	ID (INCHES)	INT AREA (SQ. INCHES)	WT/FT (POUNDS)	GPM @ 2 FPS	GPM @ 5 FPS	GPM @ 10 FPS	GPM @ 15 FPS	GPM @ 20 FPS	GPM @ 25 FPS
1/8	0.405	0.068	0.269	0.057	0.244	0.4	0.9	1.8	2.7	3.5	4.4
1/4	0.540	0.088	0.364	0.104	0.424	0.6	1.6	3.2	4.9	6.5	8.1
3/8	0.675	0.091	0.493	0.191	0.567	1.2	3.0	6.0	8.9	11.9	14.9
1/2	0.840	0.109	0.622	0.304	0.850	1.9	4.7	9.5	14.2	18.9	23.7
3/4	1.050	0.113	0.824	0.533	1.130	3.3	8.3	16.6	24.9	33.2	41.6
1	1.315	0.133	1.049	0.864	1.677	5.4	13.5	26.9	40.4	53.9	67.4
1 1/4	1.660	0.140	1.380	1.496	2.270	9.3	23.3	46.6	69.9	93.2	116.6
1 1/2	1.900	0.145	1.610	2.036	2.715	12.7	31.7	63.5	95.2	126.9	158.7
2	2.375	0.154	2.067	3.356	3.649	20.9	52.3	104.6	156.9	209.2	261.5
2 1/2	2.875	0.203	2.469	4.788	5.787	29.8	74.6	149.2	223.9	298.5	373.1
3	3.500	0.216	3.068	7.393	7.568	46.1	115.2	230.4	345.7	460.9	576.1
3 1/2	4.000	0.226	3.548	9.887	9.100	61.6	154.1	308.2	462.3	616.4	770.5
4	4.500	0.237	4.026	12.730	10.779	79.4	198.4	396.8	595.2	793.7	992.1
5	5.563	0.258	5.047	20.006	14.602	124.7	311.8	623.6	935.4	1,247.2	1,559.1
6	6.625	0.280	6.065	28.890	18.954	180.1	450.3	900.6	1,350.9	1,801.1	2,251.4
8	8.625	0.322	7.981	50.027	28.524	311.9	779.7	1,559.4	2,339.2	3,118.9	3,898.6
10	10.750	0.365	10.020	78.854	40.441	491.6	1,229.0	2,458.1	3,687.1	4,916.1	6,145.1
12[1]	12.750	0.406	11.938	111.932	53.469	697.8	1,744.6	3,489.1	5,233.7	6,978.3	8,722.9
12[2]	12.750	0.375	12.000	113.097	49.510	705.1	1,762.7	3,525.5	5,288.2	7,051.0	8,813.7

Table 1.1 - Using Tables for Sizing a Transmission Lines

1.2.4 Interactive Sizing Hydraulic Transmission Lines using Software

Figure 1.4 shows the use of *Hydraulic Components Sizing Calculator* (HCSC) to size a hydraulic line. The figure shows that the software validates the results from the chart in example 1 (upper part) and in example 2 (lower part).

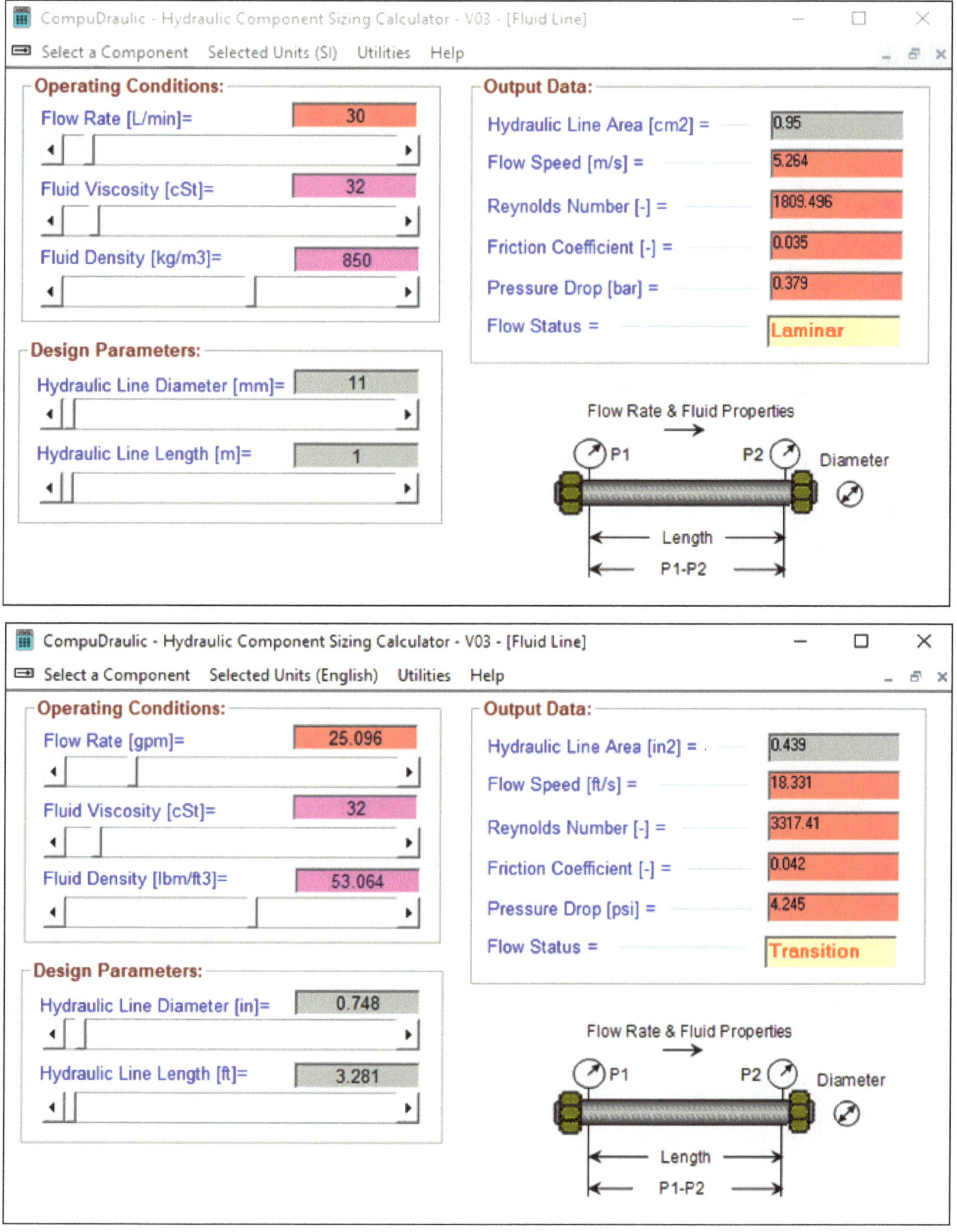

Fig. 1.4 - Using Software for Sizing a Transmission Lines (www.compudraulic.com)

1.3- Rated Pressures for Hydraulic Lines

Pressure carrying capacity is the second most important selection factor of a transmission line.

❖ **Burst Pressure:**
Burst Pressure is the maximum hydrostatic pressure above which a lines assembly fails. Burst pressure of hoses is based on the structure of the hose. Burst pressure for hard lines (i.e. tubes and pipes) depends on the material and wall thickness. Burst pressure for hydraulic transmission lines is determined by standard test procedure. However, burst pressure for hard lines (i.e. pipes and tubes) can be estimated using *Barlow's formula* as shown in Equation 1.3.

BP = (2 x t x s)/OD   **1.3**

Where:
- BP = Burst Pressure (psi)
- **t** = Wall Thickness [in]
- **s** = Ultimate Strength of Material (psi)
- **OD** = Outer Diameter of Pipe or Tube

❖ **Maximum Allowable Pressure:**
Equation 1.4 is used to calculate the *Maximum Allowable Working Pressure.*

WP = (BP/SF)   **1.4**

Where:
- WP = Allowable Working Pressure.
- BP = Burst Pressure.
- SF = Safety Factor = 4-6 (Typically, but may be different in some applications)

1.4- Hydraulic Pipes

1.4.1- Features of Hydraulic Pipes

Hydraulic *pipes* are rigid, not flexible or intend to bend. They have highest *Tensile Strength* and least expensive among all hydraulic transmission lines. They are used generally to transmit large hydraulic power (large flow and large pressure) for long distances. The weight per unit length is the largest among the hydraulic transmission lines. So, they must be clamped to a fixed frame to carry the weight of the pipe. As shown in Fig. 1.5, many industrial and mobile applications rely on hydraulic pipes in transmitting high hydraulic power.

Fig. 1.5 - Use of Hydraulic Pipes in Industrial (Left) and Mobile (right) Applications

1.4.2- Material of Hydraulic Pipes

Hydraulic pipes are produced from different grades of *Carbon Steel* depending on the pressure rating. Pipes that are used in hydraulic systems are known as black iron pipes and may be either hot or cold drawn. Galvanized pipe should never be used in a hydraulic system because some hydraulic fluids additives react with the zinc coating.

1.4.3- Sizes of Hydraulic Pipes

Sizing of pipes is identified by a *schedule number* established by **ANSI Standard**. As shown in Table 1.2, *Nominal Size* of hydraulic pipes are the *outer diameter* "**OD**" on which the fitting will be assembled. Pipe sizes are ranging between (0.5 - 12) inches (10-300 mm). Hydraulic pipes are produced for different pressure ratings. Obviously, a pipe that works at high pressure has thicker walls. A pipe with a specific OD has different *inner diameters* "**ID**" depending on the standard schedule it belongs to. So, a common mistake is that pipes are sized based on flow only. When pressure increases, moving to a higher schedule may require larger pipe size to maintain same ID, keep flow to be laminar and maintain reasonable pressure drop.

For example, a pipe of nominal size = 1 inch, it has an actual OD = 1.315 inch. This pipe has an ID = 1.049 inch in schedule 40, ID = 0.957 in schedule 80, ID = 0.815 in schedule 160, and ID = 0.599 in double extra heavy schedule. As a result, pipes that belong to large schedules have smaller ID, larger wall thickness, work at higher pressure, handle less flow, and have more weight per unit length. The table shows following schedules:

- Schedule 40, formerly called standard, is for low pressure applications.
- Schedule 80, formerly called extra strong, is for medium pressure applications.
- Schedule 160 or "XS", is for high pressure applications.
- Schedule Double Extra Strong or "XXS", is for extremely high-pressure applications.

NOMINAL SIZE	PIPE O.D.	INSIDE DIAMETER			
		SCHED. 40	SCHED. 80	SCHED. 160	DOUBLE EXTRA HEAVY
1/8	.405	.269	.215	--	--
1/4	.540	.364	.302	--	--
3/8	.675	.493	.423	--	--
1/2	.840	.622	.546	.466	.252
3/4	1.050	.824	.742	.614	.434
1	1.315	1.049	.957	.815	.599
1 1/4	1.660	1.380	1.278	1.160	.896
1 1/2	1.900	1.610	1.500	1.338	1.100
2	2.375	2.067	1.939	1.689	1.503
2 1/2	2.875	2.469	2.323	2.125	1.771
3	3.500	3.068	2.900	2.624	--
3 1/2	4.000	3.548	3.364	--	--
4	4.500	4.026	3.826	3.438	--
5	5.563	5.047	4.813	4.313	4.063

Table 1.2- ANSI Standard Schedules for Hydraulic Pipes

To get a better understanding of ANSI standard for pipes, Fig. 1.6 shows three pipes that have the same OD of one inch, and they belong to different schedules according to ANSI.

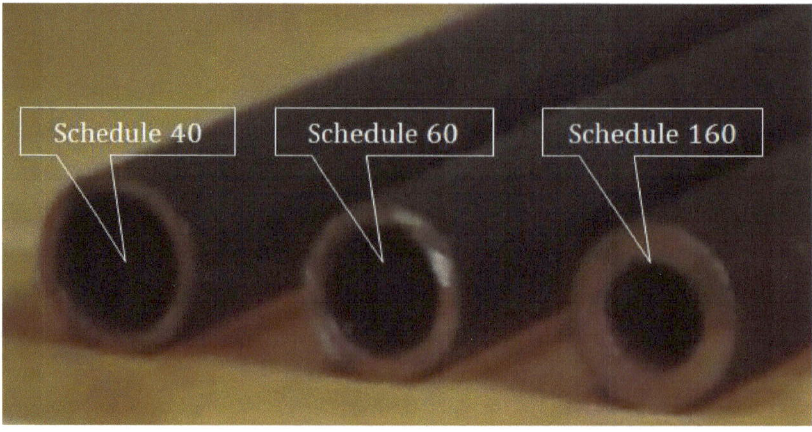

Fig. 1.6 - Hydraulic Pipes According to ANSI Standard

Example of Pipe Size Selection:

Problem Definition: A pump drives a hydraulic motor. Flow distribution analysis is conducted and shows that maximum flow in suction, pressure and return line is 40 GPM.

Required: Referring to Table 1.2 and 1.3, find the nominal and inside diameter of each line. Both the suction and return pipes shall be selected from schedule 40 and pressure pipe from schedule 80.

Suction Line (Purple): Allowable flow speed = 2 FPS, closest upper flow = 46.1 GPM → Pipe Nominal Size = 3 → ID = 3.068 in (Schedule 40).

Return Line (Blue): Allowable flow speed = 5 FPS, closest upper flow = 52.3 GPM → Pipe Nominal Size = 2 → ID = 2.067 in (Schedule 40).

Pressure Line (Red): Allowable flow speed = 10 FPS, flow = 40 GPM → Pipe Nominal Size = 1.25 → ID = 1.278 in (Schedule 80).

STANDARD PIPE – SCHEDULE 40

PIPE SIZE	OD (INCHES)	WALL (INCHES)	ID (INCHES)	INT AREA (SQ. INCHES)	WT/FT (POUNDS)	GPM @ 2 FPS	GPM @ 5 FPS	GPM @ 10 FPS	GPM @ 15 FPS	GPM @ 20 FPS	GPM @ 25 FPS
1/8	0.405	0.068	0.269	0.057	0.244	0.4	0.9	1.8	2.7	3.5	4.4
1/4	0.540	0.088	0.364	0.104	0.424	0.6	1.6	3.2	4.9	6.5	8.1
3/8	0.675	0.091	0.493	0.191	0.567	1.2	3.0	6.0	8.9	11.9	14.9
1/2	0.840	0.109	0.622	0.304	0.850	1.9	4.7	9.5	14.2	18.9	23.7
3/4	1.050	0.113	0.824	0.533	1.130	3.3	8.3	16.6	24.9	33.2	41.6
1	1.315	0.133	1.049	0.864	1.677	5.4	13.5	26.9	40.4	53.9	67.4
1 1/4	1.660	0.140	1.380	1.496	2.270	9.3	23.3	46.6	69.9	93.2	116.6
1 1/2	1.900	0.145	1.610	2.036	2.715	12.7	31.7	63.5	95.2	126.9	158.7
2	2.375	0.154	2.067	3.356	3.649	20.9	52.3	104.6	156.9	209.2	261.5
2 1/2	2.875	0.203	2.469	4.788	5.787	29.8	74.6	149.2	223.9	298.5	373.1
3	3.500	0.216	3.068	7.393	7.568	46.1	115.2	230.4	345.7	460.9	576.1
3 1/2	4.000	0.226	3.548	9.887	9.100	61.6	154.1	308.2	462.3	616.4	770.5
4	4.500	0.237	4.026	12.730	10.779	79.4	198.4	396.8	595.2	793.7	992.1
5	5.563	0.258	5.047	20.006	14.602	124.7	311.8	623.6	935.4	1,247.2	1,559.1
6	6.625	0.280	6.065	28.890	18.954	180.1	450.3	900.6	1,350.9	1,801.1	2,251.4
8	8.625	0.322	7.981	50.027	28.524	311.9	779.7	1,559.4	2,339.2	3,118.9	3,898.6
10	10.750	0.365	10.020	78.854	40.441	491.6	1,229.0	2,458.1	3,687.1	4,916.1	6,145.1
12 [1]	12.750	0.406	11.938	111.932	53.469	697.8	1,744.6	3,489.1	5,233.7	6,978.3	8,722.9
12 [2]	12.750	0.375	12.000	113.097	49.510	705.1	1,762.7	3,525.5	5,288.2	7,051.0	8,813.7

EXTRA STRONG PIPE – XS – SCHEDULE 80

PIPE SIZE	WALL (INCHES)	ID (INCHES)	INT AREA (SQ. INCHES)	WT/FT (POUNDS)	GPM @ 2 FPS	GPM @ 5 FPS	GPM @ 10 FPS	GPM @ 15 FPS	GPM @ 20 FPS	GPM @ 25 FPS
1/8	0.095	0.215	0.036	0.314	0.2	0.6	1.1	1.7	2.3	2.8
1/4	0.119	0.302	0.072	0.534	0.4	1.1	2.2	3.3	4.5	5.6
3/8	0.126	0.423	0.141	0.738	0.9	2.2	4.4	6.6	8.8	11.0
1/2	0.147	0.546	0.234	1.087	1.5	3.6	7.3	10.9	14.6	18.2
3/4	0.154	0.742	0.432	1.472	2.7	6.7	13.5	20.2	27.0	33.7
1	0.179	0.957	0.719	2.169	4.5	11.2	22.4	33.6	44.8	56.1
1 1/4	0.191	1.278	1.283	2.993	8.0	20.0	40.0	60.0	80.0	100.0
1 1/2	0.200	1.500	1.767	3.627	11.0	27.5	55.1	82.6	110.2	137.7
2	0.218	1.939	2.953	5.017	18.4	46.0	92.0	138.1	184.1	230.1
2 1/2	0.276	2.323	4.238	7.653	26.4	66.1	132.1	198.2	264.2	330.3
3	0.300	2.900	6.605	10.242	41.2	102.9	205.9	308.8	411.8	514.7
3 1/2	0.318	3.364	8.888	12.492	55.4	138.5	277.1	415.6	554.1	692.6
4	0.337	3.826	11.497	14.968	71.7	179.2	358.4	537.6	716.8	896.0
5	0.375	4.813	18.194	20.756	113.4	283.6	567.1	850.7	1,134.3	1,417.8
6	0.432	5.761	26.067	28.543	162.5	406.3	812.6	1,218.8	1,625.1	2,031.4
8	0.500	7.625	45.664	43.342	284.7	711.7	1,423.4	2,135.1	2,846.9	3,558.6
10 [3]	0.594	9.562	71.810	64.362	447.7	1,119.2	2,238.5	3,357.7	4,477.0	5,596.2
12 [3]	0.688	11.374	101.605	88.537	633.4	1,583.6	3,167.2	4,750.9	6,334.5	7,918.1

Table 1.3 - Examples of Sizing Hydraulic Pipes

1.4.4- Pressure Rating of Hydraulic Pipes

Table 1.4 presents allowable working pressure and estimated burst pressure for seamless steel pipe (material is ASTM A53 or A106 Grade B steel with *tensile strength* of 60k psi). Presented data are tabulated by ANSI B 31.3 standard based on Barlow's formula. The standard considers lower safety factor for higher schedule number.

Example: As shown in the figure, for a pipe size of nominal size =1, approximate estimated burst pressure = 6k psi in schedule 40, approximate estimated burst pressure = 10k psi in schedule 80, and approximate estimated burst pressure = 16k psi in schedule 160. The reported allowable working pressure for that pipe size indicates that safety factor for schedule $40 \approx 9$, for schedule $80 \approx 6$, and for schedule $160 \approx 5$.

Threaded Connections Pressure Ratings of Seamless Steel

1 inch OD Different schedule

PIPE SIZE & SCHEDULE	ALLOWABLE WORKING PRESSURE (PSIG)		ESTIMATED BURST VALUE P_B (PSIG)	WATER HAMMER FACTOR
	ANSI B31.1	ANSI B 31.3		
1/8-40	994	746	11,369	327.6
1/8-80	3,509	2,632	19,369	512.9
1/4-40	948	711	9,680	178.9
1/4-80	3,103	2,327	16,569	260.0
3/8-40	916	687	8,277	97.55
3/8-80	2,864	2,148	14,500	132.5
1/2-40	884	663	7,409	61.28
1/2-80	2,576	1,932	12,837	79.53
1/2-160	4,445	3,334	18,551	109.2
1/2-XXS	8,577	6,433	33,837	373.3
3/4-40	842	631	6,384	34.92
3/4-80	2,295	1,722	11,070	43.06
3/4-160	4,704	3,528	18,384	62.89
3/4-XXS	7,328	5,496	28,670	125.9
1-40	[illegible]	622	5,788	21.55
1-80	[illegible]	1,595	9,986	25.89
1-160	[illegible]	3,187	16,465	35.69
1-XXS	6,804	5,103	26,321	66.08
1 1/4-40	806	605	5,091	12.45
1 1/4-80	1,941	1,456	8,778	14.52
1 1/4-160	3,319	2,490	13,043	17.62
1 1/4-XXS	6,575	4,932	22,585	29.53
1 1/2-40	798	598	4,764	9.147
1 1/2-80	1,866	1,399	8,238	10.54
1 1/2-160	3,522	2,642	13,353	13.24
1 1/2-XXS	6,095	4,572	20,869	19.59
2-40	773	580	4,266	5.549
2-80	1,764	1,323	7,500	6.306
2-160	3,834	2,875	13,866	8.331
2-XXS	5,435	4,076	18,514	10.50
2 1/2-40	815	611	4,299	3.889
2 1/2-80	1,750	1,312	7,346	4.394
2 1/2-160	3,074	2,305	11,478	5.250
2 1/2-XXS	5,588	4,191	18,866	7.559
3-40	801	601	3,977	2.519
3-80	1,683	1,263	6,857	2.819
3-160	3,201	2,401	11,589	3.443
3-XXS	5,096	3,822	17,143	4.482
3 1/2-40	790	592	3,780	1.883
3 1/2-80	1,634	1,226	6,540	2.095
3 1/2-XXS	4,783	3,587	16,080	3.186
4-40	789	592	3,653	1.463
4-80	1,604	1,203	6,320	1.620
4-160	3,263	2,447	11,493	2.006
4-XXS	4,556	3,417	15,307	2.386
5-40	772	579	3,408	0.9308
5-80	1,543	1,157	5,932	1.023
5-160	3,270	2,453	11,325	1.275
5-XXS	4,178	3,134	14,021	1.436
6-40	766	575	3,260	0.6445
6-80	1,608	1,206	6,014	0.7144
6-160	3,275	2,456	11,212	0.8812
6-XXS	4,159	3,120	13,838	0.9887

DISCONTINUED FORMER STANDARD

Table 1.4 - ANSI B 31.3 Standard for Hydraulic Pipes Pressure Rating

1.4.5- Hydraulic Pipe Assembly

1.4.5.1- Hydraulic Pipes Threads

Pipe Connection: Pipes used in hydraulic systems are not intended to bend and should be seamless, As shown in Fig. 1.7, pipes are traditionally assembled using tapered threads. A pipe thread connection includes external threads cut on the pipe ends and internal threads cut into the openings of the pipe fittings. This thread is tapered 3/4" per foot to assure a positive seal when it is properly tightened.

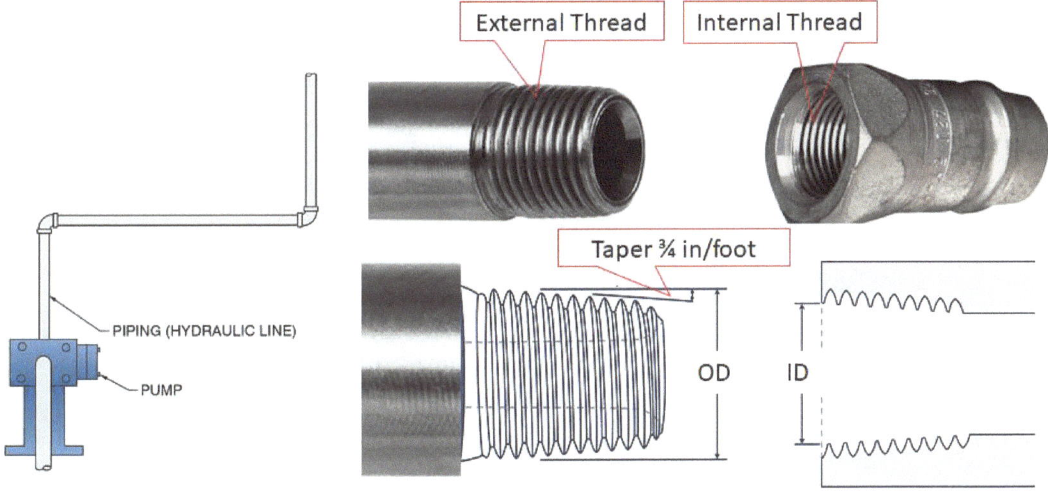

Fig. 1.7 - Pipe Connection by National Thread

Spiral vs. Dryseal Thread: Figure 1.8 shows the standard (*Spiral*) National Pipe Tapered (*NPT*) thread versus the *Dryseal* National Pipe Tapered (*NPTF*) thread. The NPT threads tend to leak under high system pressure because of a continuous internal *spiral clearance* that exists in the threads even when it is tightened. Therefore, NPTF threads are recommended over NPT threads for high pressure hydraulic lines and fuel lines. The NPT/NPTF designations are explained in the tube section.

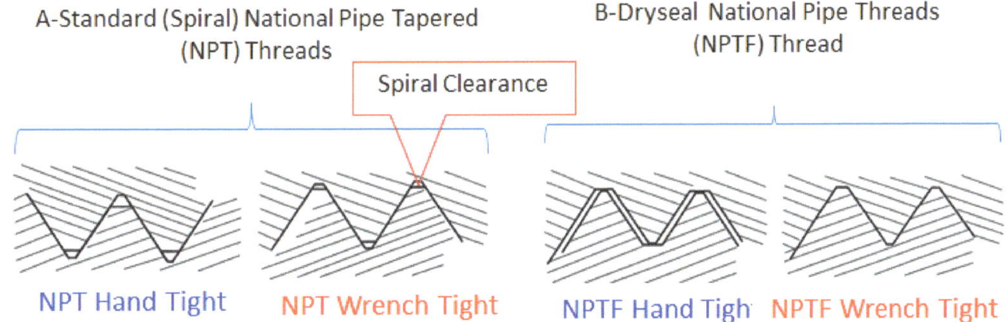

Fig. 1.8 - Standard Spiral Thread versus Dryseal Thread

1.4.5.2- Fittings for Hydraulic Pipes

As shown in Fig. 1.9, a wide variety of threaded connections are available to connect a pipe to hydraulic system. However, Fittings <u>aren't recommended for assembling pipes</u>. Alternatively, Butt & Socket Weld fittings and 4 Bolts 61/62 SAE Flanges are recommended for pipe connections.

Fig. 1.9- Hydraulic Pipe Fittings

1.5- Hydraulic Tubes

1.5.1- Features of Hydraulic Tubes

Hydraulic *Tubes* are semirigid fluid conductor that have relatively thinner and less weight per unit length as compared to hydraulic pipes. Hydraulic tubes can be bent with the minimum bend radius respected. Tubes are easier to assemble than pipes since no welding is needed. Tubes are hooked to the components by various types of fittings and flanges. Tubes have lower pressure ratings than pipes. As shown in Fig. 1.10, many industrial and mobile applications rely on hydraulic tubes in transmitting hydraulic power.

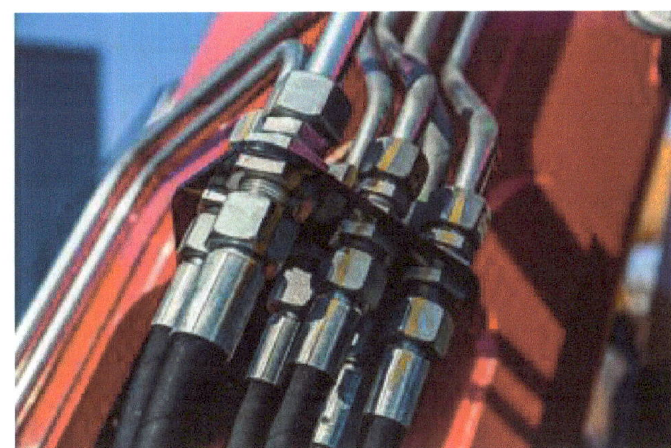

Fig. 1.10 - Use of Hydraulic Tubes in Industrial (Left) and Mobile (right) Applications

1.5.2- Material of Hydraulic Tubes

As shown in Fig. 1.11, hydraulic tubes are produced from various material as follows:
- Stainless Steel: Ultimate Strength = 60/75/100 thousands of psi.
- Low Carbon Steel Ultimate Strength = 55,000 psi.
- Cooper: Ultimate Strength = 32,000 psi.

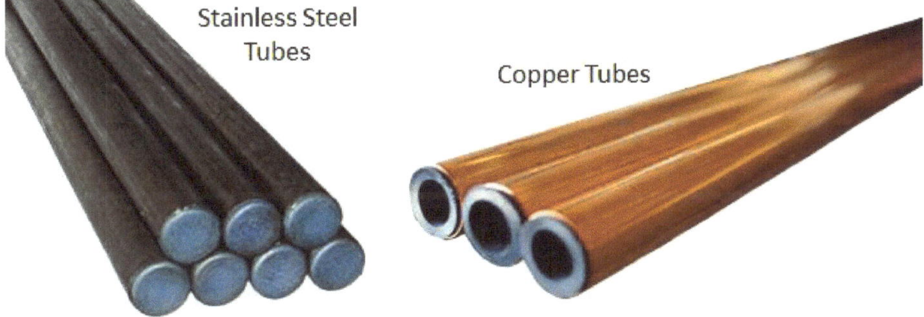

Fig. 1.11 - Material of Hydraulic Tubes

1.5.3- Sizes of Hydraulic Tubes

Like hydraulic pipes, nominal size of hydraulic tubes is the outer diameter "OD". Tube sizes are ranging between (0.125 – 2.25) inches (4-42 mm).

1.5.4- Pressure Rating of Hydraulic Tubes

As shown in Table 1.5, three values for the maximum allowable working pressure are presented based on three formulas to estimate Burt Pressure: *Barlow's Formula "1"*, *Boardman's Formula "2"*, and *Lame's Formula "3"*. In this textbook, like hydraulic pipes, Barlow's formula will only be considered in this textbook because it results in the lowest burst pressure. The values shown in the table are for *Carbon Steel* and based on design *safety factor* of approximately 4:1. Maximum allowable pressure is calculated based on an allowable fiber stress of 12,500 psi which is equivalent to 50% of the minimum yield point and approximately 28% of the *Ultimate Strength* (55000 psi) of the tube material.

Nominal Tube OD, in.		See Note*	Nominal Tube Wall Thickness, in.									
			0.028	0.035	0.049	0.065	0.083	0.095	0.109	0.120	0.134	0.148
1/8	0.125	1	5,600	7,000								
		2	6,800	9,000								
		3	6,650	8,450								
3/16	0.188	1	3,750	4,650								
		2	4,250	5,500								
		3	4,250	5,450								
1/4	0.250	1	2,800	3,500	4,900	6,500						
		2	3,100	3,950	5,800	8,200						
		3	3,100	3,950	5,750	7,800						
5/16	0.312	1	2,250	2,800	3,900	5,200						
		2	2,400	3,100	4,500	6,250						
		3	2,450	3,100	4,500	6,150						
3/8	0.375	1	1,850	2,350	3,250	4,350	5,550	6,350				
		2	2,000	2,500	3,650	5,050	6,700	7,950				
		3	2,000	2,550	3,650	5,000	6,550	7,600				
1/2	0.500	1		1,750	2,450	3,250	4,150	4,750	5,450	6,000		
		2		1,850	2,650	3,650	4,800	5,600	6,600	7,450		
		3		1,850	2,700	3,650	4,800	5,550	6,450	7,200		
5/8	0.625	1		1,400	1,950	2,600	3,300	3,800	4,350	4,800		
		2		1,450	2,100	2,850	3,700	4,350	5,050	5,650		
		3		1,500	2,100	2,850	3,750	4,350	5,050	5,600		
3/4	0.750	1		1,150	1,650	2,150	2,750	3,150	3,650	4,000		
		2		1,200	1,700	2,350	3,050	3,500	4,100	4,600		
		3		1,200	1,750	2,350	3,050	3,550	4,150	4,600		
7/8	0.875	1		1,000	1,400	1,850	2,350	2,700	3,100	3,400		
		2		1,050	1,450	1,950	2,550	2,950	3,450	3,850		
		3		1,050	1,500	2,000	2,600	3,000	3,500	3,900		
1	1.000	1		875	1,200	1,600	2,050	2,350	2,700	3,000	3,350	3,700
		2		900	1,250	1,700	2,200	2,550	3,000	3,300	3,750	4,200
		3		900	1,300	1,750	2,250	2,600	3,000	3,350	3,800	4,200

Note: Wall thicknesses having values shown to the right of the bold line are not normally considered suitable for 37° single flaring to J533.

Table 1.5 – Maximum Allowable Working Pressure (psi) for Carbon Steel Tubes

1.5.5- Hydraulic Tubes Assembly

Unlike hydraulic pipes, tubes are bent to minimize fittings and connected by fitting, or flanges.

1.5.5.1- Tube Bending

During tube bending, wall thickness of the outer surface is reduced and accordingly pressure rating too. Therefore, tube bending must respect *minimum bend radius* reported by tube manufacturer. Figure 1.12 shows that minimum bend radius is based on the inner surface or based on the center line of the tube. The figure also shows manual mandrills of different sizes for tube bending. Table 1.6 shows minimum bend radius based on the size of the tube.

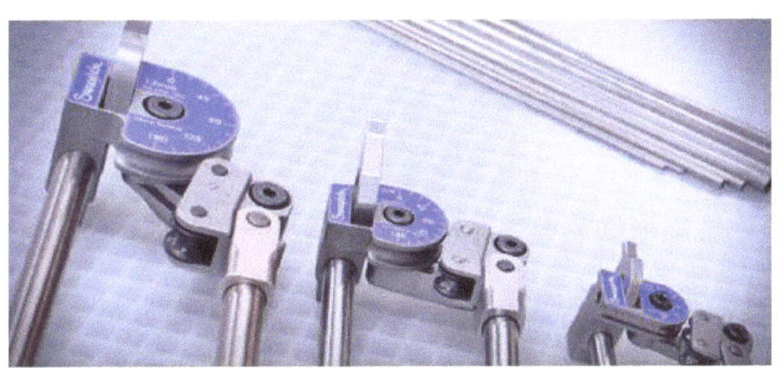
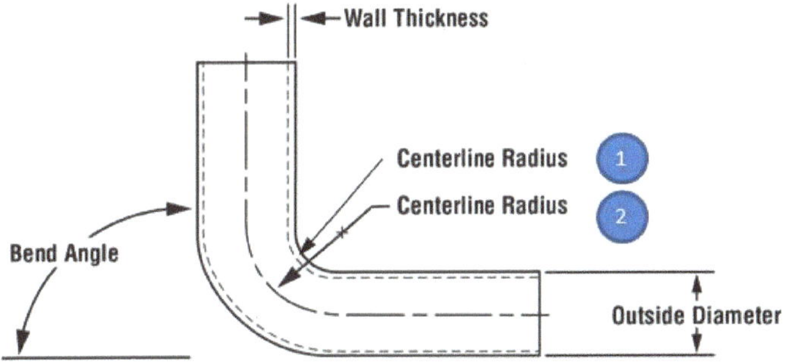

Fig. 1.12 - Manual Mandrills of Different Sizes for Tube Bending

TUBE OD (INCHES)	CLOSE① BEND (INCHES)	NORMAL BEND (INCHES)	MS33611 STANDARD	METRIC SIZES		
				TUBE OD (MM)	CLOSE① BEND (MM)	NORMAL BEND (MM)
0.125	—	0.281③	—	4	—	9
0.188	—	0.422③	—	6	—	13
0.250	0.562	0.750	—	8	—	18
0.312	0.687	1.000	—	10	32	32
0.375	0.937	1.250	1.125	12	—	32
0.500	1.250	2.000	1.500	15	38	38
0.625	1.500	2.500	1.875	16	38	38
0.750	1.750	3.000	2.250	18	44	44
0.875	2.000	3.500	2.625	20	44	44
1.000	3.000	4.000	3.000	22	—	89
1.125	3.500	4.500	3.375	25	—	100
1.250	3.750	5.000	3.750	28	—	112
1.500	5.000	6.000	4.500	30	—	128
1.750	—	7.000	—	35	—	140
2.000	—	8.000	—	38	—	152
2.250	—	9.000③	—	42	—	168

Table 1.6 - Minimum Bend Radius for Hydraulic Tubes of Different Sizes

1.5.5.2- Tube Flaring

The standard **SAE J533B** covers specifications for 37° and 45° for *tube flaring.* 37° single and double flares shall conform to the dimensions specified in Fig. 1.13 and Table 1.7. 45° Single and double flares shall conform to the dimensions specified in Fig. 1.14 and Table 1.8.

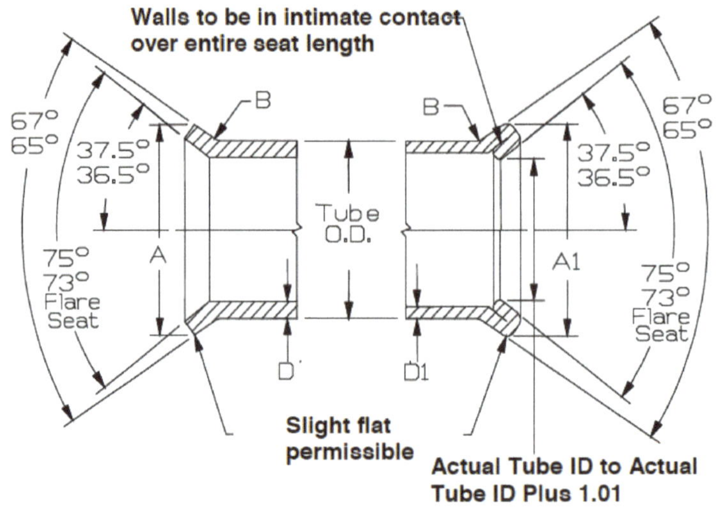

Fig. 1.13 - 37º Single and Double Flaring According to SAE J533B

Nominal Tube OD	A Single Flare Diameter		A₁ Double Flare Diameter		B Radius	Dᵇ Single Flare Wall Thickness	D₁ᵇ Double Flare Wall Thickness
in	in		in		in	in	in
	Max	Min	Max	Min	±0.02	Max	Max
1/8	0.200	0.180	0.200	0.180	0.03	0.035	0.025
3/16	0.280	0.260	0.280	0.260	0.03	0.035	0.028
1/4	0.360	0.340	0.360	0.340	0.03	0.065	0.035
5/16	0.430	0.400	0.430	0.400	0.03	0.065	0.035
3/8	0.490	0.460	0.490	0.460	0.04	0.065	0.049
1/2	0.660	0.630	0.660	0.630	0.06	0.04	0.049
5/8	0.790	0.760	0.790	0.760	0.06	0.04	0.049
3/4	0.950	0.920	0.950	0.920	0.08	0.04	0.049
7/8	0.748	1.040	1.070	1.040	0.08	0.109	0.065
1	0.916	1.170	1.200	1.170	0.09	0.120	0.065
1-1/8	1.041	1.350	1.380	1.350	0.09	0.120	0.065
1-1/4	1.157	1.480	1.510	1.480	0.09	0.120	0.065
1-1/2	1.730	1.700	1.730	1.700	0.11	0.120	0.065
1-3/4	2.110	2.080	2.110	2.080	0.11	0.120	0.065
2	2.360	2.330	2.360	2.330	0.11	0.134	0.065

Table 1.7 - 37º Single and Double Flaring Standard Dimensions According to SAE J533B

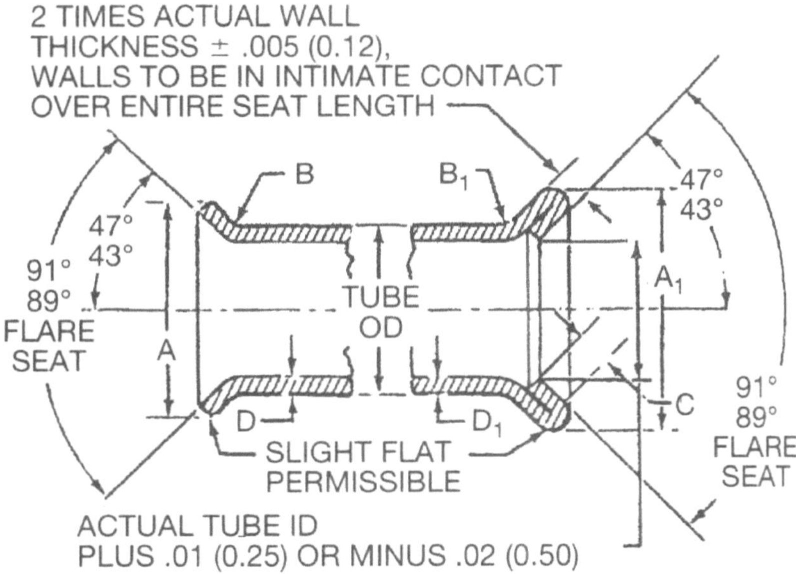

Fig. 1.14 - 45⁰ Single and Double Flaring According to SAE J533B

Nominal Tube OD	A Single Flare Diameter		A₁ Double Flare Diameter		B Single Flare Radius	B₁ Double Flare Radius	C Double Flare Coined Seat Length	Db Single Flare Wall Thickness	D₁b Double Flare Wall Thickness
in	in		in		in	in	in	in	in
	Max	Min	Max	Min	±0.01	±0.01	Min	Max	Max
1/8	0.181	0.171	0.213	0.198	0.02	0.04	0.040	0.035	0.025
3/16	0.249	0.239	0.280	0.265	0.02	0.04	0.040	0.035	0.028
1/4	0.325	0.315	0.360	0.345	0.02	0.04	0.040	0.049	0.035
5/16	0.404	0.388	0.425	0.410	0.02	0.04	0.062	0.049	0.035
3/8	0.487	0.471	0.500	0.485	0.02	0.04	0.062	0.065	0.049
7/16	0.561	0.545	0.570	0.555	0.02	0.04	0.062	0.065	0.049
1/2	0.623	0.607	0.640	0.625	0.02	0.04	0.062	0.083	0.049
9/16	0.676	0.660	0.712	0.697	0.02	0.04	0.062	0.083	0.049
5/8	0.748	0.732	0.772	0.757	0.02	0.04	0.062	0.095	0.049
3/4	0.916	0.900	0.912	0.897	0.02	0.04	0.062	0.109	0.049
7/8	1.041	1.025	—	—	0.02	—	—	0.109	—
1	1.157	1.141	—	—	0.02	—	—	0.120	—

Table 1.8 - 45⁰ Single and Double Flaring Standard Dimensions According to SAE J533B

1.5.5.3- Tube Fittings

What is a Tube Fitting? A fitting is a connection between a transmission line and a hydraulic component.

How to Identify a Fitting? Selecting proper fitting is a challenging task. Fittings are identified by Standards, Body Configuration, Seal Configuration, Thread Characteristics, Size, Material, and Interchangeability.

Fitting Body General Configurations: Fittings are provided in various configurations such as male, female, male to male, female to female, and male to female. Figure 3.15 shows that fittings could be fixed or rotating. Rotating "*Swivel*" fittings are used to connect stationary and rotating machine assemblies.

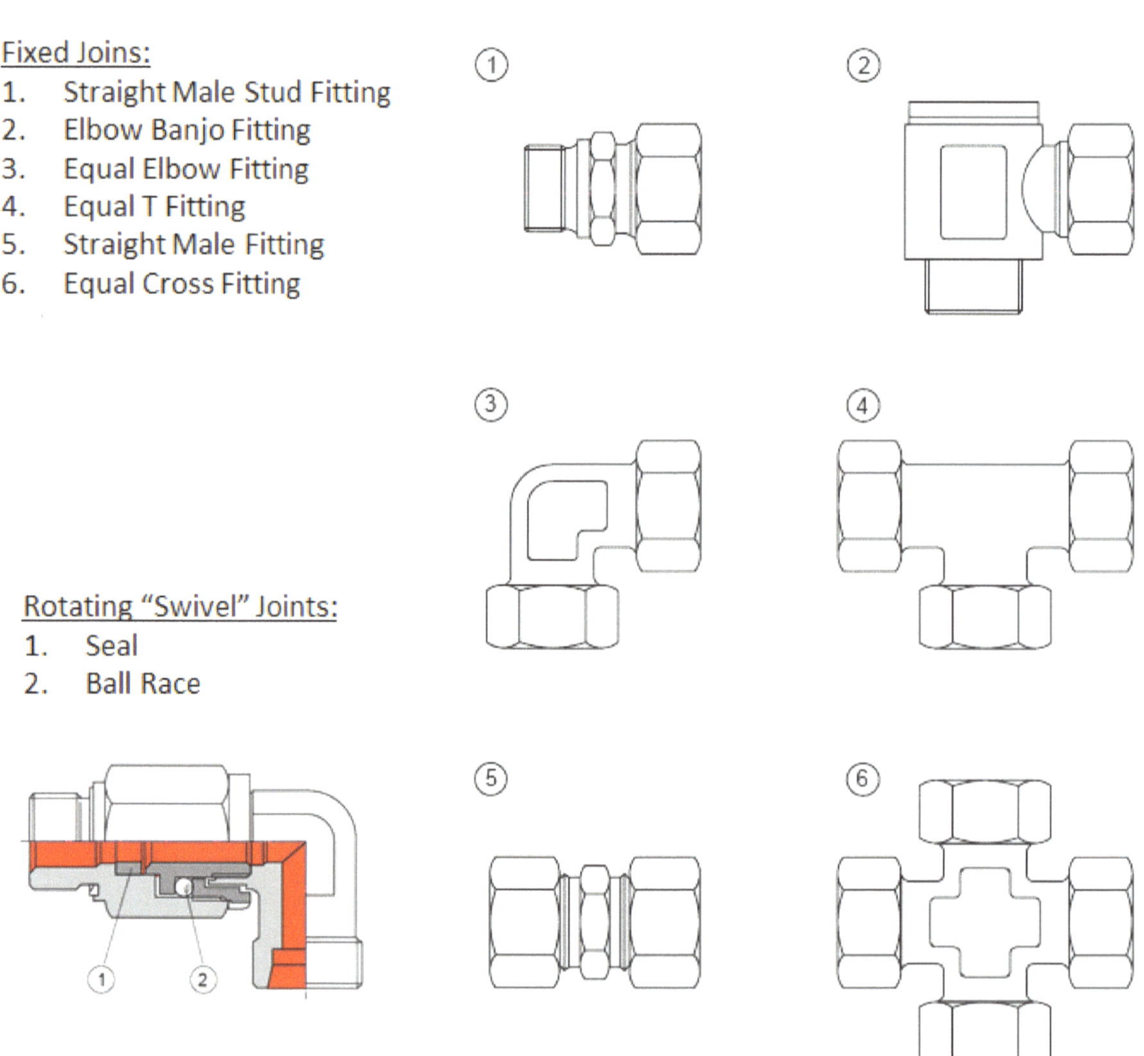

Fig. 1.15 - Fitting Body Configurations (Courtesy of Bosch Rexroth)

Fitting Seal Configuration: As shown in Fig. 1.16, sealing between the fitting and the tube is provided by several methods as follows:

- **Thread Interference Sealing:** This is the easiest type of fitting to use. A characteristic of this thread is that the male is thinner at the front than it is at the back. As the male is threaded into the female, the edges of the thread distort by flattening out. This distortion creates the seal.
- **O-Ring Sealing:** This type of seal is excellent for high-pressure applications. The O-ring is being compressed makes the seal. In both *O-Ring Face Seal* and O-*Ring Cone Seal*, the nipple is welded to the tube.
- **Mated Surfaces Sealing using Flareless Fitting:** In *Flareless* fittings, the seal takes place using a *Compression Ring*, also referred to as a *Cutting Ring*.
- **Mated Surfaces Sealing using Flared Fitting:** In *Flared* fittings, the seal takes place between male flared and female coned seats. Different seat angles are available.

Fitting Standards: There are several standards for hydraulic fittings such as: International Standard "ISO", North American "JIC/SAE/NPT", British "B", Metric French, "GAZ" German "DIN", Japanese "JIS".

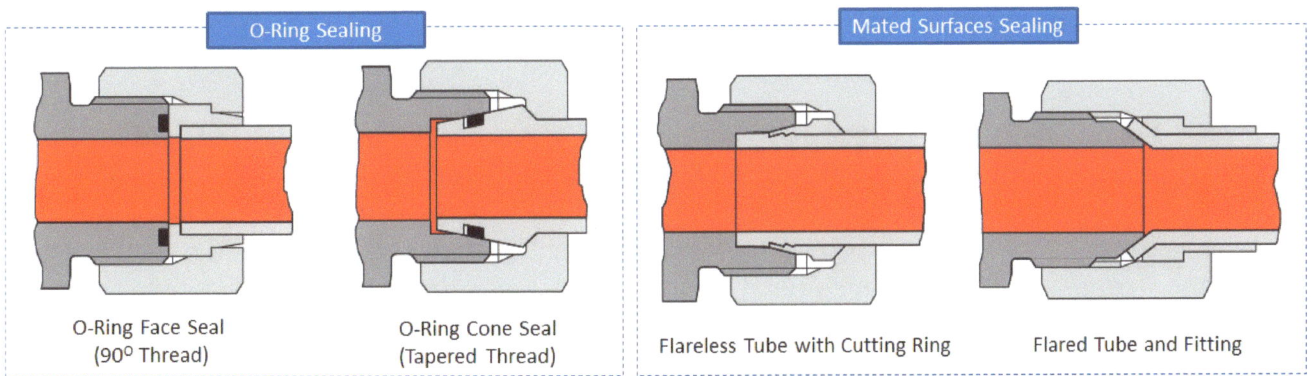

Fig. 1.16 - Fitting Seal Configurations (Courtesy of Bosch Rexroth)

1.5.5.4- Threaded Sealing Fittings

North American Standards Thread Types:

The *North American Standard* for threads is widely used for over 100 years. They are used to effectively seal pipes for fluid and gas transfer. They are available in iron or brass for low-pressure applications and carbon steel and stainless steel for high-pressure

The following designations are used to identify thread type.

N = National, **P** = Pipe, **S** = Straight, **T** = Tapered, **F** = Fuel, and **M** = Swivel Mechanical Joints.

Figure 1.17 shows the common thread sealing fittings:

- **National Pipe Tapered (NPT):** *National Pipe Tapered* thread is a *Spiral tapered* thread used commonly for low pressure application because of the spiral clearance. As shown in Fig. 1.17, the spiral clearance increases with the large size fittings and with the frequent use of the fitting.

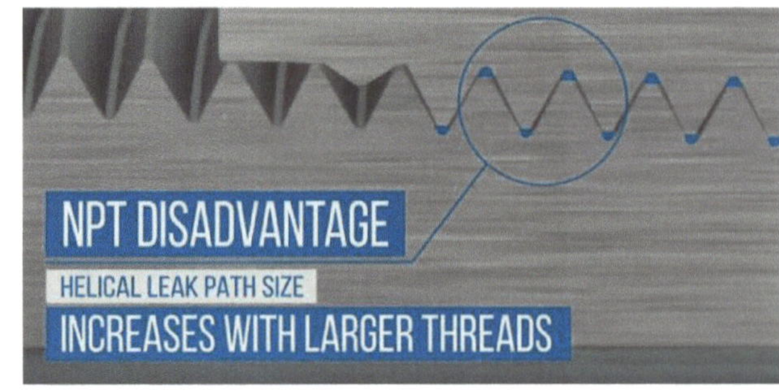

Fig. 1.17 – Spiral Clearance on NPT Fittings (Courtesy of Brennan Industries)

- **National Pipe Tapered for Fuels (NPTF):** As shown in Fig. 1.18, *National Pipe Tapered Fuel* thread is a *Dryseal* thread for Fuel and made to specifications ASME 1.20.3. It is used for both male and female ends. The NPTF male will mate with the NPTF, NPSF, or NPSM females. The NPTF male has tapered threads and a 30° inverted seat. The NPTF female has tapered threads and no seat. <u>The seal takes place by deformation of the threads.</u>

- **National Pipe Straight for Fuels (NPSF):** *National Pipe Straight* thread for Fuels is used for female ends and properly mates with the NPTF male end.

- **National Pipe Straight for Swivel Mechanical Joints (NPSM):** *National Pipe Straight* thread for Mechanical Joint has a straight thread and a 30° internal chamfer. The female half of this connection has a straight thread and an inverted 30° seat. <u>The seal takes place on the 30° seat.</u>

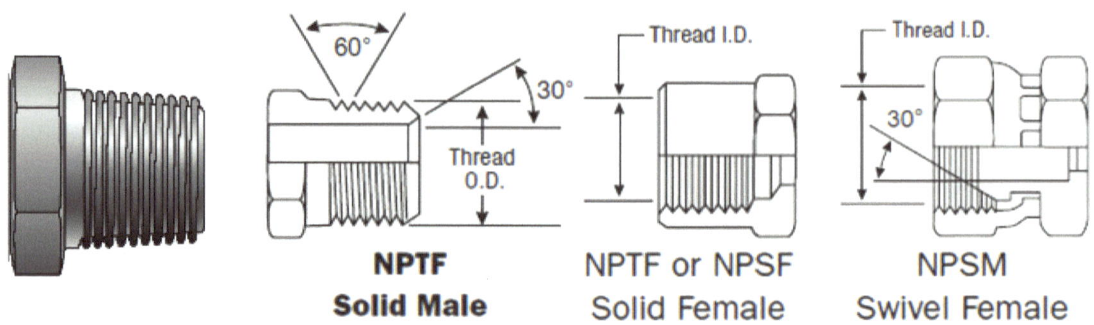

Fig. 1.18 - National Pipe Thread (Courtesy of Gates)

Figure 1.19 shows crossectional views of NPTF and NPSM fittings.

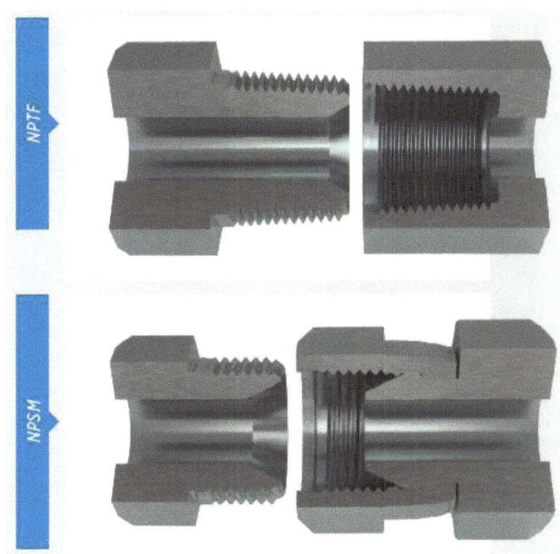

Fig. 1.19 – Crossectional Views of NPTF and NPSM Fittings (Courtesy of Brennan Industries)

North American Standards Thread Types (SAE):

- As shown in Fig. 1.20, the SAE 45° inverted flare male will mate with an SAE 42° inverted flare female only. The male has straight threads and a 45° inverted flare. The female has straight threads and a 42° inverted flare. <u>The seal is made between the 45° flare seat on the male and the 42° flare seat on the female.</u>

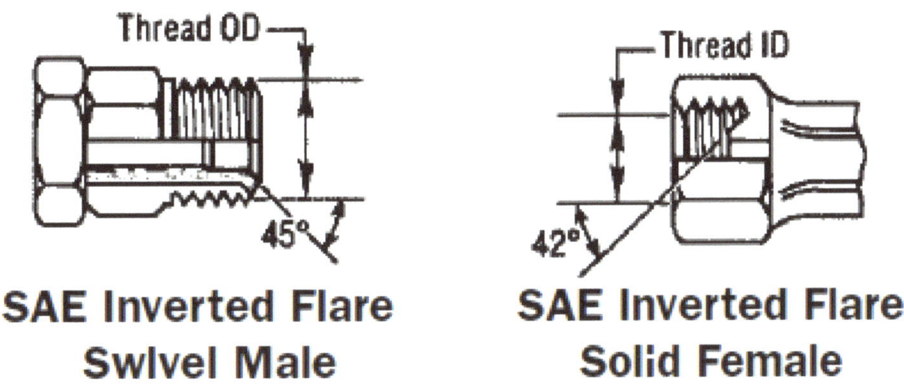

Fig. 1.20 - SAE Inverted Flare Thread (Courtesy of Gates)

British Standards Thread Types (BSP):

- **British Standard Pipe Parallel (BSPP) (Fig 1.21):** Popular couplings have *British Standard Pipe* (BSP) threads, also known as *Whitworth* threads. The BSPP male has straight threads and a 30° seat. The BSPP (parallel) male will mate with a BSPOR (parallel) female or a female port. The BSPP (parallel) connector is similar to, but not interchangeable with the NPSM connector. The thread pitch is different in most sizes, and the thread angle is 55° instead of the 60° angle on NPSM threads.

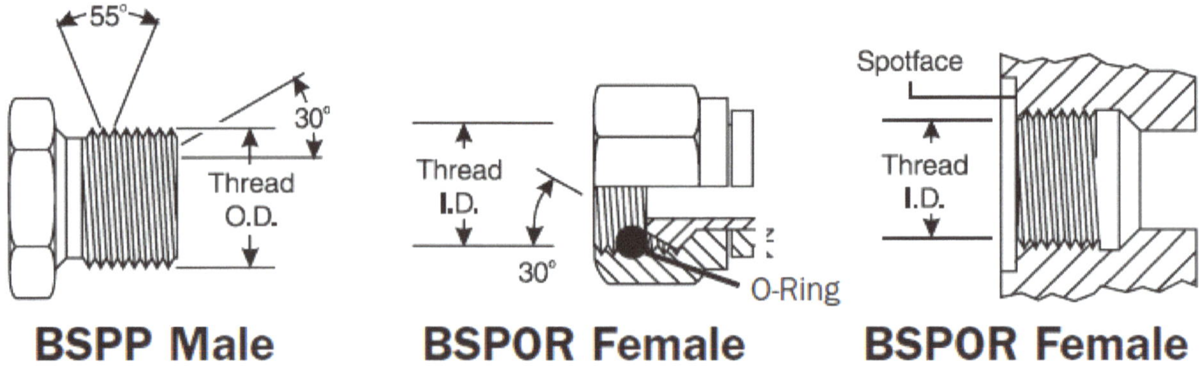

Fig. 1.21 - British Standard Pipe Parallel Thread (Courtesy of Gates)

- **British Standard Pipe Tapered (BSPT) (Fig 1.22):** It is a popular coupling. The BSPT (tapered) male will mate with a BSPT (tapered) female, or a BSPOR (parallel) female. The BSPT (parallel) connector is similar to, but not interchangeable with the NPTF connector. The thread pitch is different in most sizes, and the thread angle is 55° instead of the 60° angle on NPSM threads.

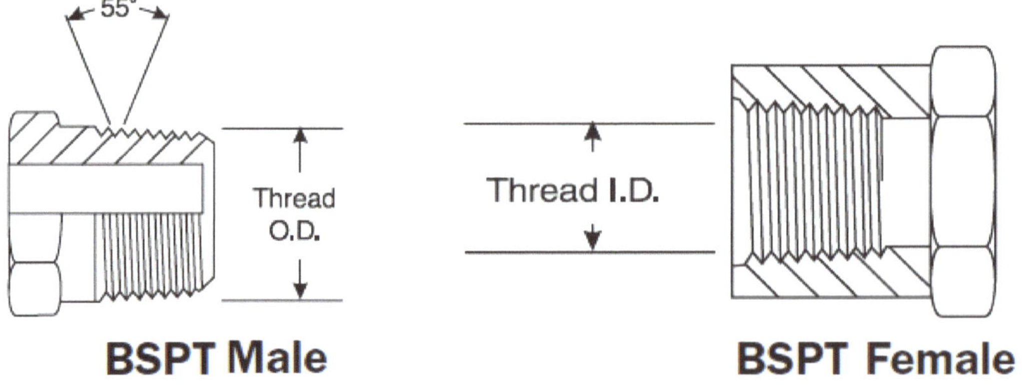

Fig. 1.22 - British Standard Pipe Tapered Thread (Courtesy of Gates)

Figure 1.23 shows sectional views for BSPT and PSPP fittings.

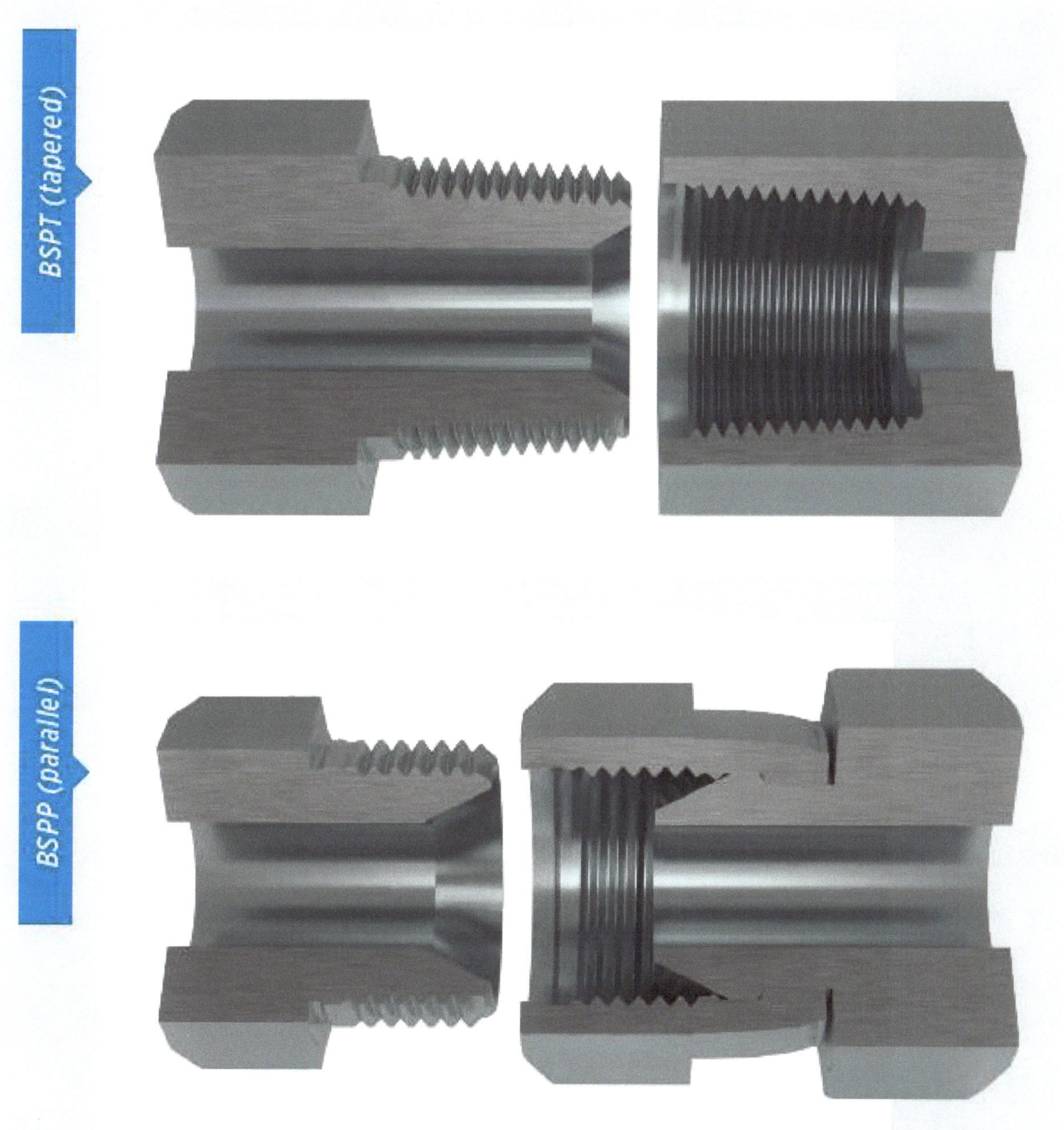

Fig. 1.23 - Sectional Views for BSPT and PSPP Fittings (Courtesy of Brennan Industries)

Metric Standards for Thread Type Couplings:

Popular couplings have metric threads. Both French "GAZ" and German "DIN" are considered metric. However, they are not the same because the threads are different in some sizes. The French GAZ couplings use fine threads in all sizes, and German DIN couplings use coarse threads in larger sizes.

Metric French Standard Thread Type (GAZ):

GAZ 24° Metric French (GAZ) (Fig. 1.24): The GAZ 24° cone male will mate with the female 24° cone or the female tube fitting. The male has a 24° seat and straight metric threads. The female has a 24° seat or a tubing sleeve. When measuring the flare angle with the seat angle gauge, use the 12° gauge. The seat angle gauge measures the angle from the connector centerline.

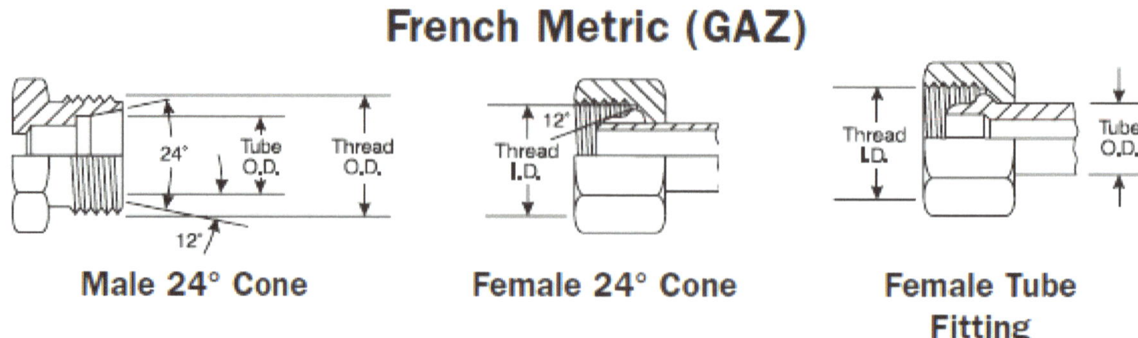

Fig. 1.24 - GAZ 24° Metric French Thread (Courtesy of Gates)

Metric German Standard Thread Type (DIN):

DIN 2353 24° Cone Metric German (Fig. 1.25): The DIN 24° cone male will mate with any of the females shown. The male has a 24° seat, straight metric threads, and a recessed counterbore which matches the tube O.D. of the coupling used with it. The mating female is a 24° cone with O-ring, a metric tube fitting or a universal 24° cone.

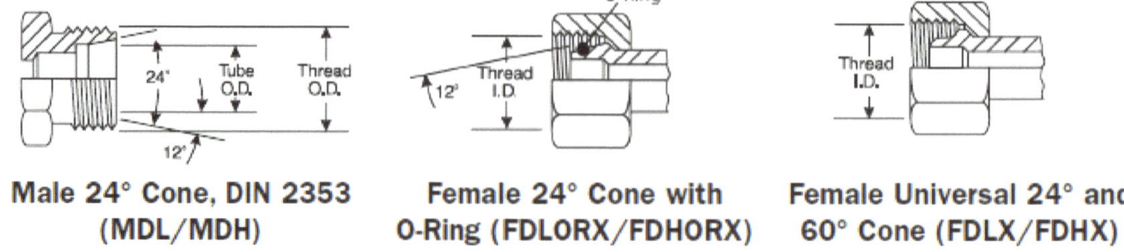

Fig. 1.25 - DIN 24° Cone Metric German Thread (Courtesy of Gates)

- **Example of DIN 2353 - Innovative Metric Fitting (EO-2) (Fig. 1.26):**
  - Description: The *EO-2* is a high-pressure tube fitting generation with elastomeric seals on all joints. EO-2 is a true metric straight thread design according to 24° bite-type standards such as: ISO 8434-1, DIN 2353 or DIN 3861. It consists of a body, a functional nut, and an elastomeric seal.

  - Operation: The elastomeric seal assures a hermetically sealed tube joint. It is located in between the inner cone of the fitting body and the tube surface, thus blocking the only possible leak path. Due to its large cross-section, the seal effectively compensates for all manufacturing tolerances on tube and fitting cone. The sealing effect is pressure supported which makes the EO-2 fitting suitable for high pressure applications. The static compression also eliminates air-ingress into the fluid system below atmospheric pressure conditions.

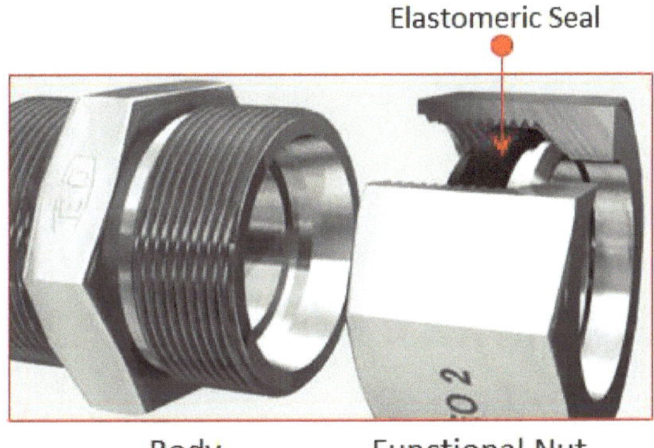

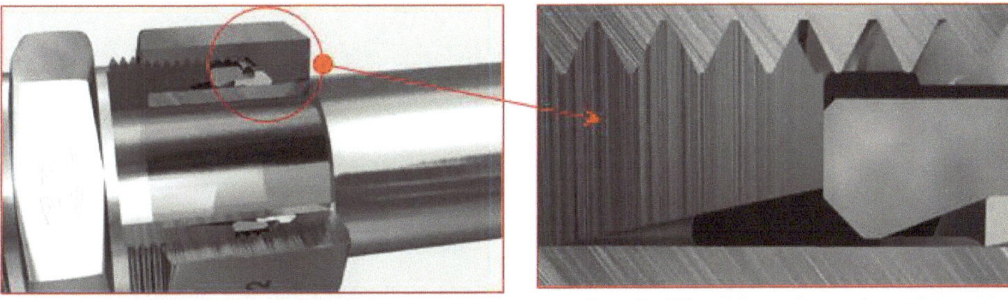

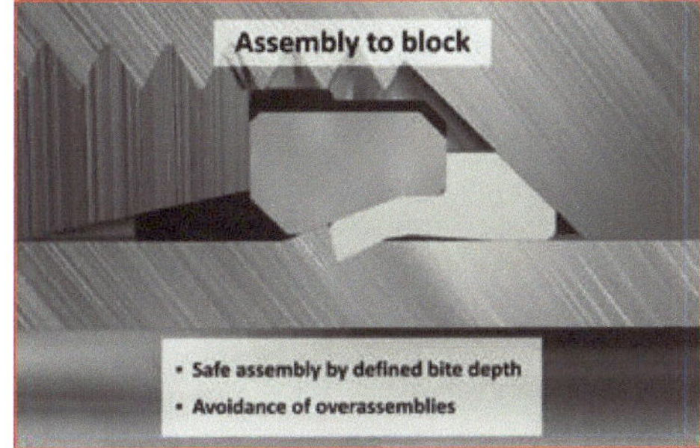

2- Tightened by Wrench
to Specified Torque

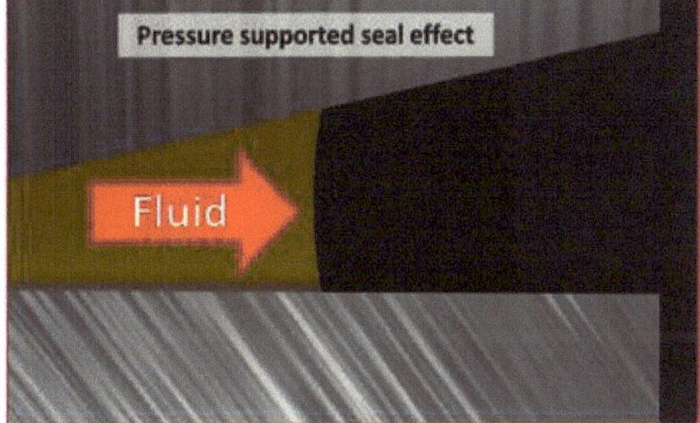

3- Fluid Pressure Applies

Fig. 1.26 - EO-2 Fitting (Courtesy of Parker)

- **DIN 6711 60° Cone Metric German (Fig. 1.27):** The DIN 60° cone male will mate with the female universal 24° or 60° cone connector only. The male has a 60° seat and straight metric threads. The female has a 60° universal seat and straight metric threads. When measuring the flare angle with the seat angle gauge, use the 30° gauge.

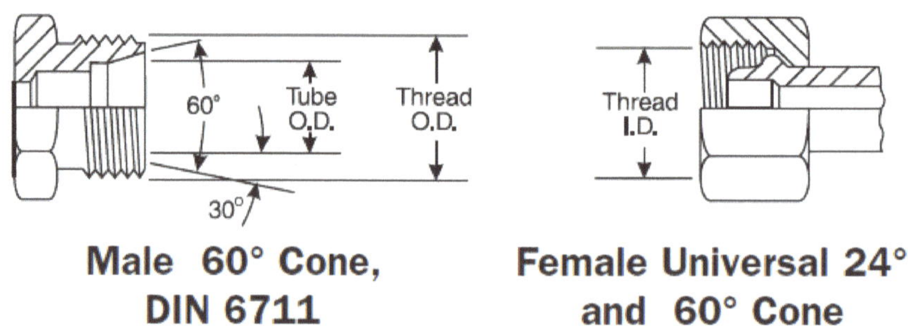

Fig. 1.27 - DIN 60° Cone Metric German Thread (Courtesy of Gates)

- **DIN 3852 Couplings Type A & B (Fig. 1.28):** The male DIN 3852 Type A & B couplings will mate with the female DIN coupling shown below. The male and female type A & B couplings have straight threads. The seal occurs when the ring seal (Type A) or the face seal (Type B) mates with the face of the female port.

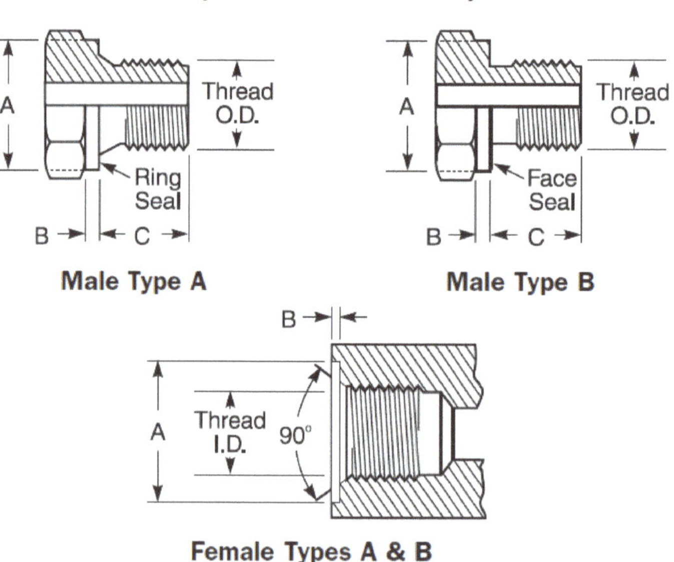

Fig. 1.28 - DIN 3852 Thread (Courtesy of Gates)

Japanese Standards Thread Types (JIS):

There are two popular types of coupling styles in Japan, Japanese Industrial Standard "JIS" and Komatsu. These couplings look similar to Male JIC and Female JIC Swivel couplings. However, there are two major differences: The threads are BSP and the seat angle is only 30° instead of 37° for JIC.

- **Japanese JIS-30° Flare Parallel Threads (Fig. 1.29):** The Japanese 30° flare male connector will mate with a Japanese 30° flare female only. The male and female have straight threads and a 30° seat. The seal is made on the 30° seat. The threads on the Japanese 30° flare connector conform to JIS B 0202, which are the same as the BSPOR threads. Both the British and Japanese connectors have a 30° seat, but they are not interchangeable because the British seat is inverted.

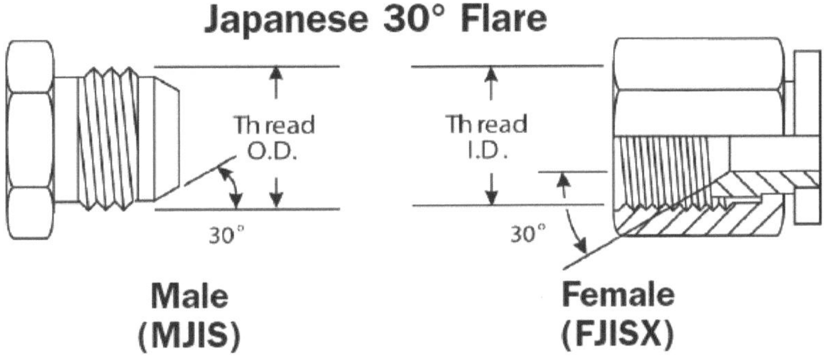

Fig. 1.29 - Japanese 30⁰ Flare Thread (Courtesy of Gates)

- **Japanese Tapered Pipe Thread (Fig. 1.30):** The Japanese tapered pipe thread connector is identical to and fully interchangeable with the BSPT (tapered) connector. The Japanese connector does not have a 30° flare and will not mate with the BSPOR female. The threads conform to JIS B 0203, which are the same as BSPT threads. The seal on the Japanese tapered pipe thread connector is made on the threads.

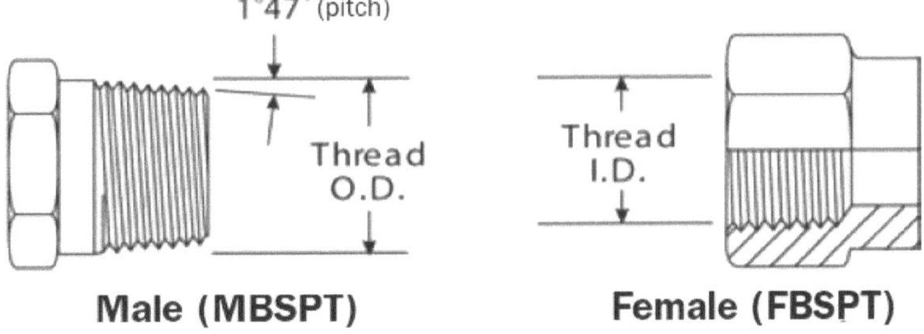

Fig. 1.30 - Japanese Tapered Pipe Thread (Courtesy of Gates)

1.5.5.5- O-Ring Sealing Fittings

North American Standards (SAE):

- **SAE-JI926 Straight Thread O-Ring Boss (ORB):** Figure 1.31 shows the *O-Ring Boss* male type. The male will only mate with an O-ring boss female, and the female is generally found on ports. The male has straight threads and an O-ring. The female has straight threads and a sealing face. The seal is made at the O-ring on the male and the sealing chamfer on the female. It can be configured as a straight or in elbow coupling to ease routing a line. Figure 1.32 shows typical O-Ring Boss fittings of different configurations.

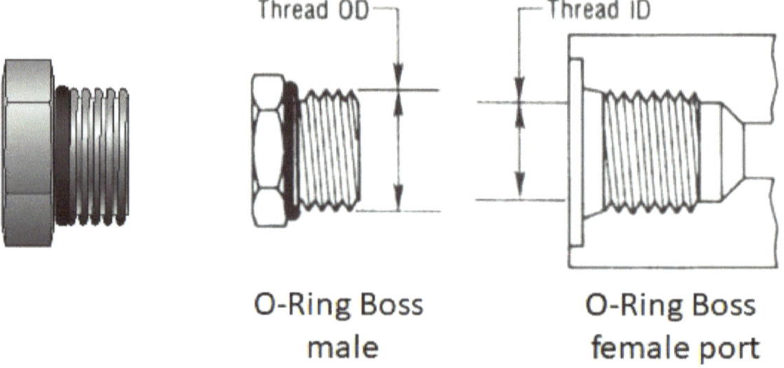

Fig. 1.31 - SAE JI926 Straight Thread O-Ring Boss (ORB)

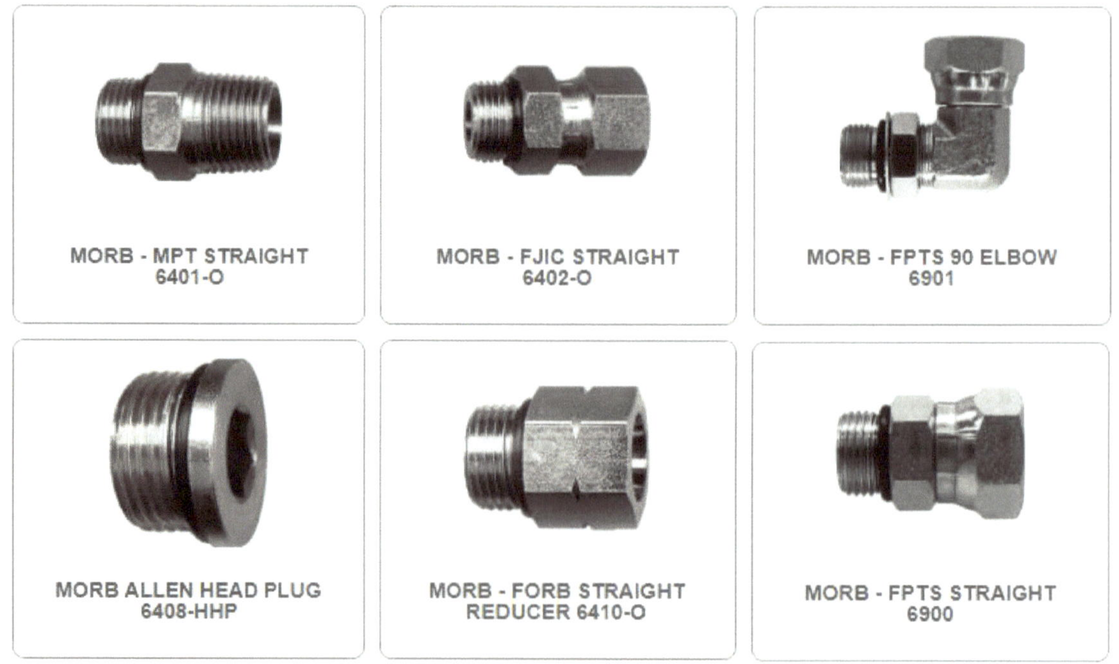

Fig. 1.32 - Typical O-Ring Boss Fittings of Different Configurations (www.redl.com)

ORB is recommended by the *National Fire Protection Association* (N.F.P.A.) for leak prevention in medium and high-pressure hydraulic systems. Figure 1.33 shows a sectional view of an ORB fitting.

Fig. 1.33 – Sectional View of an ORB Fitting (Curtesy of Brennan Industries)

- **O-Ring Face Seal (ORFS) – SAE-J1453 (Fig. 1.34):** *In the O-Ring Face Seal*, the O-Ring is placed in a captive groove on the face of the fitting. The solid male O-ring face seal will mate with a swivel female O-ring face seal SAE J1453 fitting only. An O-ring rests in the O-ring groove in the male coupling. The seal is made when the O-ring in the male contacts the flat face on the female coupling. The limitation on using such fittings comes from the maximum operating temperature under which the O-Ring performs properly. Figure 1.35 shows typical O-Ring Face Seal fittings of different configurations.

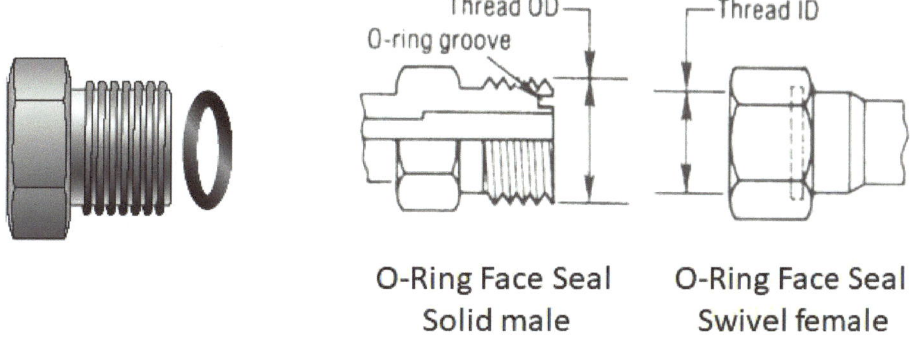

Fig. 1.34 - O-Ring Face Seal SAE J1453 (Courtesy of Gates)

**Fig. 1.35 - Typical O-Ring Face Seal Fittings of Different Configurations
(Courtesy of Parker)**

British Standards:
- Same as the SAE (ORB and ORFS) but the thread characteristic is different.

Metric Standard:
- ISO/DIS 8434-3 same as SAE (ORB and ORFS) but the thread characteristic is different.

Example of O-Ring Face Seal (O-Lok) (Fig. 1.36):

Description: *O-Lok* fittings are O-ring face seal (ORFS) fittings developed to eliminate leakage in high pressure hydraulic systems. The O-Lok system is designed to meet the needs of mobile equipment, mining, agriculture and other heavy equipment.

Features:
- Working pressure up to 630 bar.
- Available as standard in steel and stainless steel.
- Available in sizes from 1/4" to 2".
- Can be connected to both tubes and hoses.
- Can be used with tubes in different materials.
- Corrosion resistant.
- Easy to assemble.
- Best choice if you work with construction machinery.
- Meets international standards SAE J1453 and ISO8434-3.

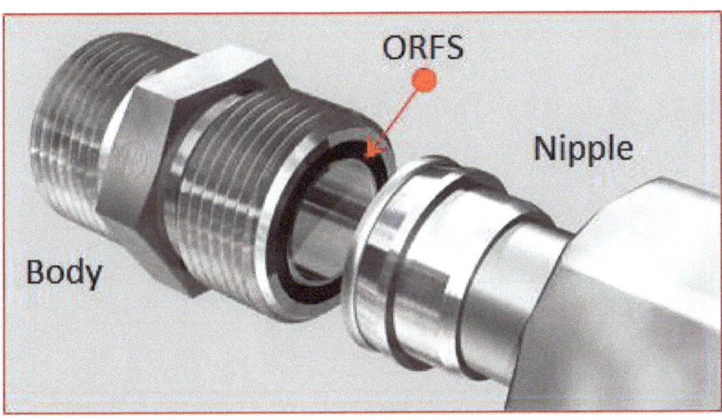

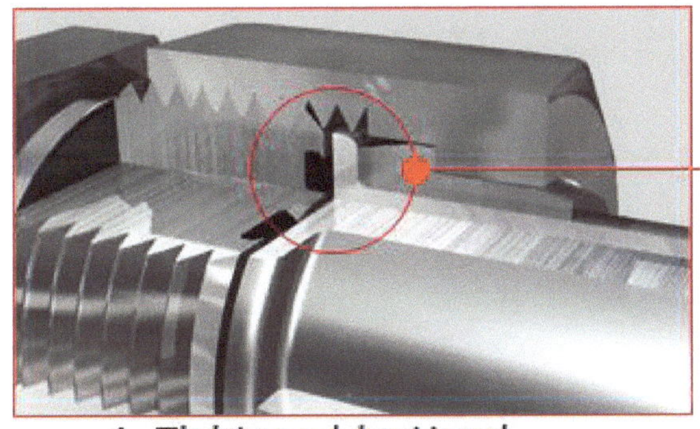
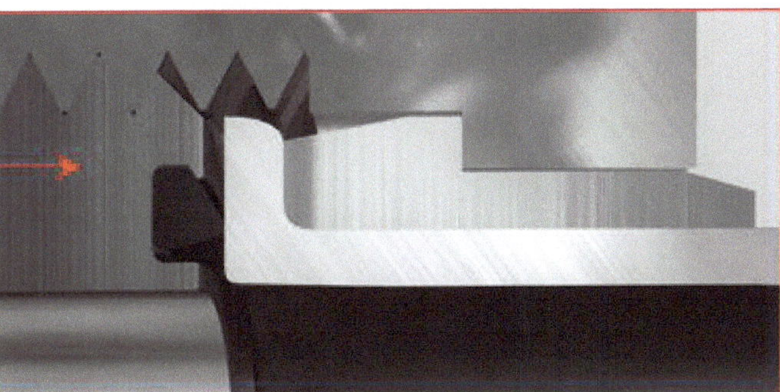

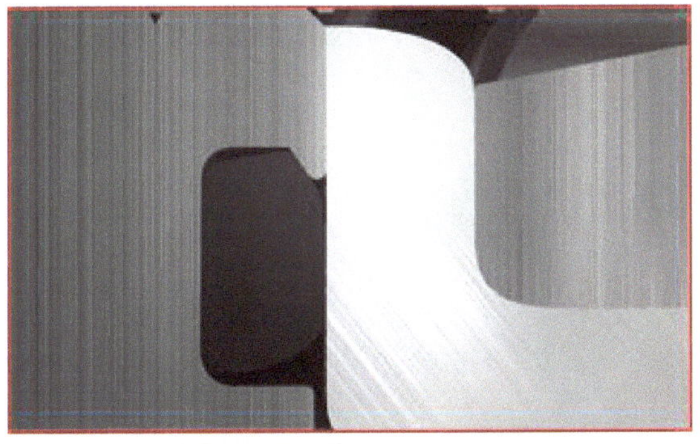
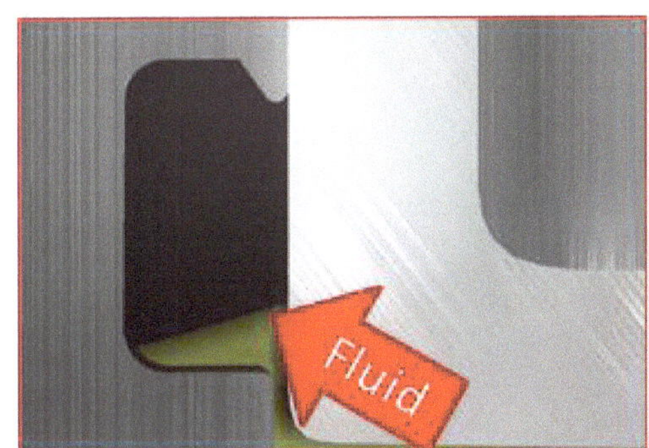

2- Tightened by Wrench
to Specified Torque

3- Fluid Pressure Applies

Fig. 1.36 - O-Lok Fitting (Courtesy of Parker)

1.5.5.6- Flareless Fittings with Cutting Rings

Flareless Tube Fittings (Fig. 1.37): This assembly is referred to as the *Standpipe* assembly. It could be North American (NASP) or Metric (MSP). This assembly consists of flareless solid male that mates only with a female flareless nut and cutting ring (compression sleeve). The male has straight threads and a 24° seat. The female has straight threads and a cutting ring for a sealing surface. <u>The seal is made between the cutting ring and the 24° seat on the male, and between the cutting ring and the tube on the female.</u> Figure 1.38 shows typical standpipe flareless assemblies.

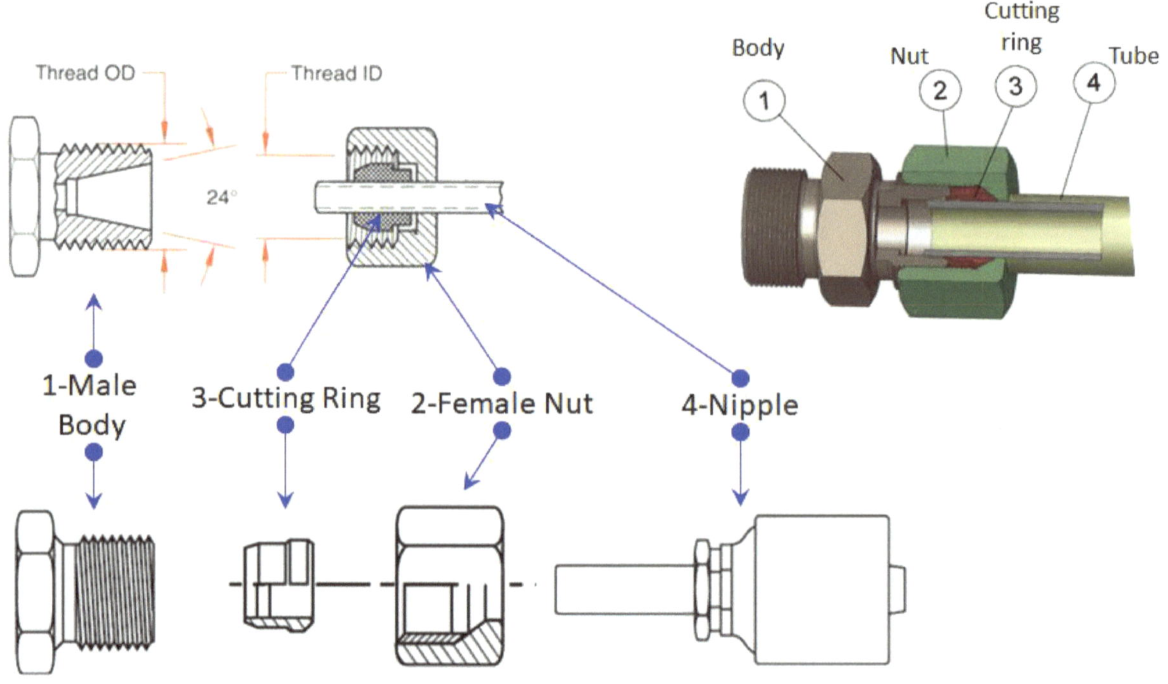

Fig. 1.37 - North American or Metric Flareless Assembly (Courtesy of Gates)

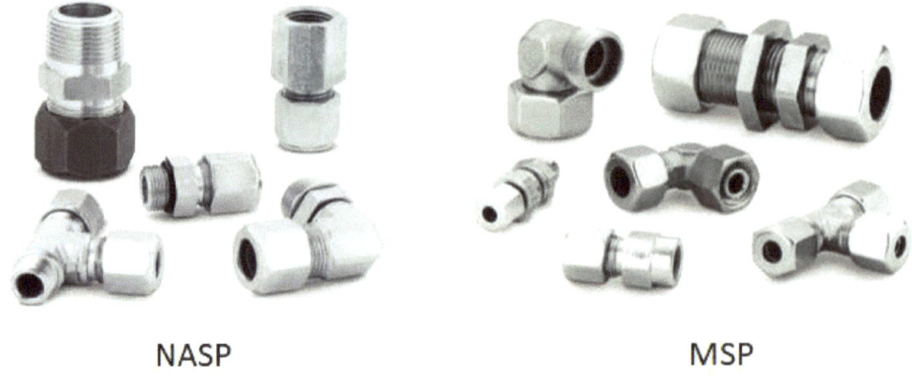

Fig. 1.38 - Typical North American or Metric Flareless Assembly (Courtesy of Parker)

How Cutting Rings Work (Fig. 1.39): As shown in the figure, after tightening to specified torque, the two cutting edges on the inner surface of the cutting ring carve into the outer surface of the tube, hence ensuring the necessary holding power and sealing for high operating pressures. Simultaneously, the outer surface of the cutting ring mates with the inner surface of the body.

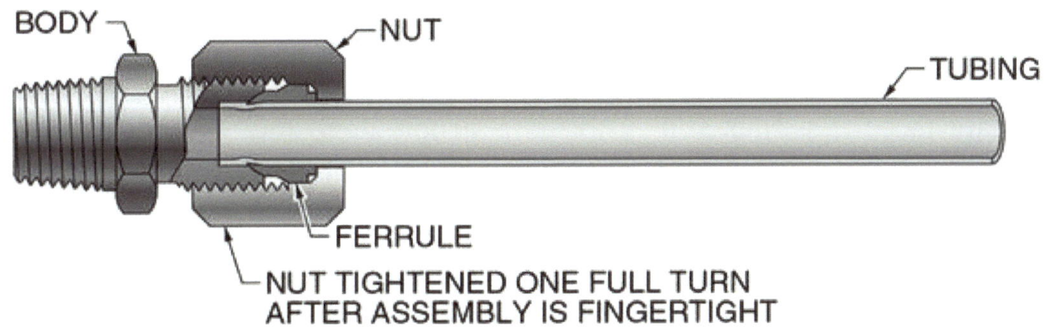

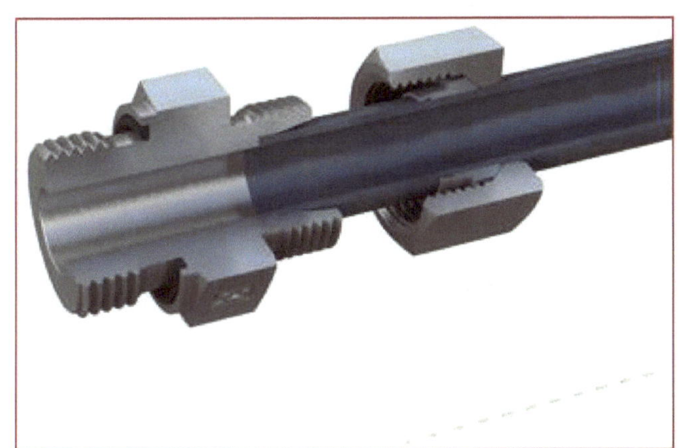

1- Untightened

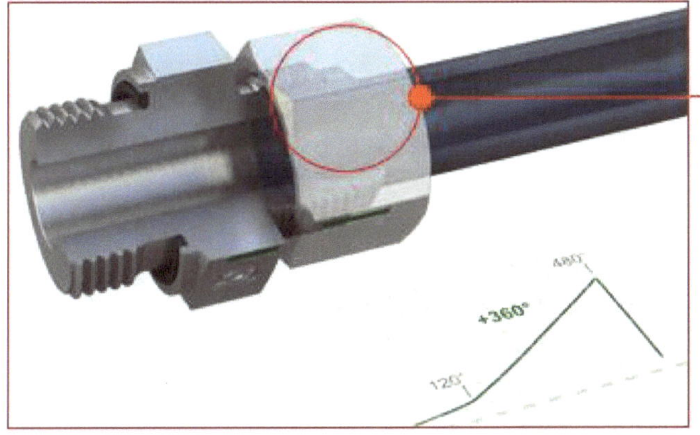

3- Tightened by Wrench to Specified Torque

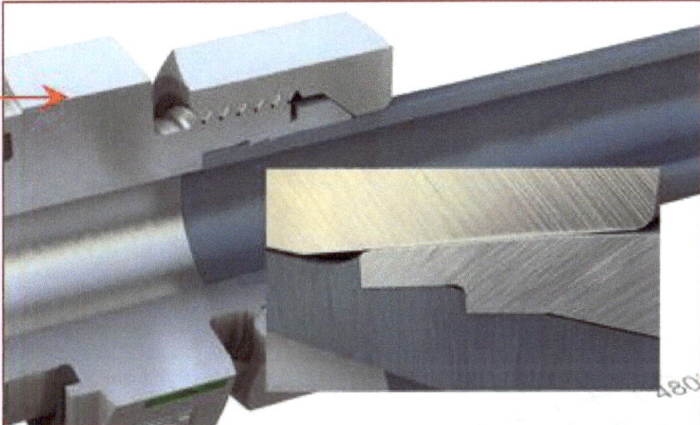

2- Tightened by Hand

Fig. 1.39 - How Cutting Rings Work (www.stauffusa.com)

1.5.5.7- Flared Fittings for Flared Tubes

<u>North American Standards:</u>

- **SAE-J514 37⁰ Flared Fittings (Fig. 1.40):** The *Joint Industrial Council* (JIC) is now defunct, and this standard is included as a part of SAE J5I4. Both male and female have straight threads and a 37⁰ flare seat. Seat angle must be carefully measured to differentiate between the two 37⁰ and 45⁰. Figure 1.41 shows typical JIC 37⁰ fittings. They are common in most fluid power systems. They are available in various material such as Nickel Alloys, Brass, Carbon Steel, and Stainless Steel. The absence of O-Ring makes these fittings recommended for high-temperature applications.

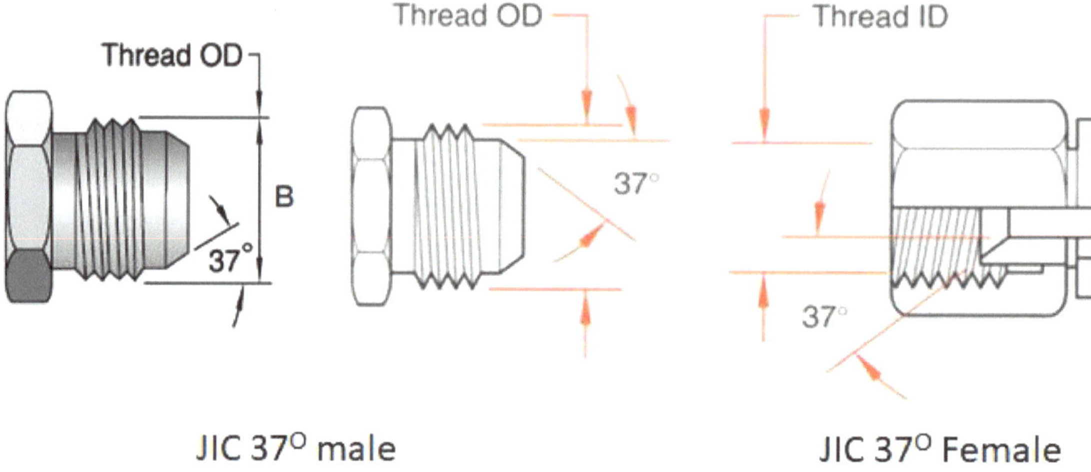

Fig. 1.40 – Characteristics of 37⁰ Flare Fitting (Courtesy of Gates)

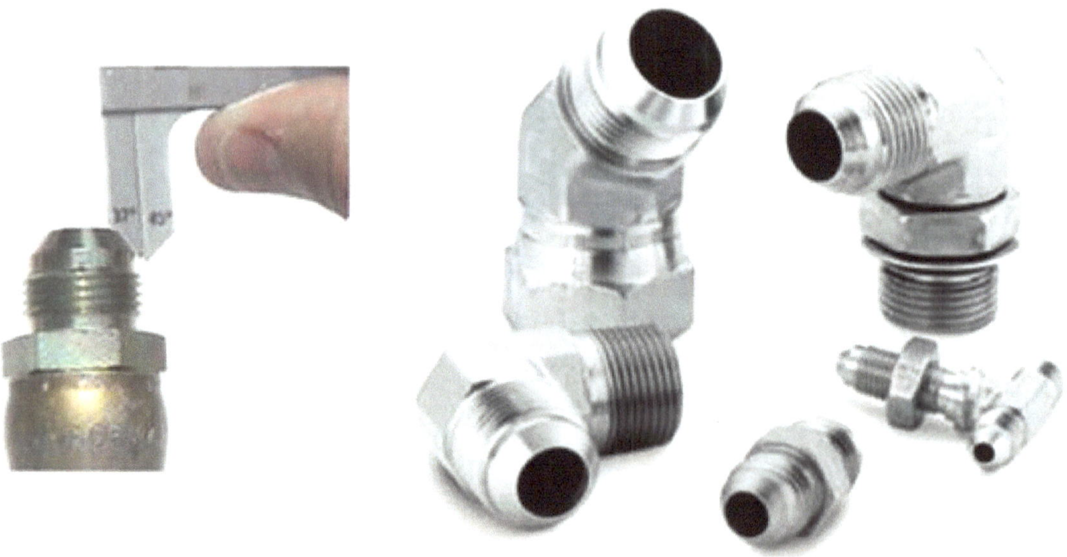

Fig. 1.41 – Typical 37⁰ Flared Fitting Configurations (Courtesy of Parker)

Figure 1.42 shows a sectional view of a sectional view of a 37⁰ flared fitting.

**Fig. 1.42 – Sectional View of a 37⁰ Flare Fitting Configurations
(Courtesy of Brennan Industries)**

Example of SAE-514 37⁰ Flare (Triple-Lok) (Fig. 1.43):

Description: *Triple-Lok* 37° flare fittings are the widely known and used.

Features as reported by manufacturer:
- Reliable and compact design. Easy to assembly.
- Meet the strict requirements of SAE-J514 and ISO 8434-2 industry standards.
- Adaptable to inch or metric tubes, worldwide available and accepted.
- Used for both tube and hose adapters.
- Working pressure up to 600 bar (9000 psi) depends on size.
- Available in sizes from 1/4" size to 2" size.

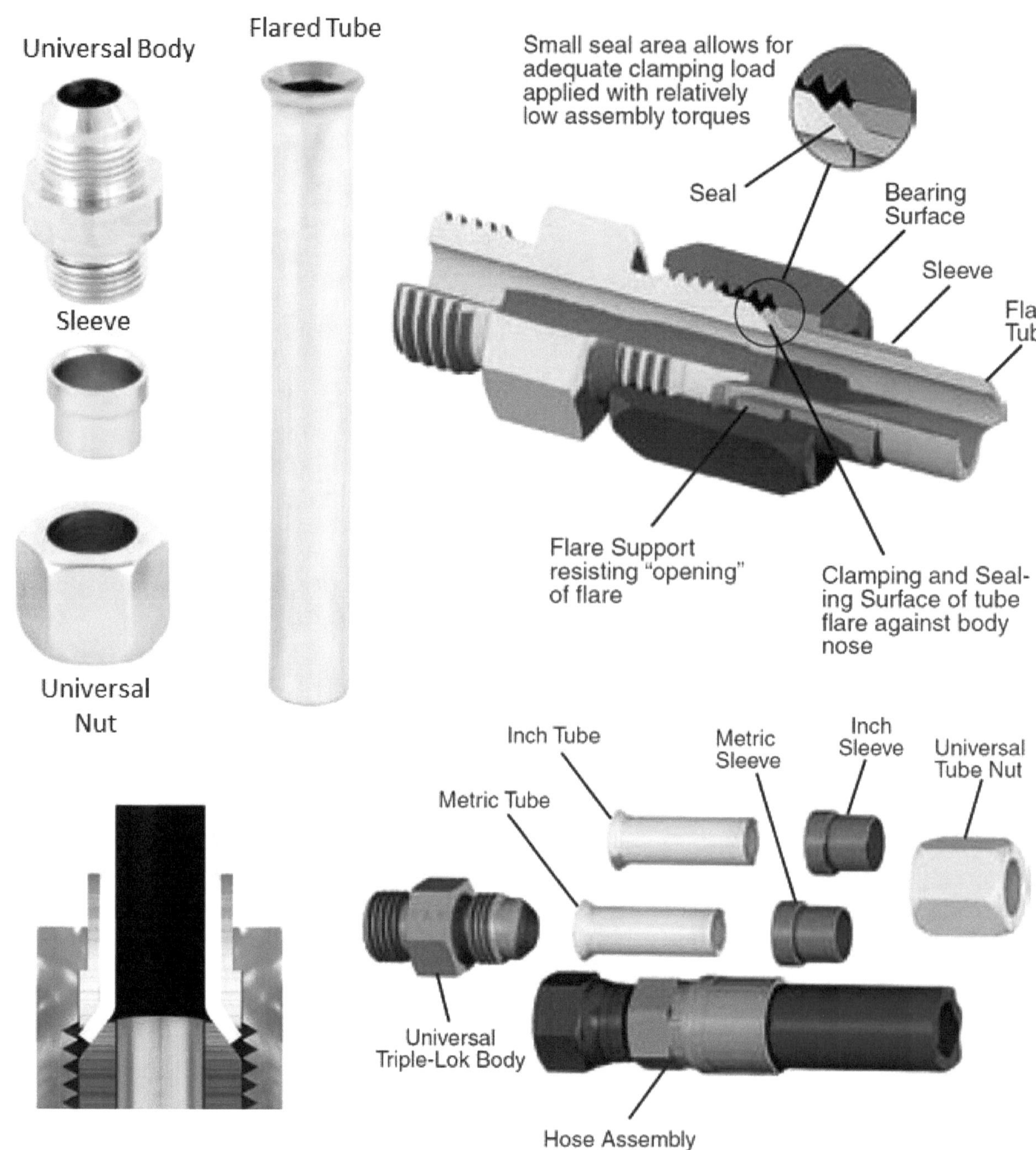

Fig. 1.43 - Triple-Lok Fitting (Courtesy of Parker)

- **SAE-J512 45⁰ Flare (Fig. 1.44):** The SAE 45° flare will only mate with a SAE 45° flare female. Both male and female couplings have straight threads and a 45° flare seat. The seal is made on the 45° flare seat. Once again, because some sizes of this coupling have the same threads as the JIC 37° flare, carefully measure the seat angle before use. Figure 1.45 shows typical JIC 45⁰ fittings. Such fittings are typically brass material and are not used in high-pressure hydraulic systems. They rather used in low-pressure fluid systems such as fuel systems, brakes, air conditioning systems, power steering, transmission, refrigerators, and oil coolers.

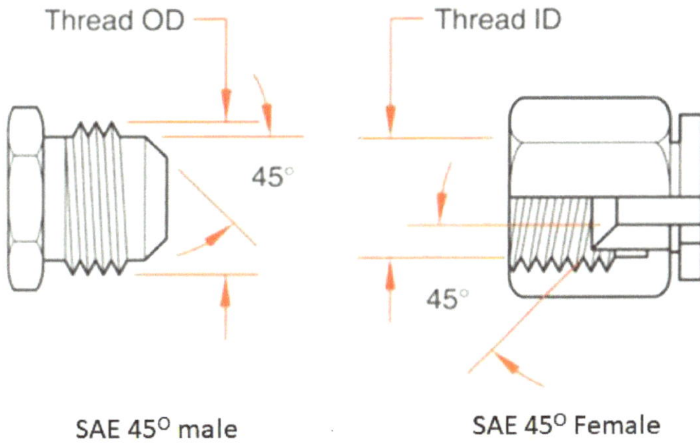

Fig. 1.44 - SAE 45⁰ Flare (Courtesy of Gates)

Fig. 1.45 - Typical JIC 45⁰ Flare (Courtesy of Parker)

Metric Standard:

- 037° flare fitting ISO 8434-2 ® 90° thread fitting with O-ring.

Japanese Standard:

- **Japanese 30° Flare Parallel Threads (Fig. 1.46):** The Japanese 30° flare male connector will mate with a Japanese 30° flare female only. The male and female have straight threads and a 30° seat. The seal is made on the 30° seat. The threads on the Japanese 30° flare connector conform to JIS B 0202, which are the same as the BSPOR threads. Both the British and Japanese connectors have a 30° seat, but they are not interchangeable because the British seat is inverted.

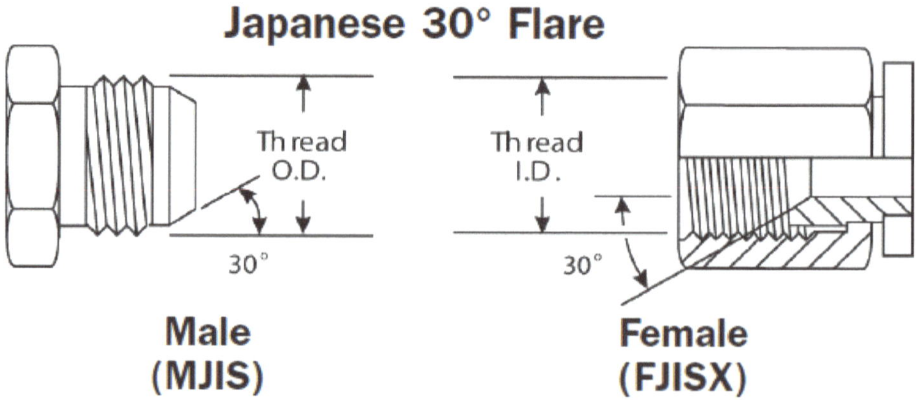

Fig. 1.46 - Japanese 30º Flare Thread (Courtesy of Gates)

1.5.5.8- Adaptors

To connect fittings from foreign (British or Metric) standards to a fitting from North American National Standard, *Adaptors* are used. Figure 1.47 shows examples of available adaptors:
1. Male Foreign Standard Pipe Tapered Thread to Male JIC 37° Flare.
2. Male Foreign Standard Pipe Parallel to Male JIC 37° Flare.
3. Male Foreign Standard Pipe Parallel to Male Pipe NPTF.
4. Male Foreign Standard Pipe Parallel with O-Ring Boss to Male JIC 37° Flare.
5. Male Foreign Standard Pipe Parallel with O-Ring Boss to Male Flat-Face O-Ring.
6. Female Foreign Standard Pipe Parallel to Male Pipe NPTF.

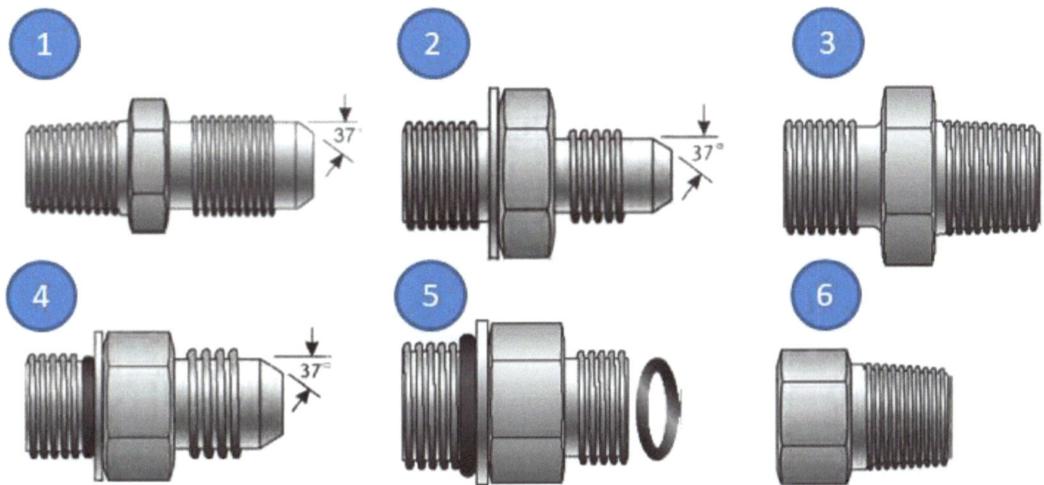

Fig. 1.47 - Adaptors to Connect Fittings from Foreign Standards to SAE Standards (Courtesy of Gates)

Figure 1.48 shows typical adapters.

Fig. 1.48 - Typical Adapters (Courtesy of Parker)

1.5.5.9- Thread Characteristics Identification

❖ **Identifying Fittings by Measurements (Fig. 1.49):**

Step 1- measure the Thread Diameter: use a combination O.D./I.D. *Caliper* to measure the thread diameter. It is to be noted that the threads of a used fitting can become worn and distorted, so the measurements may not be exact.

Step 2- Measure the Thread Pitch: use a *Thread Pitch Gauge* to identify the threads. Place the gauge on the threads until it fits snugly. For British and other European threads the thread pitch gauge measures the threads per inch. For metric connections, measure the distance between threads.

Step 3: Measure the Seat Angle: If the port is angled, determine the seat angle by using a *Seat Angle Gauge* on the sealing surface. The centerline of the fitting and the gauge must be parallel.

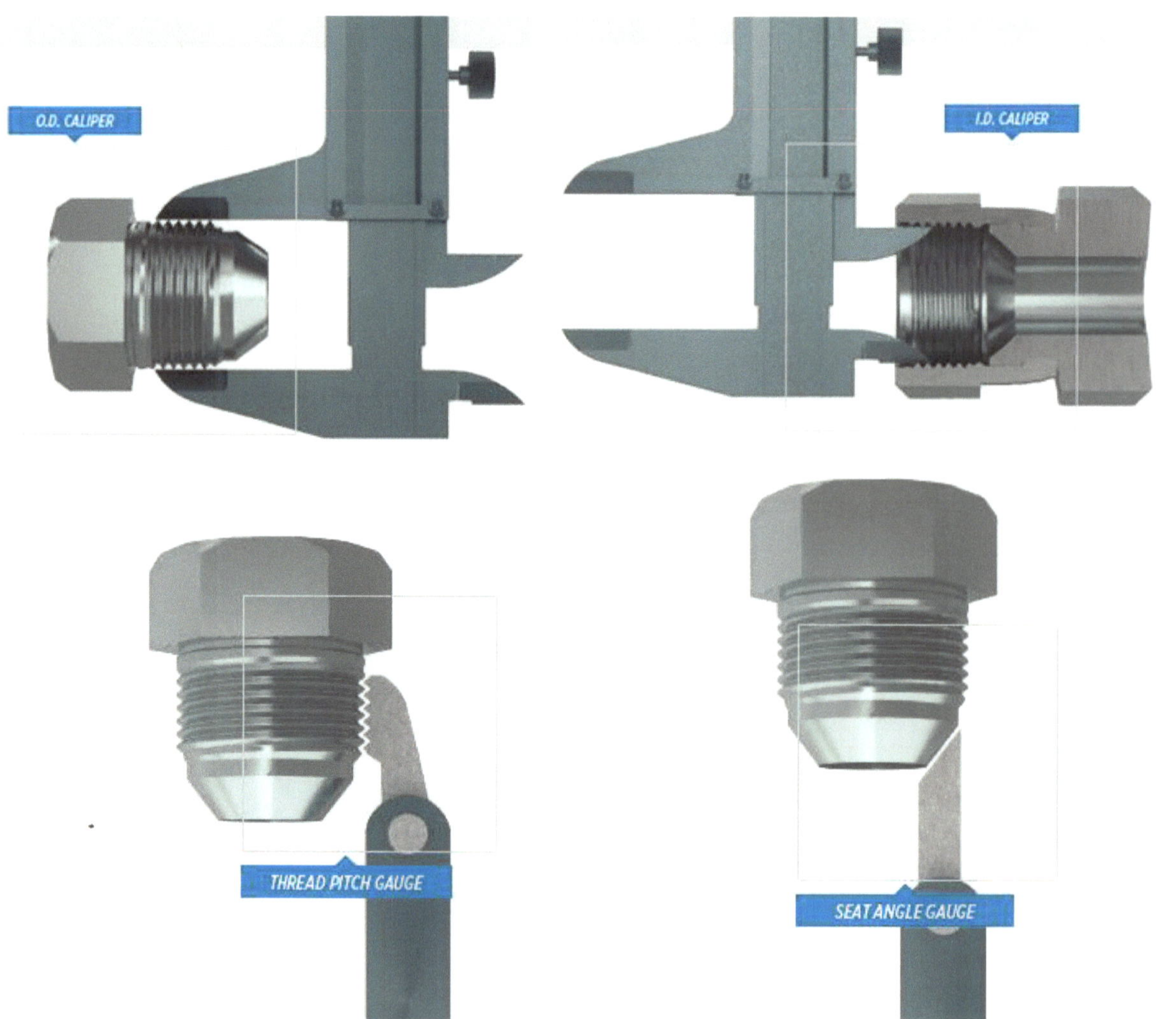

Fig. 1.49 – Identifying Fittings by Measurements (Courtesy of Brennan Industries)

❖ **Identifying Fittings by Thread ID Kit (Fig. 1.50):**

A *Thread ID Kit* is used for faster and easier identification of fittings. For ease of use, each fitting is color coded, and the size of the fitting is imprinted on the side. Each fitting has a female connector in one end and a male connector in the other end.

Fig. 1.50 – Identifying Fittings by Thread ID Kit (Courtesy of Brennan Industries)

1.5.5.10- Swivel Joints

Conventional Swivel Joints: In some systems, there might be a need for a component to rotate through an angular displacement, up to 360⁰ in some cases. In such systems, *Swivel Joints* are usable with hoses and rigid lines. As shown in Fig. 1.51, straight swivel joints provide rotation in one axis only, and elbow swivel joints provide rotation in two axes. The figure shows the construction of the swivel joint that guarantees rotary motion, sealing, and protection from introduced contamination.

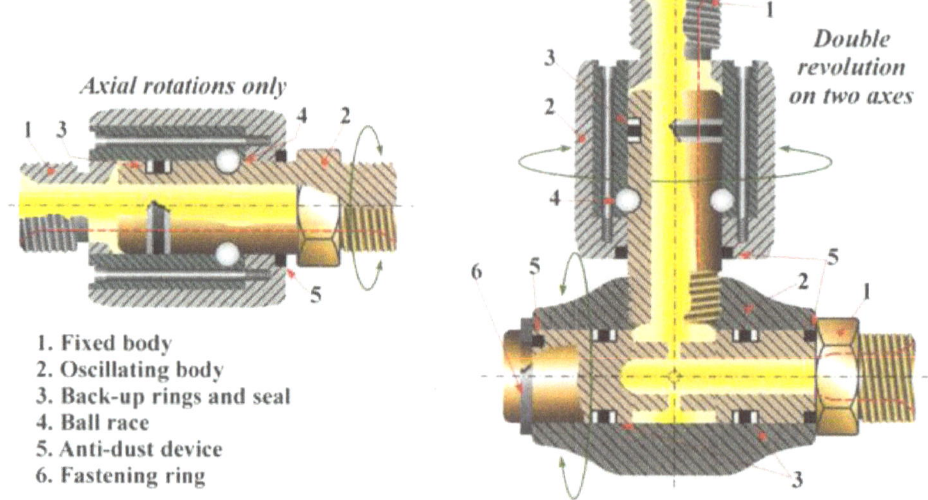

Fig. 1.51 - Construction of Conventional Swivel Joints (Courtesy of Assufluid)

Figure 1.52 shows various styles of swivel joints with the following technical characteristics:
- Pressure capabilities up to 350 bar (5000 psi).
- Variety of port options with size range is 1/4" through 2".
- Full flow design minimizes pressure drop for optimum system performance.
- Nickel plating and a wide range of seal options.
- Hardened bearing races for extended service life.
- Sealed bearing design isolates bearing race from media and environment.

Fig. 1.52 - Typical Conventional Swivel Joints (Courtesy of Parker)

Innovative Swivel Joints - WEO Plug-In hose fittings (Fig. 1.53):

<u>Description:</u> Over the time, more applications industry-wide are being converted from threaded to thread-less connectors, *WEO Plug-In* hose fittings are good choice for leak-free innovative swivel joints. As shown in the figure, such fittings are available in various configurations and sizes.

<u>Features:</u> WEO Plug-In quick swivel connection has the following features:
- Easy to install where no follow-up tightening needed.
- Minimum space requirement enables new system designs.
- Easy to service leak-free connection.
- Work injuries associated with connection/disconnection are eliminated.
- Longer hose life lower overall cost.

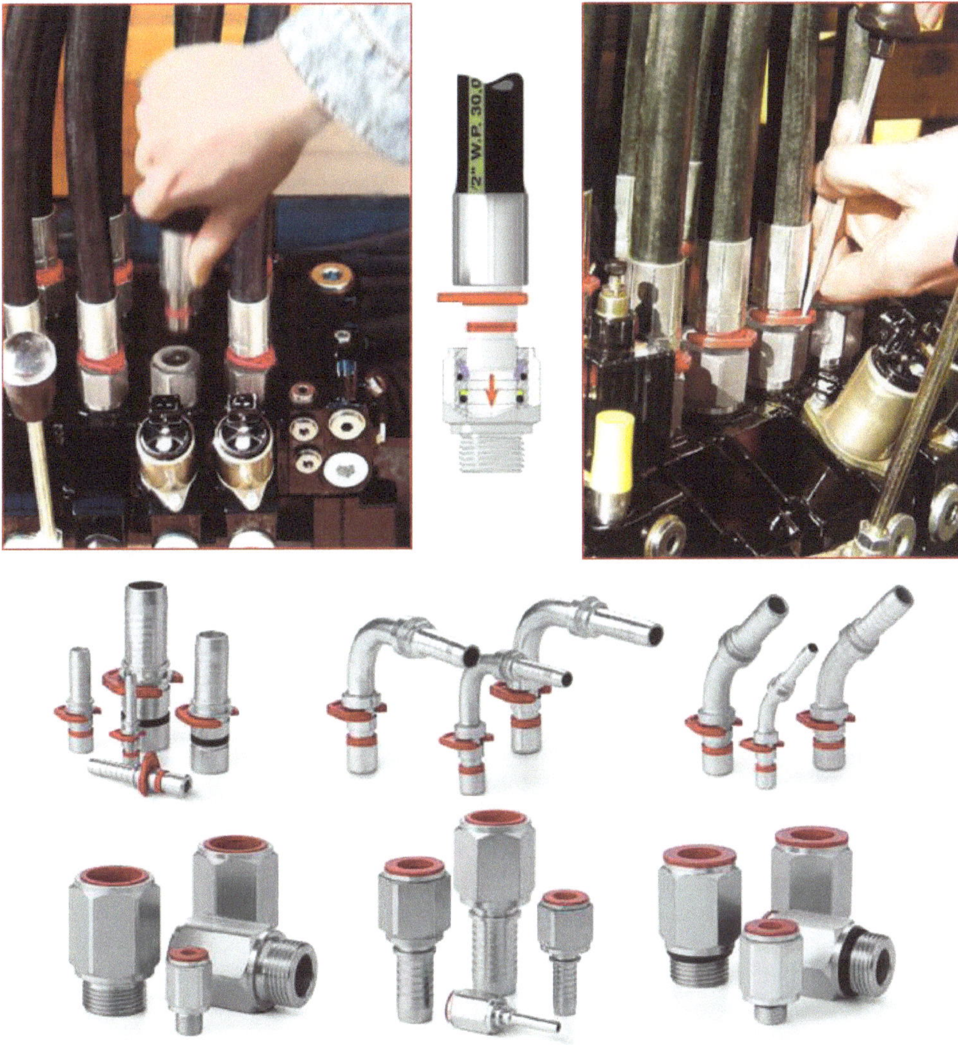

Fig. 1.53 - WEO Plug-In Swivel Joints (Courtesy of CEJN)

<u>Technical Characteristics</u>:
- <u>Working Pressure:</u>
  - Sizes (1/4 – 3/4) inch have a maximum working pressure of 350 bar (5075 psi).
  - Size 1-inch has a maximum working pressure of 250 bar (3625 psi).
- <u>Burst Pressure:</u> A minimum safety factor of 4:1. So, burst pressure is a minimum of four times the working pressure.
- <u>Temperature:</u> -30°C to +100°C (-22°F to +212°F).

<u>Construction:</u> Figure 1.54 shows the construction of the WEO Plug-in swivel joints.

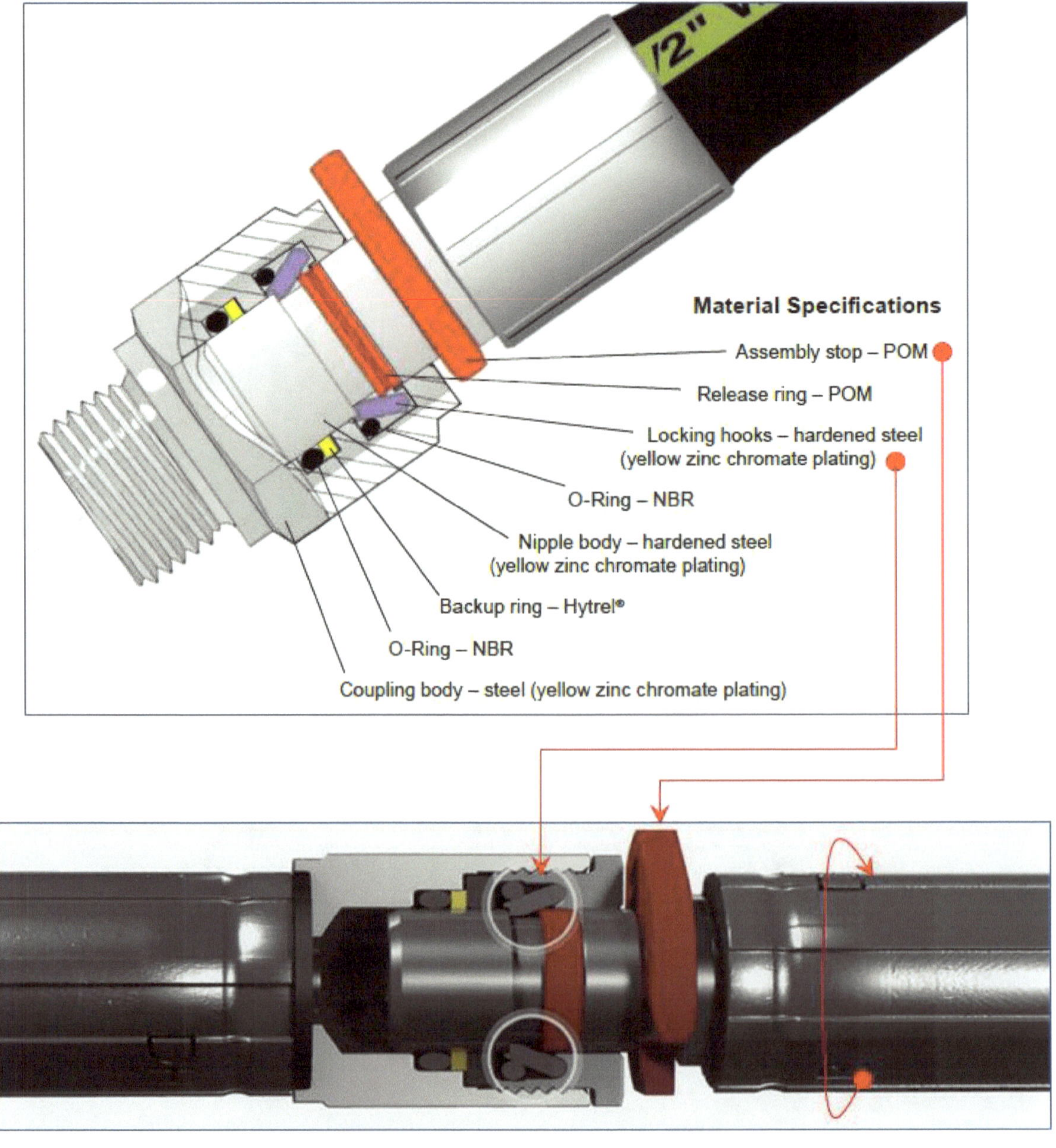

Fig. 1.54 - Construction of WEO Plug-In Swivel Joints (Courtesy of CEJN)

<u>Operation:</u> As shown in Fig. 1.55, WEO swivel joint fittings are designed with an innovative click-to-connect feature that provides easy engagement of the product's male and female halves, without the aid of tools or wrenches.

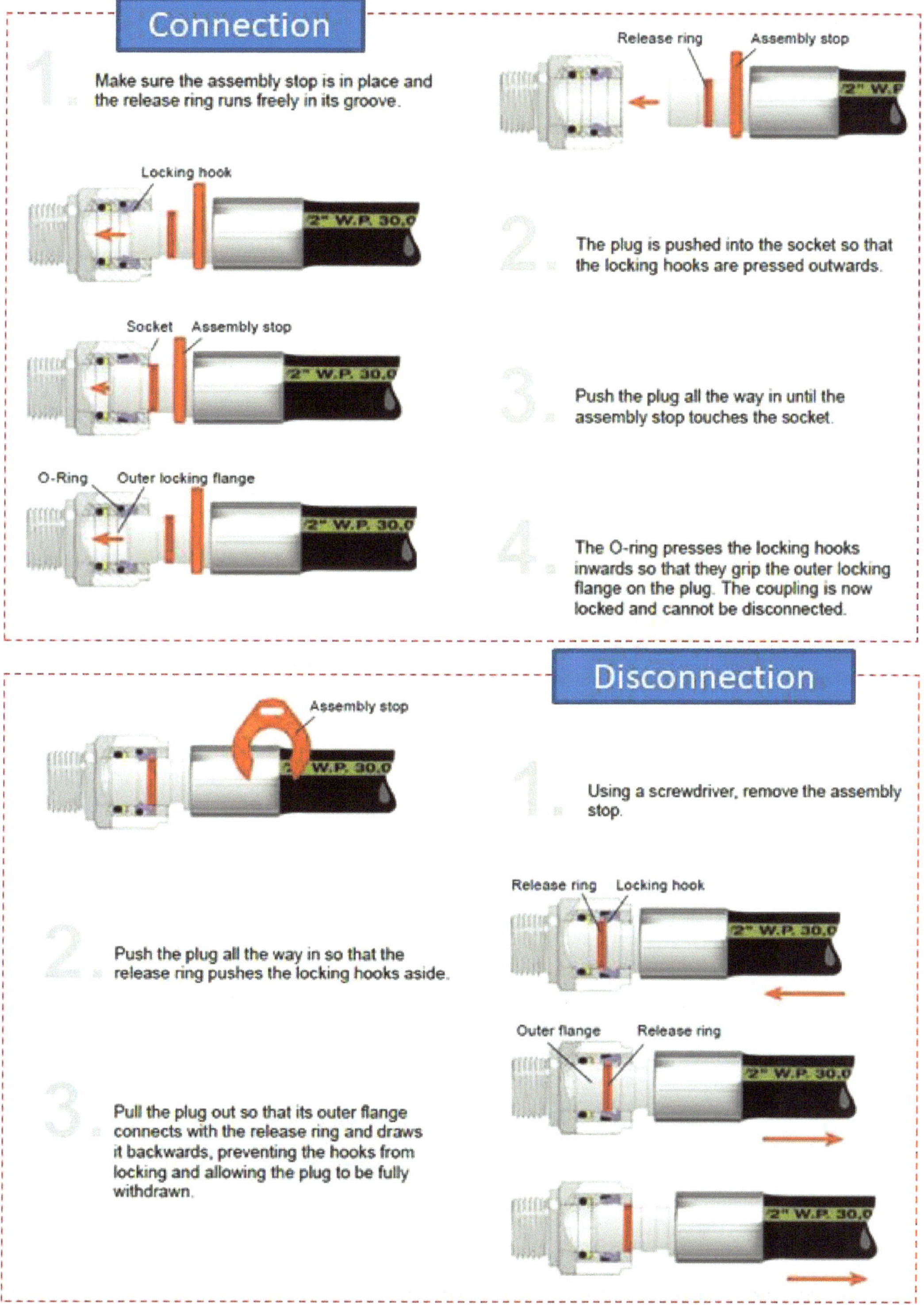

Fig. 1.55 - Operation of WEO Plug-In Swivel Joints (Courtesy of CEJN)

1.6- Hydraulic Hoses

1.6.1- Features of Hydraulic Hoses

Hydraulic hoses have the following features:

- Easy installation and flexible routing.
- Permit connections to components on moving parts.
- Permit misalignment between components.
- Require compatibility with system hydraulic fluid.
- Low equivalent bulk modulus due to the flexible walls.
- Isolate noise, shocks and vibrations.
- Reduce hydraulic stiffness and consequently system response.
- Usually more expensive than tubes or pipes.

Countless number of applications both in mobile and industrial side rely on hydraulic hoses. Figure 1.56 shows hydraulic hoses connecting hydraulic cylinders to the hydraulic system. The lower photo shows hydraulic hoses in stationary hydraulic power unit.

Fig. 1.56 - Examples of using Hydraulic Hoses in Mobile and Industrial Applications

1.6.2- Material and Construction of Hydraulic Hoses

As shown in Fig. 1.57, hydraulic hoses are generally constructed from the following main three parts:

Central Tube: The *Central Tube* is a seamless tube of synthetic rubber or plastic that must be compatible with the hydraulic fluid. This layer works as leakage barrier and must withstand the operating temperature range (low to high).

Reinforcement Layers: *Reinforcement Layers* consist of braided or spiraled textile strings or metal wires. The number of reinforcement layers is proportional to the maximum allowable pressure and inversely proportional to the hose flexibility. Figure 1.58 shows braded versus spiraled reinforcement layers. Hoses with braded reinforcement layers are for low to medium pressures. Hoses with spiraled reinforcement layers are for medium to high pressure. Braided hose is generally more flexible than spiral hose.

Outer Protective Cover: The *Outer Protective Cover* is to protect the hose from environmental conditions, hose abrasion, etc. Various materials are used such as plastic, cloth, etc. Hoses for special applications are covered by abrasion resistant covers.

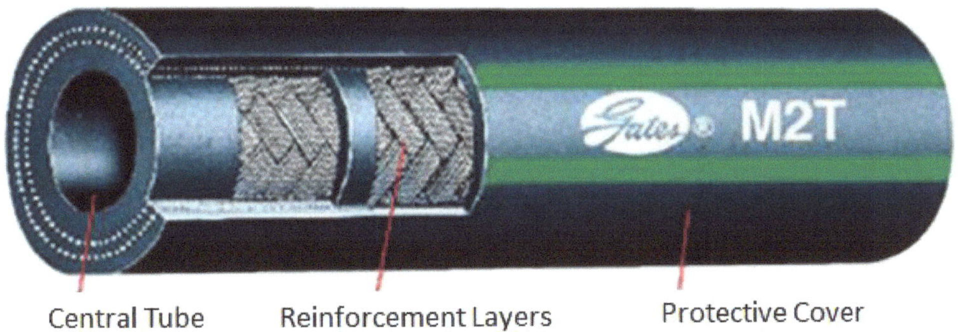

Fig. 1.57 - Basic Construction of Hydraulic Hoes (Courtesy of Gates)

Fig. 1.58 - Spiraled versus Braded Reinforcement Layers in Hydraulic Hoses

1.6.3- Sizes of Hydraulic Hoses

As shown in Fig. 1.59, unlike hydraulic pipes and tubes, nominal size of hydraulic hose is based on the inner diameter "ID". That is because the outer diameter depends on the hose structure and its pressure rating. As shown in Table 1.9, hoses have size range (3/16 – 4) inches (4 – 100) mm. The table shows that the dimension of hydraulic hoses is designated in 16th of an inch using the (-#) format. An example is -8 means 8/16, i.e. half inch.

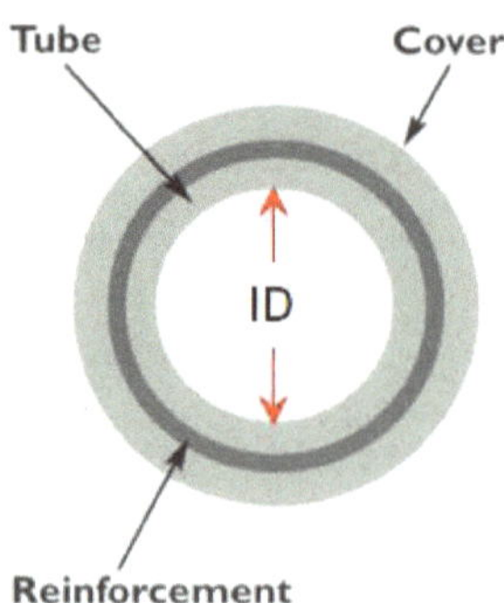

Fig. 1.59 - Nominal Size of Hydraulic Hoses (Courtesy of Gates)

Dash No.	Hose I.D.	
	Inches	Millimeters
-3	3/16	4.8
-4	1/4	6.4
-5	5/16	7.9
-6	3/8	9.5
-8	1/2	12.7
-10	5/8	15.9
-12	3/4	19.0
-14	7/8	22.2
-16	1	25.4
-20	1-1/4	31.8
-24	1-1/2	38.1
-32	2	50.8
-40	2-1/2	63.5
-48	3	76.2
-56	3-1/2	88.9
-64	4	101.6

Table 1.9- Size Range of Hydraulic Hoses (Courtesy of Gates)

1.6.4- Pressure Rating of Hydraulic Hoses

Hydraulic hose construction and performance is covered by various national and international standards including ISO, SAE, DIN, etc. The SAE standards being the most widely followed in the U.S. SAE standards provide general, dimensional, and performance specifications for the most common hoses used in hydraulic systems in mobile and stationary equipment.

SAE J517 Standard: It provides guidelines for a series of hydraulic hoses. These manufacturer-driven SAE standards have been based on design, construction, and pressure ratings to ensure that hydraulic hoses meet minimum construction and performance requirements.

Designations of hydraulic hoses follows **SAE-J517 (May 1989)** industry standard. An example of a hydraulic hose designation based on SAE standard is as **SAE100R2AT-16.** The interpretation of this designation is as follows:

- **SAE100RX:** This part describes basic construction and pressure rating.
- **A or B:** A is for braided wire and B is for spiral wire reinforcement layers.
- **T:** Appears only if the cover does not need to be removed to attach the fitting.
- **Dash number:** is the inside diameter in sixteenths of an inch. (e.g. -16 means ID = 1 in).

Table 1.10 presents SAE hose data. As shown in the table, a safety factor of 4 is used to calculate the maximum allowable working pressure as function of the burst pressure. The table also shows that, for a specific nominal size, with the increase of the maximum operating pressure, both outer diameter and minimum bend radius increases. The following examples show interpretation for the presented data.

Referring to Table 1.10:
- **SAE 100R1A:** A hose with single layer of braided wire.
- **SAE 100R2A:** A hose with double layers of braided wire.
- **SAE 100R2B:** Two spiral plies and one braid of wire.
- **SAE 100R4:** Wire Inserted Hydraulic Suction Hose.
- **SAE 100R5:** Single Wire Braid, Textile Covered Hydraulic Hose.
- **SAE 100R6:** Single Fiber Braid, (Nonmetallic), Rubber Covered Hydraulic Hose.
- **SAE 100R7:** Thermoplastic Hydraulic Hose.
- **SAE 100R8:** High Pressure Thermoplastic Hydraulic Hose.
- **SAE 100R9:** High Pressure, 4-Spiral Steel Wire, Rubber Covered Hydraulic Hose.
- **SAE 100R10:** Heavy Duty, 4-Spiral Steel Wire, Rubber Covered Hydraulic Hose.
- **SAE 100R11:** Heavy Duty, 6-Spiral Steel Wire Rubber Covered Hydraulic Hose.
- **SAE lOOR12:** Heavy Duty, High Impulse, 4-Spiral Wire, Rubber Cover Hydraulic Hose.
- **SAE 100R13:** Heavy Duty, High Impulse, Multiple Spiral Wire, Rubber Covered Hose.
- **SAE lOOR14:** This specification covers hose for use with petroleum, synthetic, and water base hydraulic fluids.

[3]Minimum burst pressure is 4 times maximum operating pressure.

I.D. INCHES	DASH NO. REF.	SAE NO. & TYPE SPEC.	O.D. MAX. INCHES	MIN. BEND (INTERNAL) RAD. IN. AT MAX. OPERATING PRESSURE	[1]MAX. OPERATING PRESSURE PSIG
1/8	-3	100R14	0.268	1.5	1,500
3/16	-3	100R1-A	0.531	3.5	3,000
3/16	-3	100R1-AT	0.494	3.5	3,000
3/16	-3	100R2-A&B	0.656	3.5	5,000
3/16	-3	100R2-AT&BT	0.557	3.5	5,000
3/16	-3	100R3	0.531	3	1,500
3/16	-4	100R5	0.539	3	3,000
3/16	-3	100R6	0.469	2	500
3/16	-3	100R7	0.450	3.5	3,000
3/16	-3	100R8	0.575	3.5	5,000
3/16	-3	100R10-A	0.781	4	10,000
3/16	-3	100R11	0.906	4	12,500
3/16	-4	100R14	0.324	2	1,500
1/4	-4	100R1-A	0.656	4	2,750
1/4	-4	100R1-AT	0.557	4	2,750
1/4	-4	100R2-A&B	0.719	4	5,000
1/4	-4	100R2-AT&BT	0.619	4	5,000
1/4	-4	100R3	0.594	3	1,250
1/4	-5	100R5	0.601	3.38	3,000
1/4	-4	100R6	0.531	2.5	400
1/4	-4	100R7	0.538	4	2,750
1/4	-4	100R8	0.660	4	5,000
1/4	-4	100R10-A	0.844	5	8,750
1/4	-4	100R11	0.969	5	11,250
1/4	-5	100R14	0.397	3	1,500
5/16	-5	100R1-A	0.719	4.5	2,500
5/16	-5	100R1-AT	0.619	4.5	2,500
5/16	-5	100R2-A&B	0.781	4.5	4,250
5/16	-5	100R2-AT&BT	0.682	4.5	4,250
5/16	-5	100R3	0.719	4	1,200
5/16	-6	100R5	0.695	4	2,250
5/16	-5	100R6	0.594	3	400
5/16	-5	100R7	0.615	4.5	2,500
5/16	-6	100R14	0.458	4	1,500
3/8	-6	100R1-A	0.812	5	2,250
3/8	-6	100R1-AT	0.713	5	2,250
3/8	-6	100R2-A&B	0.875	5	4,000
3/8	-6	100R2-AT&BT	0.777	5	4,000
3/8	-6	100R3	0.781	4	1,125
3/8	-6	100R6	0.656	3	400
3/8	-6	100R7	0.725	5	2,250
3/8	-6	100R8	0.800	5	4,000
3/8	-6	100R9-A	0.875	5	4,500
3/8	-6	100R9-AT	0.831	5	4,500
3/8	-6	100R10-A	0.969	6	7,500
3/8	-6	100R11	1.094	6	10,000
3/8	-6	100R12	0.828	5	4,000
3/8	-7	100R14	0.526	5	1,500
13/32	-6.5	100R1-A	0.844	5.5	2,250
13/32	-6.5	100R1-AT	0.744	5.5	2,250
13/32	-8	100R5	0.789	4.62	2,000
13/32	-8	100R14	0.562	5.25	1,000
1/2	-8	100R1-A	0.938	7	2,000
1/2	-8	100R1-AT	0.846	7	2,000
1/2	-8	100R2-A&B	1.000	7	3,500
1/2	-8	100R2-AT&BT	0.908	7	3,500
1/2	-8	100R3	0.969	5	1,000
1/2	-10	100R5	0.945	5.5	1,750
1/2	-8	100R6	0.812	4	400
1/2	-8	100R7	0.885	7	2,000
1/2	-8	100R8	0.970	7	3,500
1/2	-8	100R9-A	1.000	7	4,000
1/2	-8	100R9-AT	0.958	7	4,000
1/2	-8	100R10-A	1.125	8	6,250
1/2	-8	100R11	1.250	8	7,500
1/2	-8	100R12	0.966	7	4,000
1/2	-10	100R14	0.663	6.5	800
5/8	-10	100R1-A	1.062	8	1,500
5/8	-10	100R1-AT	0.971	8	1,500
5/8	-10	100R2-A&B	1.125	8	2,750
5/8	-10	100R2-AT&BT	1.034	8	2,750
5/8	-10	100R3	1.094	5.5	875
5/8	-12	100R5	1.101	6.5	1,500
5/8	-10	100R6	0.938	5	350
5/8	-10	100R7	1.015	8	1,500
5/8	-10	100R8	1.175	8	2,750
5/8	-10	100R12	1.111	8	4,000
5/8	-12	100R14	0.793	7.75	800
3/4	-12	100R1-A	1.219	9.5	1,250
3/4	-12	100R1-AT	1.127	9.5	1,250
3/4	-12	100R2-A&B	1.281	9.5	2,250
3/4	-12	100R2-AT&BT	1.190	9.5	2,250
3/4	-12	100R3	1.281	6	750
3/4	-12	100R4	1.375	5	300
3/4	-12	100R6	1.000	6	300
3/4	-12	100R7	1.125	9.5	1,250
3/4	-12	100R8	1.300	9.5	2,250
3/4	-12	100R9-A	1.266	9.5	3,000
3/4	-12	100R9-AT	1.255	9.5	3,000
3/4	-12	100R10-A	1.469	11	5,000
3/4	-12	100R10-AT	1.450	11	5,000
3/4	-12	100R11	1.594	11	6,250
3/4	-12	100R12	1.241	9.5	4,000
3/4	-12	100R13	1.306	9.5	5,000
3/4	-14	100R14	0.917	9	800
7/8	-14	100R1-A	1.344	11	1,125
7/8	-14	100R1-AT	1.252	11	1,125
7/8	-14	100R2-A&B	1.406	11	2,000
7/8	-14	100R2-AT&AB	1.315	11	2,000
7/8	-16	100R5	1.266	7.38	800
7/8	-16	100R14	1.061	9	800
1	-16	100R1-A	1.547	12	1,000
1	-16	100R1-AT	1.440	12	1,000
1	-16	100R2-A&B	1.609	12	2,000
1	-16	100R2-AT&BT	1.531	12	2,000
1	-16	100R3	1.547	8	565
1	-16	100R4	1.625	6	250
1	-16	100R7	1.445	12	1,000
1	-16	100R8	1.520	12	2,000
1	-16	100R9-A	1.609	12	3,000
1	-16	100R9-AT	1.594	12	3,000
1	-16	100R10-A	1.797	14	4,000
1	-16	100R10-AT	1.790	14	4,000
1	-16	100R11	1.953	14	5,000
1	-16	100R12	1.542	12	4,000
1	-16	100R13	1.567	12	5,000
1	-18	100R14	1.175	12	800
1 1/8	-20	100R5	1.531	9	625
1 1/8	-20	100R14	1.320	16	600
1 1/4	-20	100R1-A	1.875	16.5	625
1 1/4	-20	100R1-AT	1.766	16.5	625
1 1/4	-20	100R2-A&B	2.062	16.5	1,625
1 1/4	-20	100R2-AT&BT	1.953	16.5	1,625
1 1/4	-20	100R3	1.812	10	375
1 1/4	-20	100R4	2.000	8	200
1 1/4	-20	100R9-A	2.062	16.5	2,500
1 1/4	-20	100R9-AT	1.997	16.5	2,500
1 1/4	-20	100R10-A	2.062	18	3,000
1 1/4	-20	100R10-AT	2.060	18	3,000
1 1/4	-20	100R11	2.219	18	3,500
1 1/4	-20	100R12	1.912	16.5	3,000
1 1/4	-20	100R13	2.021	16.5	5,000
1 3/8	-24	100R5	1.781	10.5	500
1 1/2	-24	100R1-A	2.125	20	500
1 1/2	-24	100R1-AT	2.047	20	500
1 1/2	-24	100R2-A&B	2.312	20	1,250
1 1/2	-24	100R2-AT&BT	2.203	20	1,250
1 1/2	-24	100R4	2.250	10	150
1 1/2	-24	100R9-A	2.312	20	2,000
1 1/2	-24	100R10-A	2.312	22	2,500
1 1/2	-24	100R10-AT	2.310	22	2500
1 1/2	-24	100R11	2.469	22	3,000
1 1/2	-24	100R12	2.167	20	2,500
1 1/2	-24	100R13	2.315	20	5,000
1 13/16	-32	100R5	2.266	13.25	350
2	-32	100R1-A	2.688	25	375
2	-32	100R1-AT	2.594	25	375
2	-32	100R2-A&B	2.812	25	1,125
2	-32	100R2-AT&BT	2.703	25	1,125
2	-32	100R4	2.750	12	100
2	-32	100R9-A	2.875	26	2,000
2	-32	100R10-A	2.844	28	2,500
2	-32	100R10-AT	2.840	28	2,500
2	-32	100R11	3.031	28	3,000
2	-32	100R12	2.688	25	2,500
2	-32	100R13	2.862	25	5,000
2 3/8	-40	100R5	2.922	24	350
2 1/2	-40	100R2-A&B	3.312	30	1000
2 1/2	-40	100R11	3.656	36	2500
2 1/2	-40	100R4	3.250	14	62
3	-48	100R4	3.750	18	56
3	-48	100R5	3.609	33	200
3 1/2	-56	100R4	4.250	21	45
4	-64	100R4	4.750	24	35

Table 1.10 – Hoses Specifications in accordance with SAE Standard

Table 1.11 presents more data about fluid compatibility and temperature range in accordance with the SAE standard.

SAE DIMENSIONAL AND PERFORMANCE STANDARDS FOR HYDRAULIC HOSE							
SAE standard hydraulic hose type/application	Compatible hydraulic fluids	Temperature range, °F	ID range, in.	Maximum operating pressure, psi	Proof pressure range, psi	Minimum burst pressure range, psi	Minimum bend radius, in.
100R1—Steel wire reinforced, rubber coated	Petroleum & water based	-40 to 212	$3/16$ to 2	575 to 3,250	1,150 to 6,500	2,300 to 13,000	3.5 to 25
100R2—High-pressure steel wire, reinforced rubber cover	Petroleum & water based	-40 to 212	$3/16$ to 2	1,150 to 6,000	2250 to 12,000	4,500 to 24,000	3.5 to 25
100R3—Double fiber, braid rubber cover – High-temp, low-pressure	Petroleum & water based	-40 to 212	$3/16$ to 1-1	375 to 1,500	750 to 3,000	1,500 to 6,000	3 to 10
100R4—Wire inserted, hydraulic suction and return	Petroleum & water based	-40 to 212	3 to 4	35 to 300	70 to 600	140 to 1,200	5 to 24
100R5—Single wire braid, textile cover; Transportation/DOT	Petroleum & water based	-40 to 212	$3/16$ to 3-$1/16$	200 to 3,000	400 to 6,000	800 to 12,000	3 to 33
100R6—Single fiber braid, rubber cover–Transportation	Petroleum & water based	-40 to 212	$3/16$ to 3	300 to 500	600 to 1,000	1,200 to 2,000	2 to 6
100R7—Single fiber braid, thermoplastic–Hydraulic	Petroleum, water based & synthetic	-40 to +212	$1/8$ to 1	1,000 to 3,000	2,000 to 6,000	4,000 to 12,000	1 to 12
100R8—High pressure, thermoplastic-Hydraulic	Petroleum, water based & synthetic	-40 to 212	$1/8$ to 1	2,000 to 6,000	4,000 to 12,000	8,000 to 24,000	1 to 12
100R9	No longer part of the SAE standard.						
100R10	No longer part of the SAE standard.						
100R11	No longer part of the SAE standard.						
100R12—Heavy duty, high impulse, four-spiral wire reinforced, rubber cover–Hydraulic	Petroleum & water based	-40 to 250	$3/8$ to 2	2,500 to 4,000	5,000 to 8,000	10,000 to 16,000	5 to 25
100R13—Heavy duty, high impulse, 4- & 6-spiral steel wire reinforced, rubber cover; Hydraulic	Petroleum & water based	-40 to 250	3 to 2	5,000	10,000	20,000	9.5 to 25
100R14—High temperature, corrosive fluids, PTFE-lined hydraulic hose, single-stainless steel braid	Petroleum, water based & synthetic	-65 to +400	$3/16$ to 1-1	600 to 1,500	1,200 to 6,000	2,500 to 12,000	1.5 to 16
100R15—Heavy duty, ultra-high-pressure, 6-spiral steel wire reinforced, rubber cover–Hydraulic	Petroleum based	-40 to 250	$3/8$ to 1-1	6,000	12,000	24,000	6 to 21
100R16—Compact, high-pressure, 2-braided wire reinforced rubber cover; Hydraulic	Petroleum & water based	-40 to 212	1 to 1-1	1,800 to 5,800	3,600 to 11,600	7,200 to 23,200	2 to 8
100R17—Compact, maximum operating pressure, 1- and 2-steel braided wire reinforced rubber cover; Hydraulic	Petroleum & water based	-40 to 212	$3/16$ to 1	3,000	6,000	12,000	2 to 6
100R18—Thermoplastic, synthetic-fiber reinforcement, thermoplastic cover; Hydraulic	Petroleum, water-based, synthetic	-40 to 212	$1/8$ to 1	3,000	6,000	12,000	1 to 10
100R19—Compact, maximum operating pressure, 1- and 2-braided steel wire, reinforced rubber cover; Hydraulic	Petroleum & water based	-40 to 212	$3/16$ to 1	4,000	8,000	16,000	2 to 6

Table 1.11 - SAE Standard for Hydraulic Hoses (Hydraulic & Pneumatic Magazine)

ISO 18752 Standard Pressure Classes: More recently, many large OEMs switched to ISO standards in their design and manufacturing process to ensure the sale and service of their equipment globally. ISO Standard 18752, released in 2006, takes a different approach centered around the design practices of users who typically design hydraulic systems based on performance and pressure requirements. Table 1.12 shows the ISO pressure class specifications.

Class		35	70	140	210	250	280	350	420	560
MWP[a] (bar)		35	70	140	210	250	280	350	420	560
MWP[a] (MPa)		3.5	7	14	21	25	28	35	42	56
MWP[a] (psi)		500	1000	2000	3000	3500	4000	5000	6000	8000
Nominal Size ISO	Inch									
5	-3	•	•	•	•	•	•	•	•	N/A
6.3	-4	•	•	•	•	•	•	•	•	N/A
8	-5	•	•	•	•	•	•	•	•	N/A
10	-6	•	•	•	•	•	•	•	•	N/A
12.5	-8	•	•	•	•	•	•	•	•	N/A
16	-10	•	•	•	•	•	•	•	•	•
19	-12	•	•	•	•	•	•	•	•	•
25	-16	•	•	•	•	•	•	•	•	•
31.5	-20	•	•	•	•	•	•	•	•	•
38	-24	•	•	•	•	•	•	•	•	N/A
51	-32	•	•	•	•	•	•	•	•	N/A
63	-40	•	•	N/A	N/A	N/A	N/A	N/A	N/A	N/A
76	-48	•	•	N/A	N/A	N/A	N/A	N/A	N/A	N/A
102	-64	•	N/A	N/A	N/A	N/A	N/A	N/A	N/A	N/A

Note: • = Applicable   N/A = Not Applicable   [a] = Maximum Working Pressure

Table 1.12 - ISO 18752 Standard Pressure Classes for Hydraulic Hoses (Courtesy of Parker)

ISO 18752 Standard Resistance to Impulse: As shown in Table 1.13, in ISO 18752, hydraulic hoses specifications are classified according to their resistance to impulse pressure in four grades: A, B, C, and D. Each grade requires a specific number of impulse cycles at a certain temperature and impulse pressure in order to meet the standard. In addition, each grade is then classified by the OD of the hose into standard types (AS, BS, CS) or compact types (AC, BC, CC, DC). Compact types have a smaller OD and bend radius than the standard types.

Grades and Types				
		Resistance to Impulse		
Grade	Type[a]	Temperature °C	Impulse Pressure (% of MWP[b])	Minimum Number of Cycles
A	AS	100	133%	200,000
	AC			
B	BS	100	133%	500,000
	BC			
C	CS	120	133% and 120%[c]	500,000
	CC			
D	DC	120	133%	1,000,000

[a] Standard or compact, e.g. CS is grade C and standard type.

Standard types have larger outside diameters and larger bend radii and compact types have smaller outside diameters and smaller bend radii.

[b] Maximum working pressure.

[c] 120% of the MWP shall be used for classes 350, 420 and 560 instead of 133%.

Table 1.13 - ISO 18752 Standard Pressure Classes for Hydraulic Hoses (Courtesy of Parker)

1.6.5- Hydraulic Hose Bend Radius

When application space is tight, bend radius is another important consideration. Otherwise, Installation of a hose at less than the minimum listed bend radius kinks the hose, reduce the hose life, and may result in premature failure.

Minimum bend radius of a hose depends on hose construction and is reported by the hose manufacturer. Table 1.14 shows an example of typical minimum bend radius reported by the hose manufacture

387
Hydraulic – Constant Working Pressure
ISO 18752

# Part Number 387	Standard Cover 387	Tough Cover 387TC	Super Tough 387ST	Hose I.D.		Hose O.D.		Working Pressure		Minimum Bend Radius		Weight		Vacuum Rating inches		Parkrimp 43 Series	Parkrimp 77 Series
	ISO 18752 Performance			inch	mm	inch	mm	psi	MPa	inch	mm	bs/ft	kg/m	of Hg	kPa		
387-4	AC	AC	AC	1/4	6,3	0.53	13,4	3000	21,0	2	50	0.16	0,24	24	80	●	
387-6	AC	AC	AC	3/8	10	0.69	17,4	3000	21,0	2-1/2	65	0.23	0,34	24	80	●	
387-8	AC	AC	AC	1/2	12,5	0.82	20,7	3000	21,0	3-1/2	90	0.29	0,43	24	80	●	
387-10	AC	AC	AC	5/8	16	0.94	23,9	3000	21,0	4	100	0.33	0,49	24	80	●	
387-12	AC	AC	AC	3/4	19	1.10	27,8	3000	21,0	4-3/4	120	0.58	0,86	24	80	●	
387-16	AC	AC	AC	1	25	1.40	35,4	3000	21,0	6	150	0.79	1,17	24	80	●	
387-20	BC	CC	CC	1-1/4	31,5	1.82	46,3	3000	21,0	8-1/4	210	1.74	2,59	18	60	●	●
387-24	BC	CC	CC	1-1/2	38	2.08	52,8	3000	21,0	10	250	2.01	2,99	18	60		●
387-32	BC	CC	CC	2	51	2.61	66,2	3000	21,0	12-1/2	320	2.75	4,09	18	60		●

Table 1.14 - Example of Minimum Bend Radius Reported by the Hose Manufacturer (Courtesy of Parker)

As shown in Fig. 1.60, when considering the bend radius of a hose assembly, a minimum straight length of twice the hose's outside diameter should be allowed between the hose fitting and the point at which the bend starts.

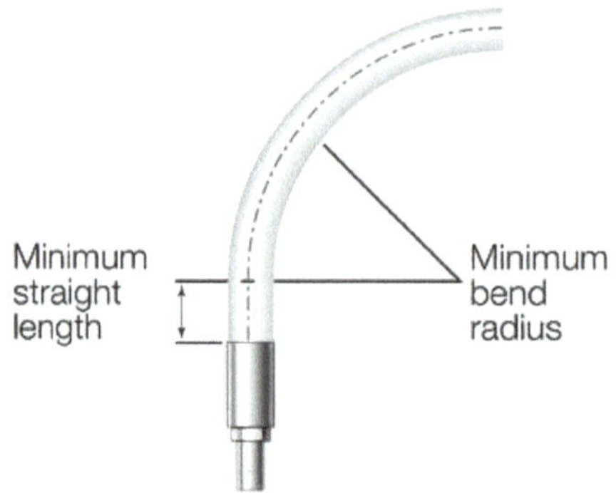

Fig. 1.60 - Minimum Straight Length for Bent Hoses

1.6.6- Hydraulic Hose Assembly

Before cutting a hose, make sure you understand the difference between "*Cut Hose Length*" and "*Assembly Overall Length*". As shown in Fig. 1.61. the overall assembly length (OAL) for a straight hose includes the hose cut length (HCL) of the hose plus the length of the two hose couplings C1 and C2. The overall assembly length of bended hoses is calculated as shown in Fig. 1.62.

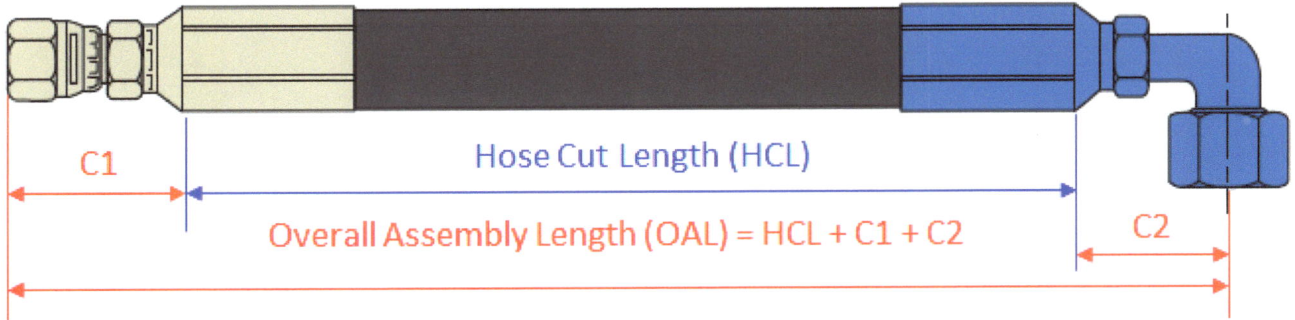

Fig. 1.61 - Straight Hose Assembly Length

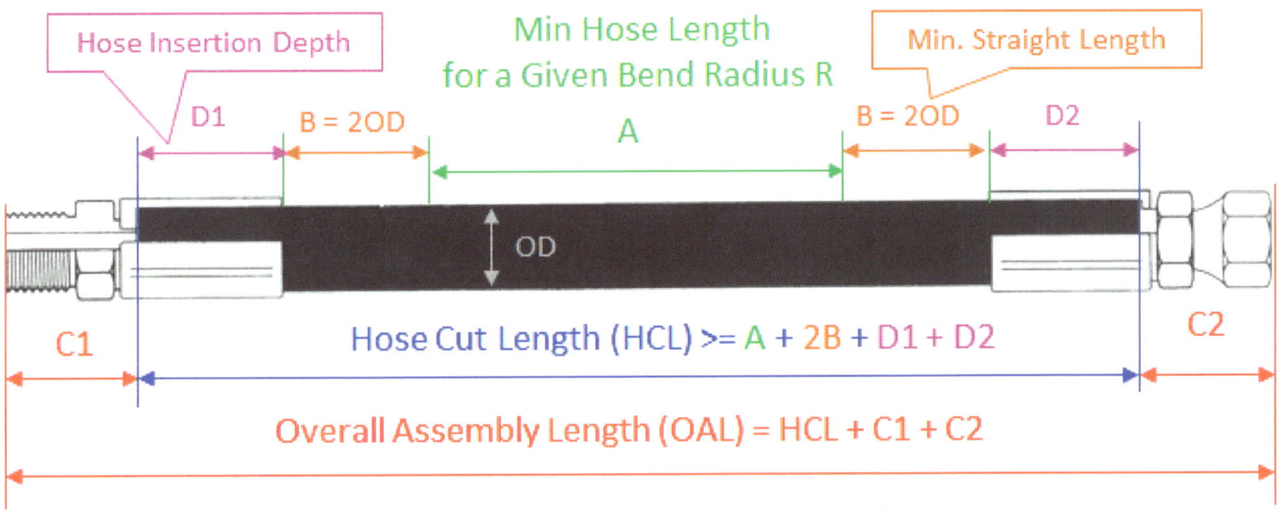

Fig. 1.62 - Bent Hose Assembly Length

Equation 1.5 shows a general Formula for Minimum Hose Cut Length for a Given Bend Radius:

A = (Angle of Bend/360) x 2 π R   **1.5**

Example: to make a 90⁰ bend for hose that has a minimum bend radius r = 4.5 inches, the minimum hose length is **A = (90/360) x 2 x π x 4.5 = 7 inches.**

Figure 1.63 shows the overall assembly length according to DIN 20066 for various hose configurations.

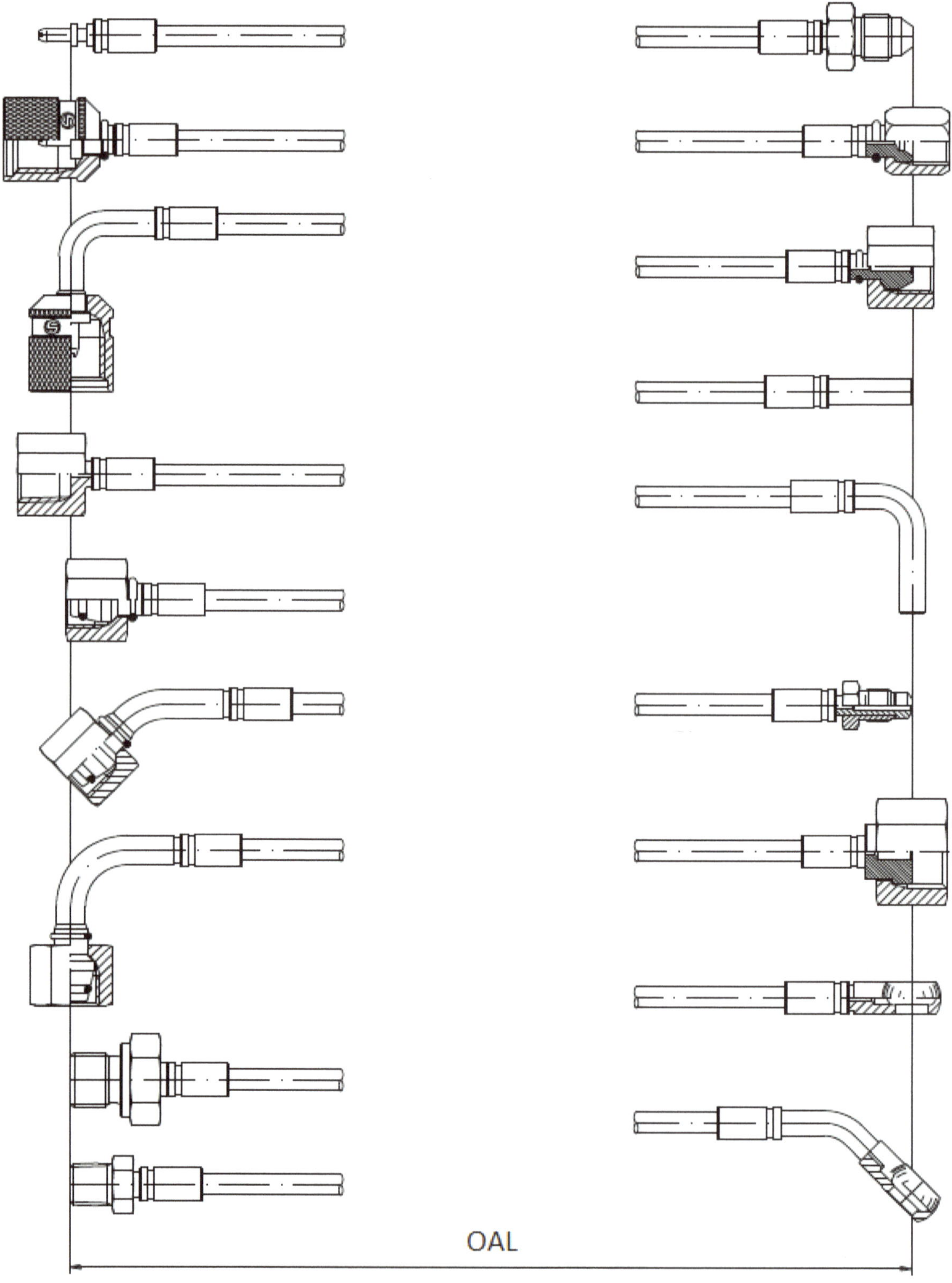

Fig. 1.63 - Hose Overall Assembly Length according to DIN 20066

1.6.7- Hose Couplings

1.6.7.1- Identification of Hose Couplings

A hydraulic hose requires two end joints to connect the hose from both ends. These end joints are referred to *Hose Couplings*. As shown in Fig. 1.64, a coupling is identified as follows:
- The *hose end* for hose attachment. The hose end is identified by the hose size and type to which it is attached.
- The *thread end* (or adaptor) for port attachment. The thread end, as shown in the figure, are available in different configurations with sizes depending on the port to which it will be attached.

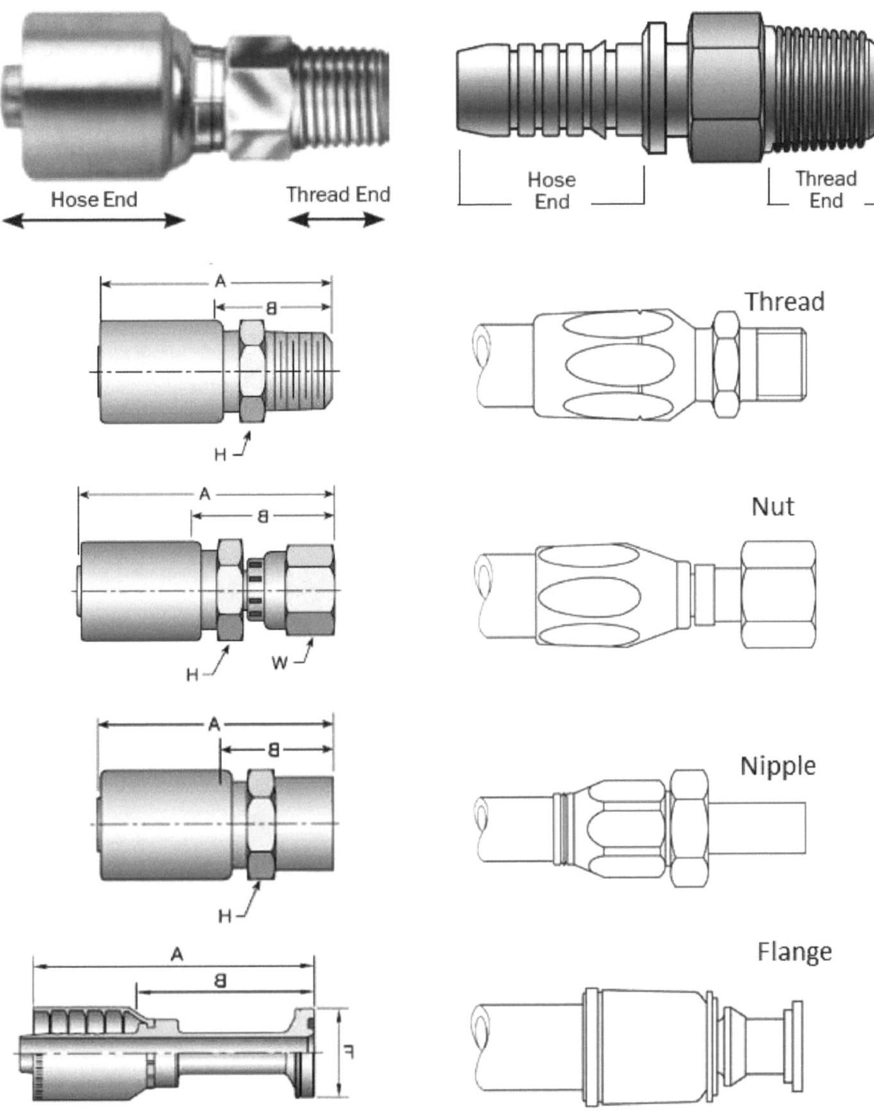

Fig. 1.64 - Identification of Hose Couplings (Courtesy of Gates)

1.6.7.2- Reusable versus Permanent Couplings

As shown in Fig. 1.65, there are two types of hydraulic couplings *Reusable Couplings* and *Permanent Couplings* for one-time use.

Reusable Couplings: As shown in Fig. 1.66, reusable (*field Attachments or Screw in Connectors*) hose fittings are an excellent choice for emergency repairs. Assembly can be accomplished with hand wrenches; there is no crimper required. However, these connectors aren't recommended if there are high pulsating pressure.

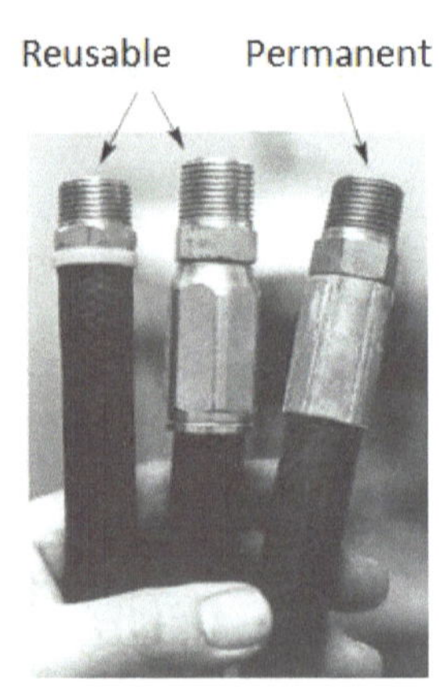

Fig. 1.65 - Permanent versus Reusable Hose Couplings (Courtesy of Gates)

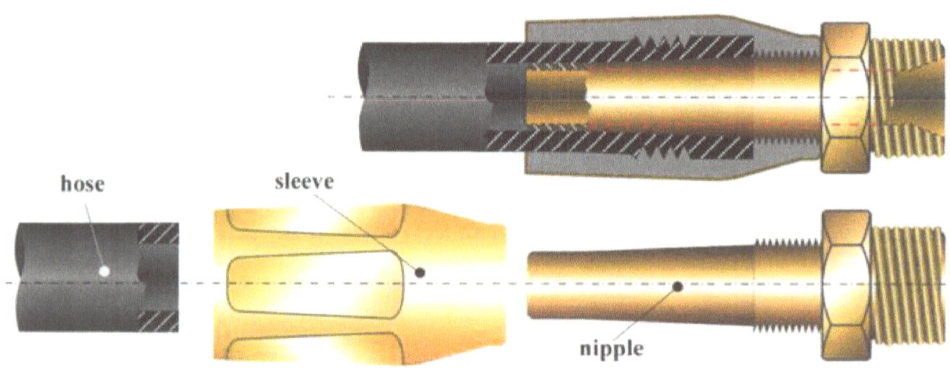

Fig. 1.66 - Reusable Hose Couplings (Courtesy of Assufluid)

Permanent Couplings: As shown in Fig. 1.67, the nipple ensures seizing and prevent hose narrowing. They require *crimping equipment* to assemble to a hose.

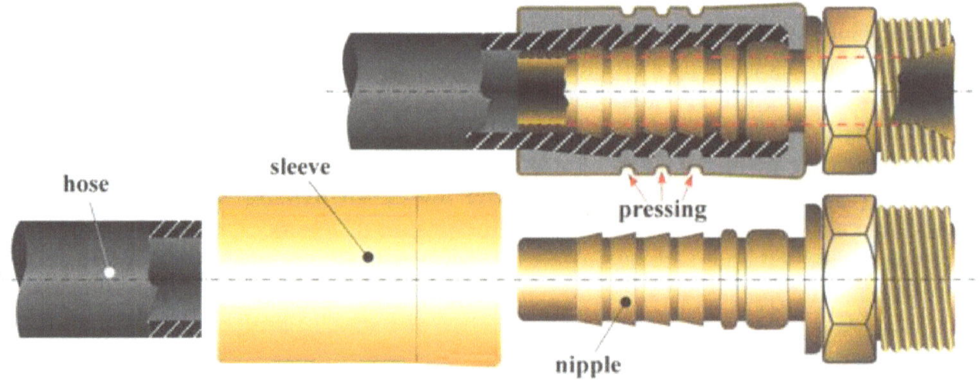

Fig. 1.67 - Permanent Hose Couplings (Courtesy of Assufluid)

As shown in Fig. 1.68, permanent hose couplings are available in either preassembled or two-piece configurations.

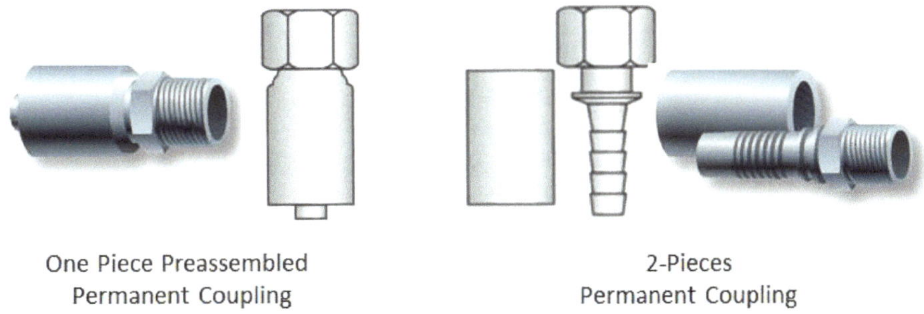

Fig. 1.68 - Permanent Hose Couplings (Courtesy of Gates)

1.6.7.3- Hose Couplings Configurations

Figure 1.69 shows the common configuration of hose couplings. However, Fig. 1.70 shows typical configurations for high and medium pressure.

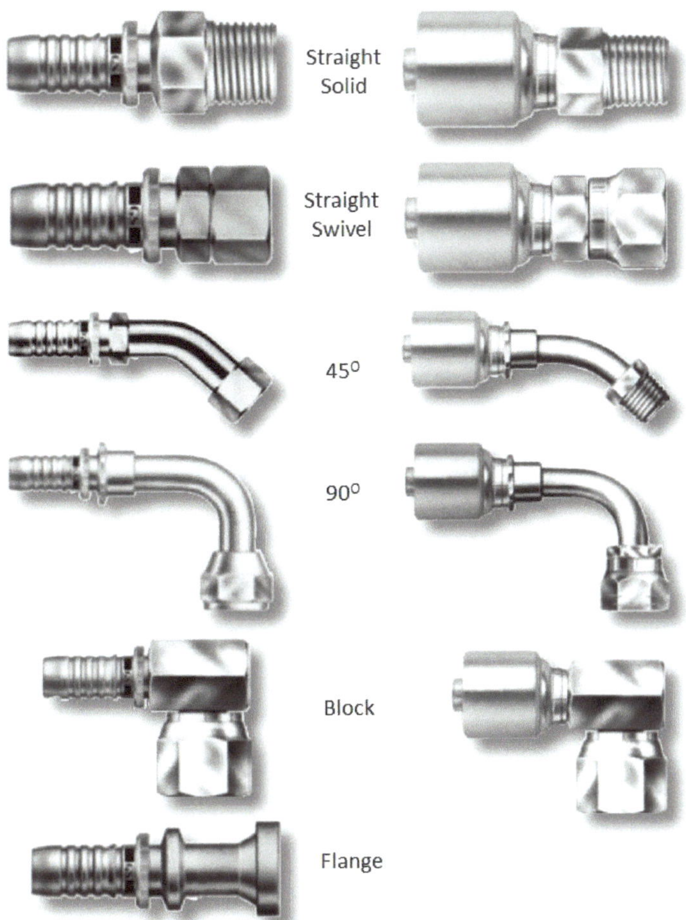

Fig. 1.69 - Hose Coupling Common Configurations (Courtesy of Gates)

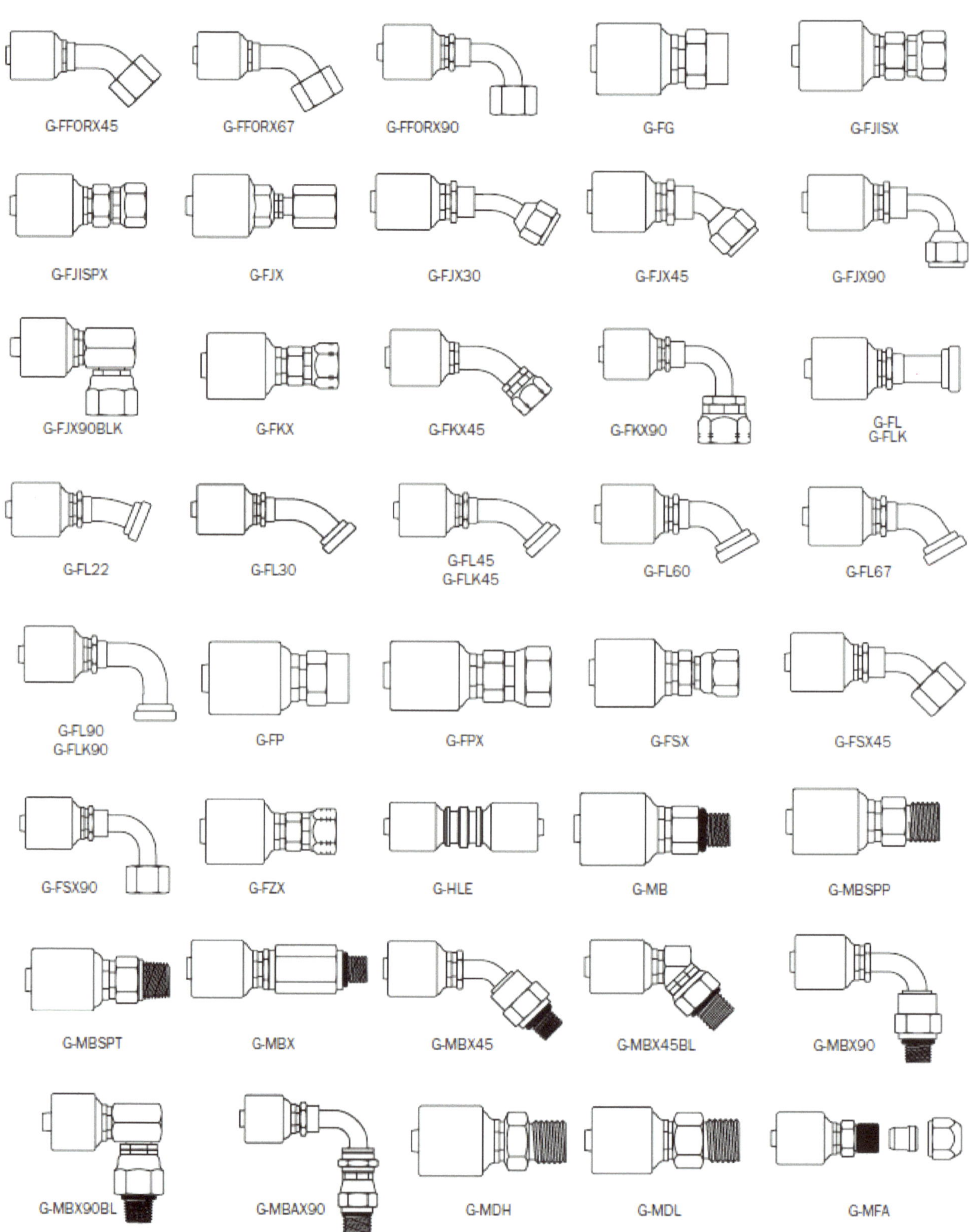

**Fig. 1.70 - Hose Coupling Configurations for Medium and High Pressure
(Courtesy of Gates)**

1.6.7.4- Hose Crimping

Crimpable Coupling: Figure 1.71 shows a typical design of one-piece crimpable hose coupling. The teeth in the crimpable fittings bite down to the hose wire for a metal-to-metal grip with maximum integrity. One-piece fittings can be combined with many types of hose covers.

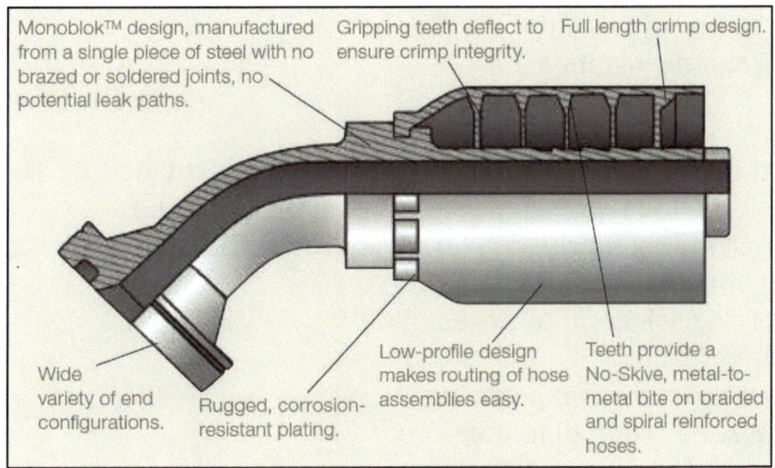

**Fig. 1.71 - Typical Design of Preassembled Cimpable Hose Coupling
(Courtesy of Parker)**

Crimping Machines: As shown in Fig. 1.72, a crimper is a hydraulic ram that uses fluid under pressure to extend the ram and crimp the fittings. They are available in various styles and power capacity.

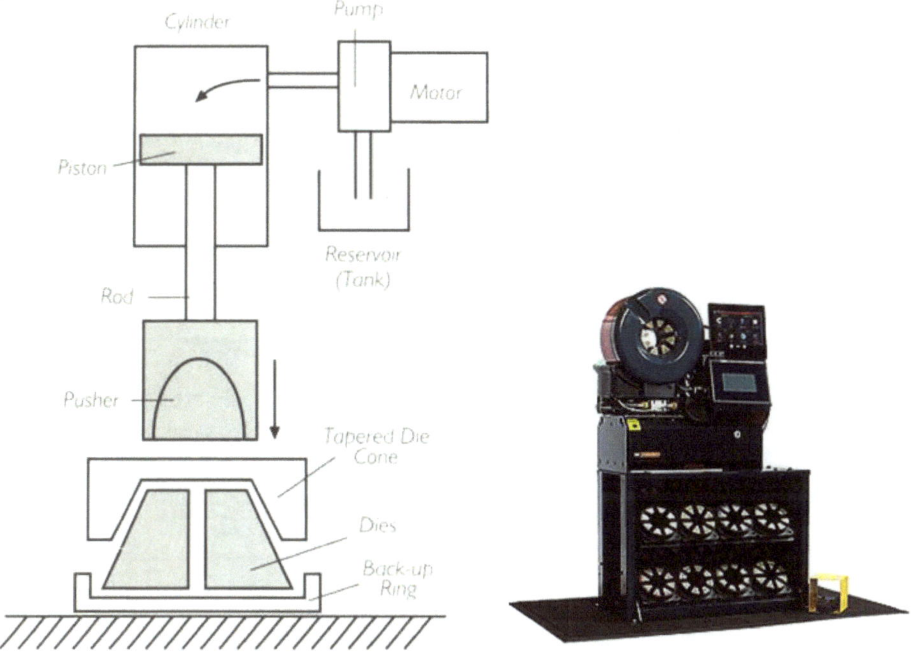

Fig. 1.72 - Hose Crimping Machine (Courtesy of Gates)

Hose Crimping Method:

Before crimping a hose, be aware of the following:

- Never re-crimp used hose with permanent or field attachable couplings.
- As shown in Fig. 1.73, some hose design requires that part of the rubber cover be removed from the end so that the fitting can be installed.

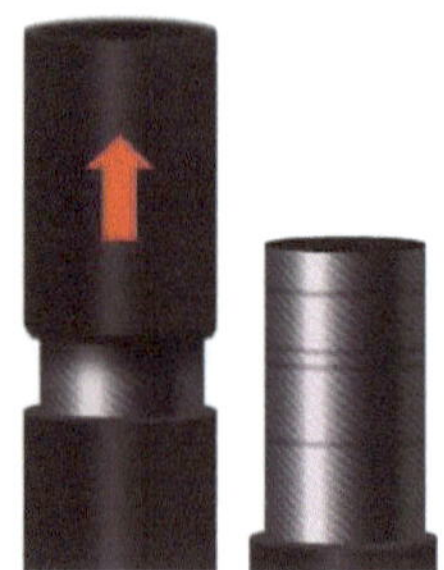

Fig. 1.73 - Prepare a Hose for Crimping

For a specific crimping machine, refer to instructions provided by the appropriate operating manual. However, the following steps shows general guidelines:

1. Setup the crimping machine referring to the manufacturer' instructions identifying the following information:
    A. Ferrule selection (if necessary).
    B. Die selection.
    C. Approximate crimp setting.
    D. Finished crimp diameter.
2. Load the selected die into the crimper.
3. Adjust the machine to the proper crimp setting.
4. Insert the hose assembly and properly align the coupling with the die fingers, as required by the particular model crimper that you're using.
5. Install die cone, if used.
6. Always wear safety glasses and keep hands and clothing away from the moving parts of the crimper.
7. Activate the crimping mechanism.
8. Remove crimped assembly from dies and measure final crimp diameter.

Measuring Crimp Diameters: As shown in Fig. 1.74, *Crimp Diameter* is measured between the ridges, halfway between the top and bottom of the coupling. When using calipers, be sure the caliper fingers do not touch the ridges.

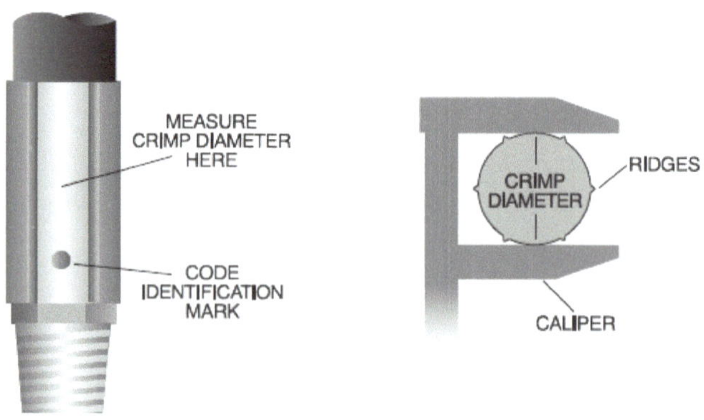

Fig. 1.74 – Measuring Crimp Diameter (Courtesy of Gates)

1.6.8- Quick Connect Couplings

Conventional Quick Connect Coupling: A *Quick* Connect Coupling is a component which can quickly connect and disconnect hoses without the use of tools or special devices. As shown in Fig. 1.75, according to ISO 16028, a *Flat Face* quick connect coupling consists of the following two pieces:

- **Coupler:** Other nomenclature are "Female", "Socket", "Body".
- **Nipple:** Other nomenclature are "Male", "Plug", "Adapter".

Connection is performed by pushing the movable part (usually the coupler) towards the fixed part (usually the nipple). Disconnection is performed by pulling the nut (2) in the coupler backward. Such couplings are used for lines where no residual pressure or fluid leakage is found upon connection or disconnection.

Caution: These types of couplings can exhibit high-pressure drop (5-10 bar) or more if not sized properly. These couplings should be sized based on the internal flow area of the coupling, which is typically significantly less than the internal area of the connection part.

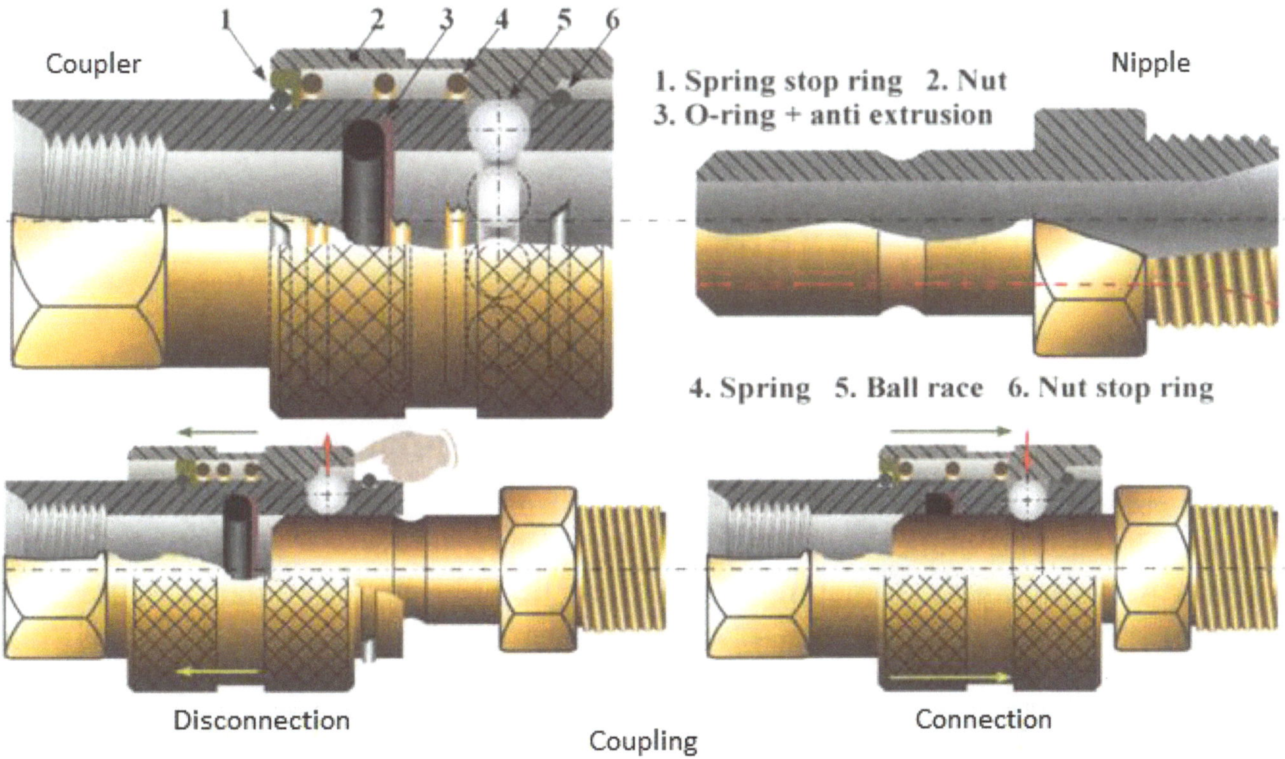

Fig. 1.75 - Flat Face Quick Connect Coupling (Courtesy of Assufluid)

Quick Connect Coupling with Nonreturn Valve: Figure 1.76 shows a <u>single non-return valve</u> (7) is included in the female coupler design. When the coupling is disconnected, if there is pressure, the fluid itself presses on the surface ensuring perfect sealing.

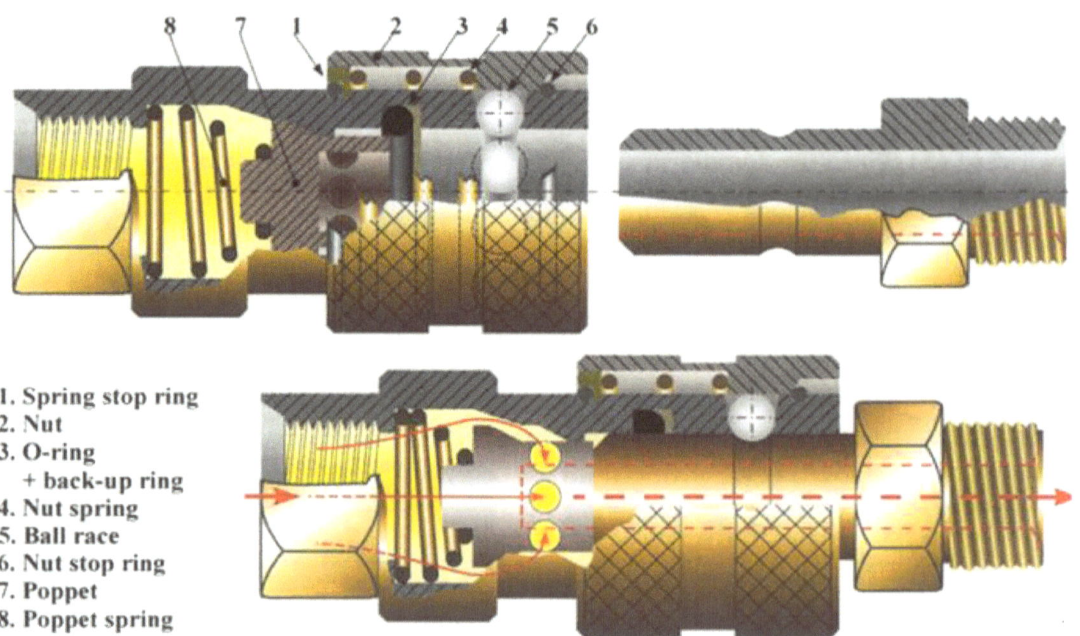

Fig. 1.76 - Quick Connect Coupling with Single Non-Return Valve (Courtesy of Assufluid)

Figure 1.77 shows another configuration where <u>two non-return valves</u> are included, one in the female side (7) and one in the male side (9). Such quick connect coupling is recommended for connecting two pressurized lines. Couplings with non-return valves work under pressure 400-600 bar (6000-9000 psi) depends on the design of the couplings.

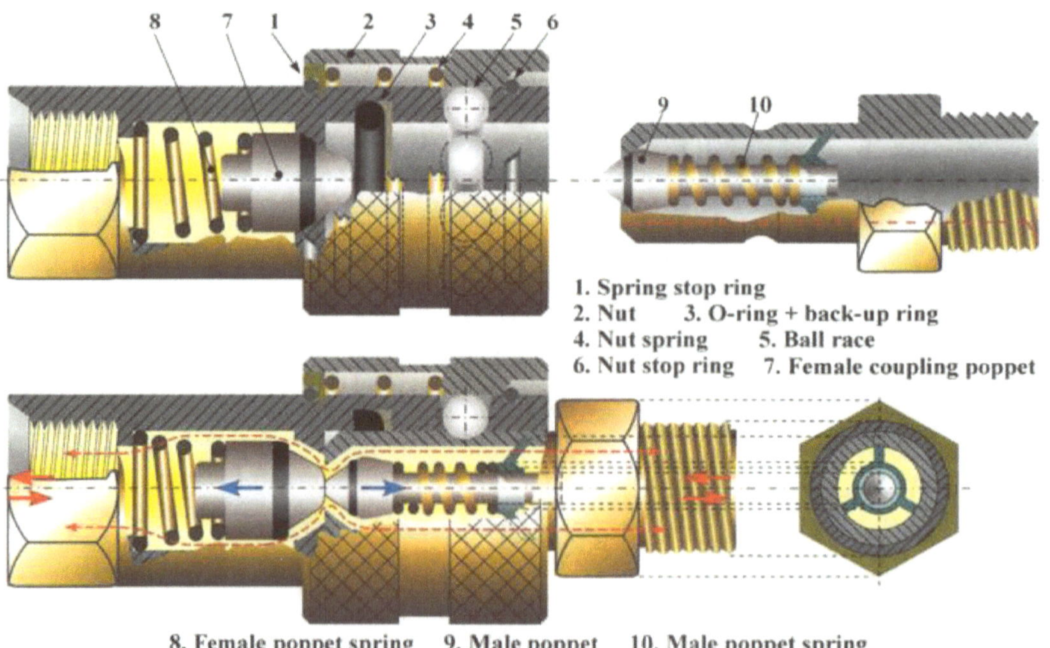

Fig. 1.77 - Quick Connect Coupling with Double Non-Return Valves (Courtesy of Assufluid)

Innovative iLok™ Coupling: In some applications, such as mining machines, workers often are involved in dangerous practices to free the couplings when it's time to move the equipment. They use hammers to beat on the coupling ferrule or the hose itself. The result is often a cracked ferrule or damaged hose, leading to a leak or catastrophic burst when lines are re-pressurized. The *iLok™ Coupling* is an innovative quick connect coupling. Figure 1.78 shows the male and the female side of such a connector that has the following characteristics:

- Easy to connect and disconnect by hand.
- Withstands high impulse applications.
- Use a secure, visible locking system.
- Release residual pressure away from workers during disconnection.
- Compact to fit in tight spaces and smooth to prevent abrasion against adjoining hoses.

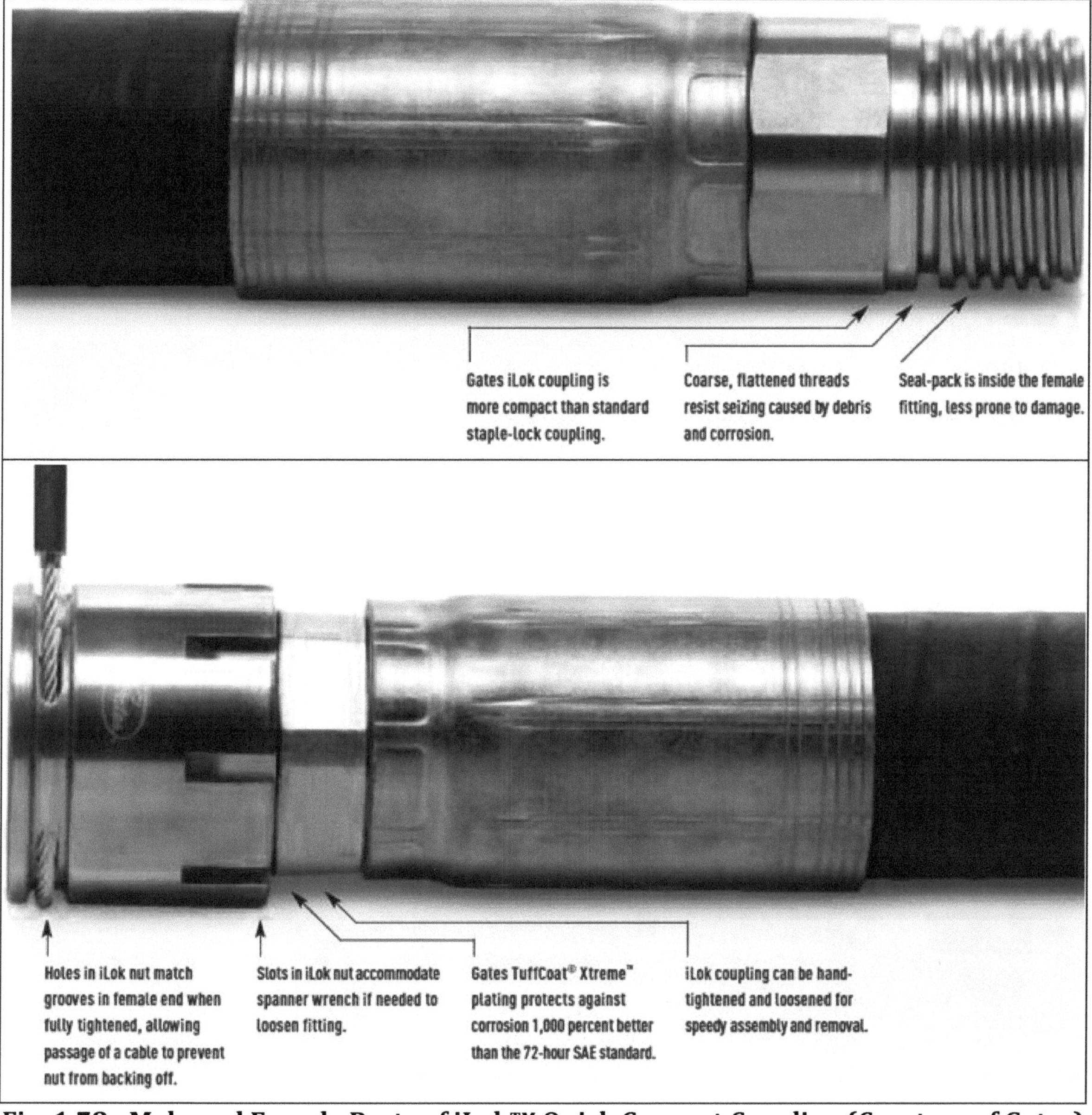

Fig. 1.78 - Male and Female Parts of iLok™ Quick Connect Coupling (Courtesy of Gates)

As shown in Fig. 1.79, when fully connected, openings in the iLok nut align with a groove in the female fitting to accommodate a cable lock. The cable won't pass through the grooves unless the nut is completely tightened, providing positive proof of a secure connection. The cable lock "flags" indicates that a connection has been made. The iLock has the following technical specifications:

- Impulse tested to 133% of operating pressure at 250°F (+121°C) for 1 million impulse cycles with no leaks or failures.
- Exceeds Code 62 of SAE J518 flange specifications.
- 4:1 design factor (burst pressure to working pressure ratio).
- TuffCoat® Xtreme™ plating provides red rust protection that exceeds the 72-hour SAE standard by 1000 percent.
- MSHA-approved for underground mining applications.

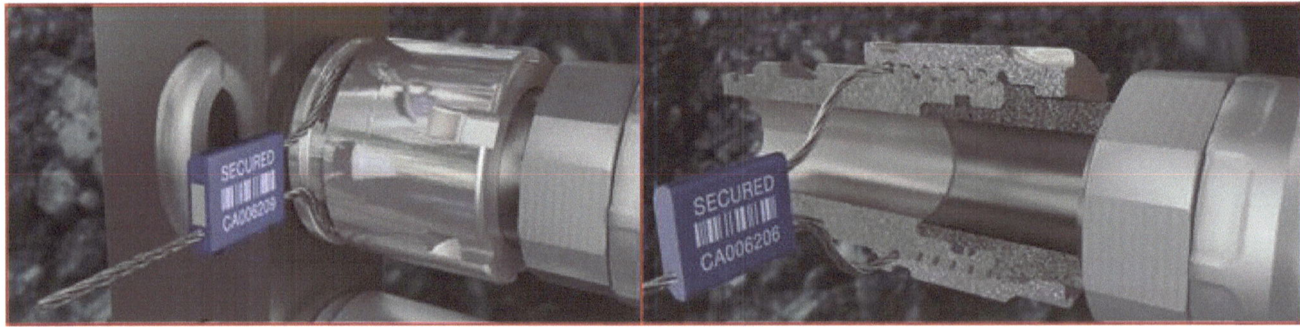

Fig. 1.79 - Locking Mechanism of iLok™ Quick Connect Coupling (Courtesy of Gates)

Innovative TLX Coupling: Figure 1.80 shows same concept has been developed by another manufacturer under a brand name of *TLX Quick Connect Coupling*.

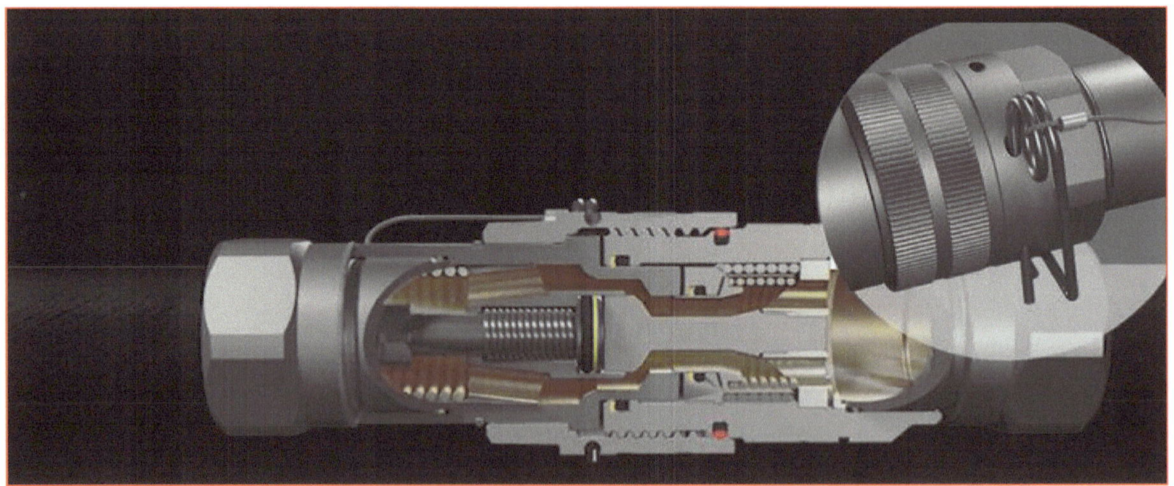

Fig. 1.80 - Locking Mechanism of TLX Quick Connect Coupling (Courtesy of CEJN)

Universal Push-to-Connect (UPTC) Assembly:

<u>Description (Fig. 1.81)</u>: *Universal Push-to-Connect* (*UPTC*) assembly is designed to achieve reliable leak-free connections for hose and tube assembly. UPTC assembly utilizes standard O-Ring Face Seal (ORFS) or EO (24º DIN cone) fitting bodies, and it is suited for hydraulic hose (rubber or thermoplastic) and tube (inch or metric) assemblies.

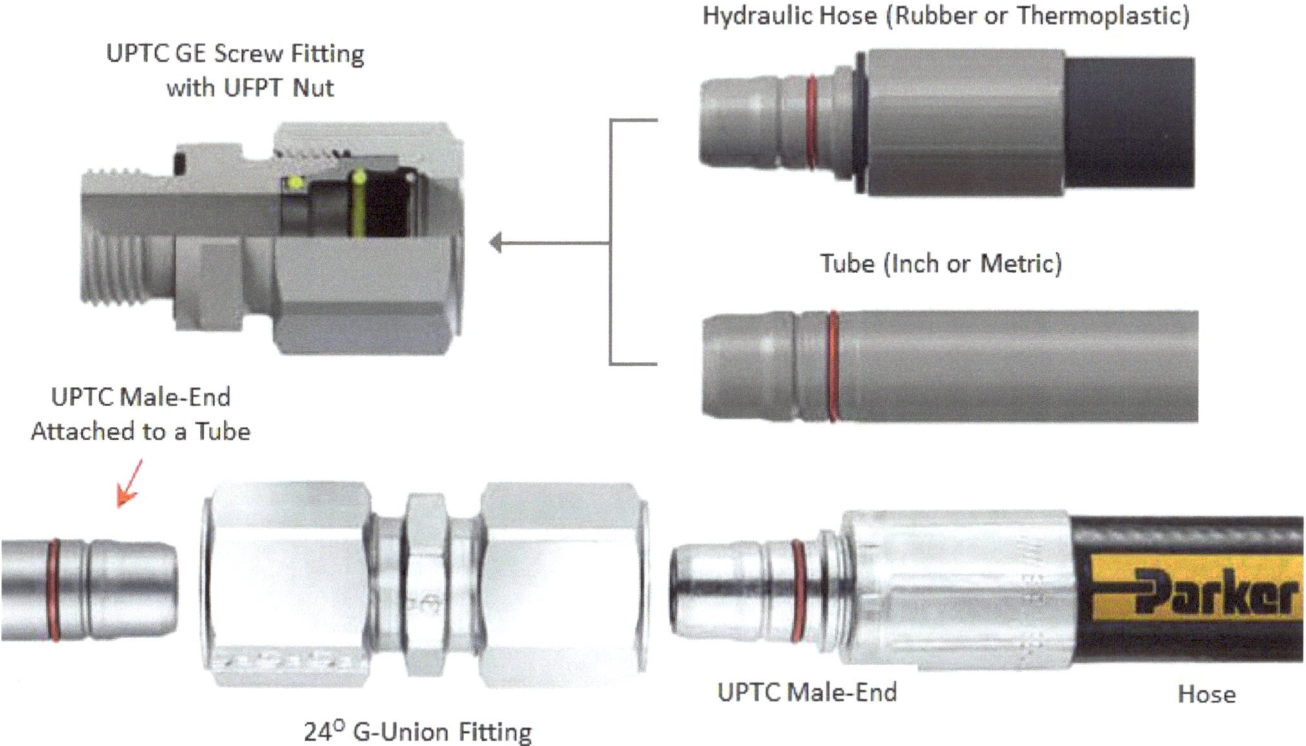

Fig. 1.81 - Universal Push-to-Connect (UPTC) (Courtesy of Parker)

<u>Features:</u>
- <u>Assembly Time Savings:</u> Simple push-to-connect design requires low pushing force with no special assembly tools.
- <u>Space Saving:</u> Easy to use in hard-to-reach areas and allows for a more compact system.
- <u>Assured Proper Connection:</u> Visual and tactile installation indicators easily help to assure proper connection, assembly, inspection and diagnosis to minimize error and its costly consequences.
- <u>Elimination of Hose Twist:</u> Self-aligning nipple eliminates hose twisting during assembly for longer service life.
- <u>Leak-Free Performance:</u> Leak-free performance is assured by mathematical simulation, rigorous lab tests and significant field-tests with leading equipment manufacturers.
- <u>Easy Design Implementation:</u> UPTC allows for incorporation into any current hydraulic system with virtually no adverse effect on other components, or the system.
- <u>Excellent Field Serviceability:</u> Allowing standard wrench disassembly and damaged hose replacement.

<u>Assembling (Fig. 1.82)</u>: As shown in the figure, assembly process is as easy as pushing the push-in male part (1) inside the system nut (2) with low pushing force as low as 60N. When a click is heard, and the red ring (6) isn't visible, this means that the stainless-steel clip ring (3) is snapped correctly, and the connection is done properly.

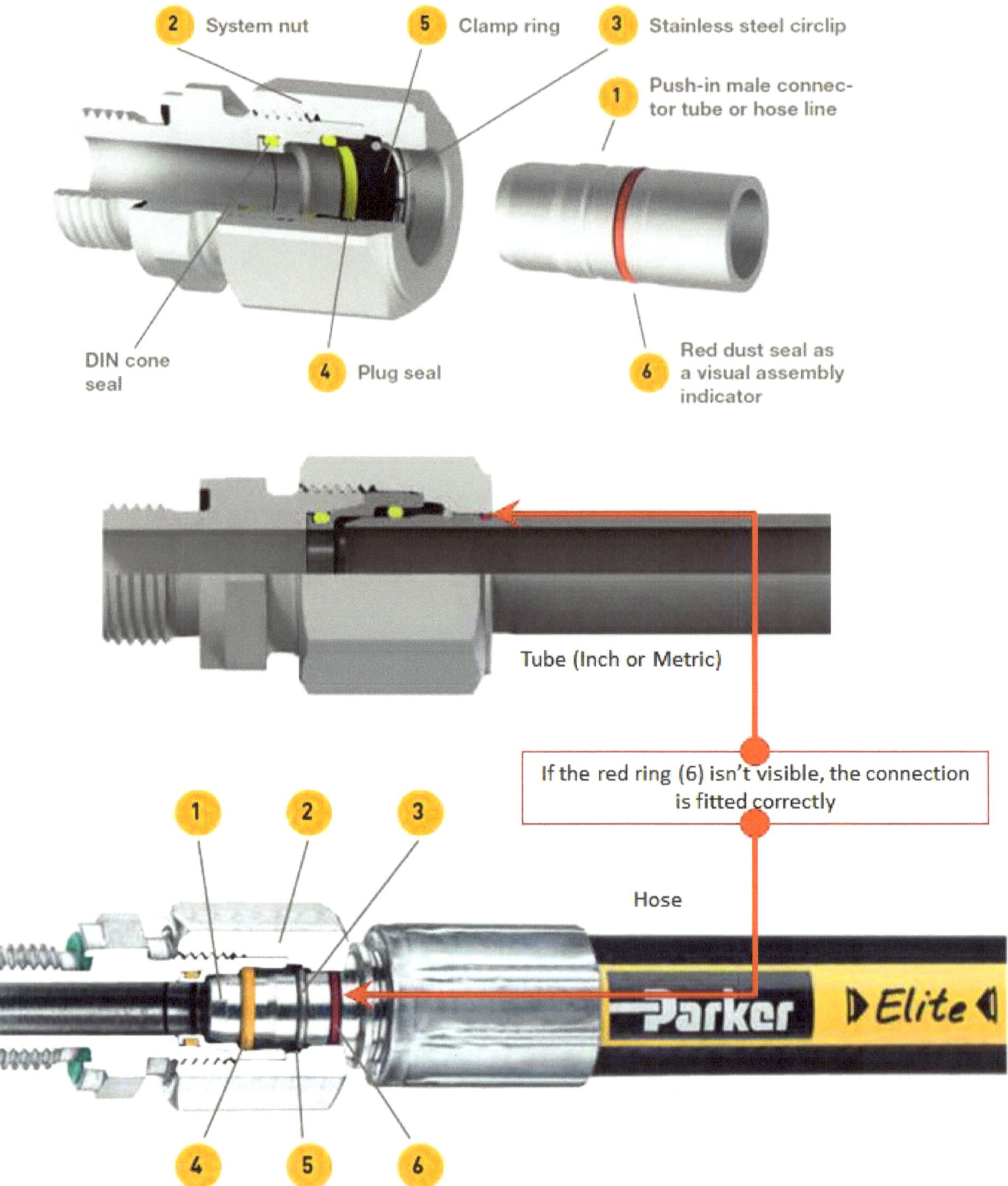

Fig. 1.82 - Assembling of Universal Push-to-Connect (UPTC) (Courtesy of Parker)

<u>Sealing (Fig. 1.83)</u>: after proper assembling, sealing is taken place by series of sealing defense lines; the cone seal, the plug seal (4), and the clamp ring (5).

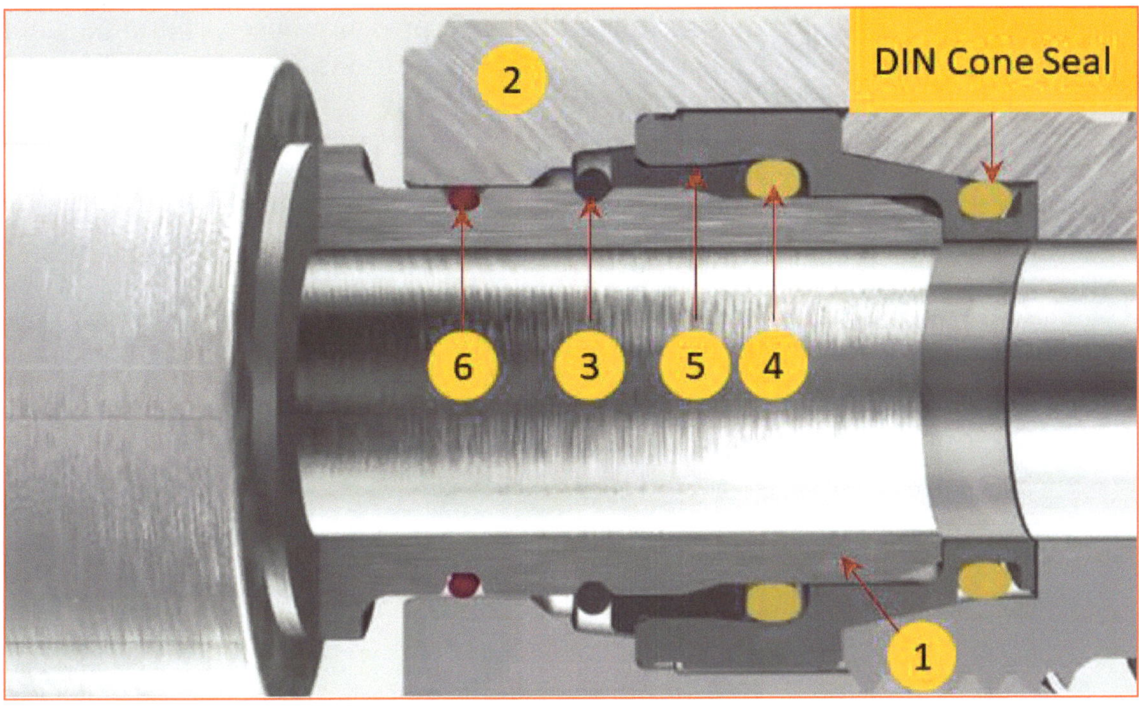

Fig. 1.83- Sealing of Universal Push-to-Connect (UPTC) (Courtesy of Parker)

<u>Disassembling (Fig. 1.84)</u>: As shown in the figure, during disassembling, the system nut (2) will be removed from the screw socket. The male end and UPTC system nut remain as a single unit.

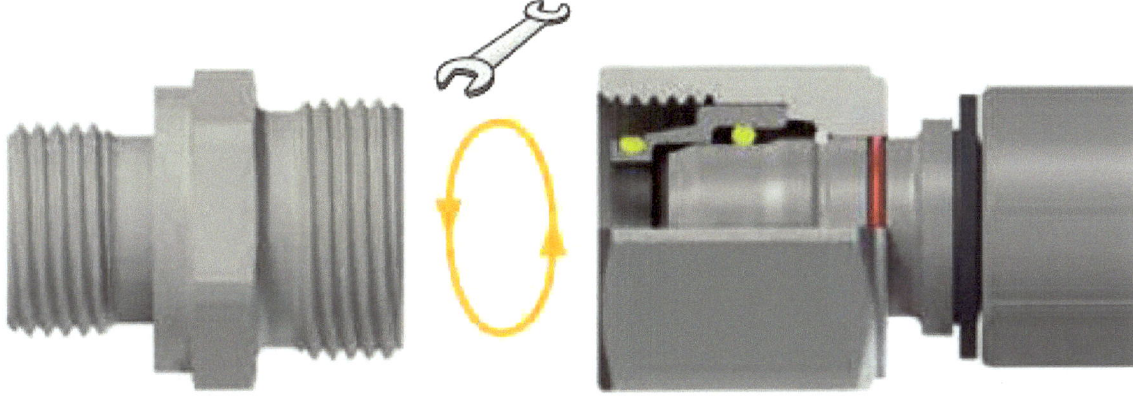

Fig. 1.84 - Disassembling of Universal Push-to-Connect (UPTC) (Courtesy of Parker)

Push-Lok Assembly for Low Pressure:

<u>Description (Fig. 1.85)</u>: *Push-Lok* is an easy-to-use complete line of hose and fittings for low pressure applications. They are structured, like high pressure hoses, containing an inner tube, reinforcement layer, and a protective cover. In applications where several hose lines carry different media, Push-Lok colors reduce timely "tracing" of lines, preventing connection/disconnection of wrong line and unnecessary, downtime.

<u>Features:</u>
- Easy assembly with no clamps or special tools is required during installation.
- Time and cost saving.
- Ensures reliable, durable, leak-free service.
- High functional safety with a design factor of 4.
- Wide range of hose and fittings for a wide range of applications.
- Wide range of fluid compatibility that is color-coded.

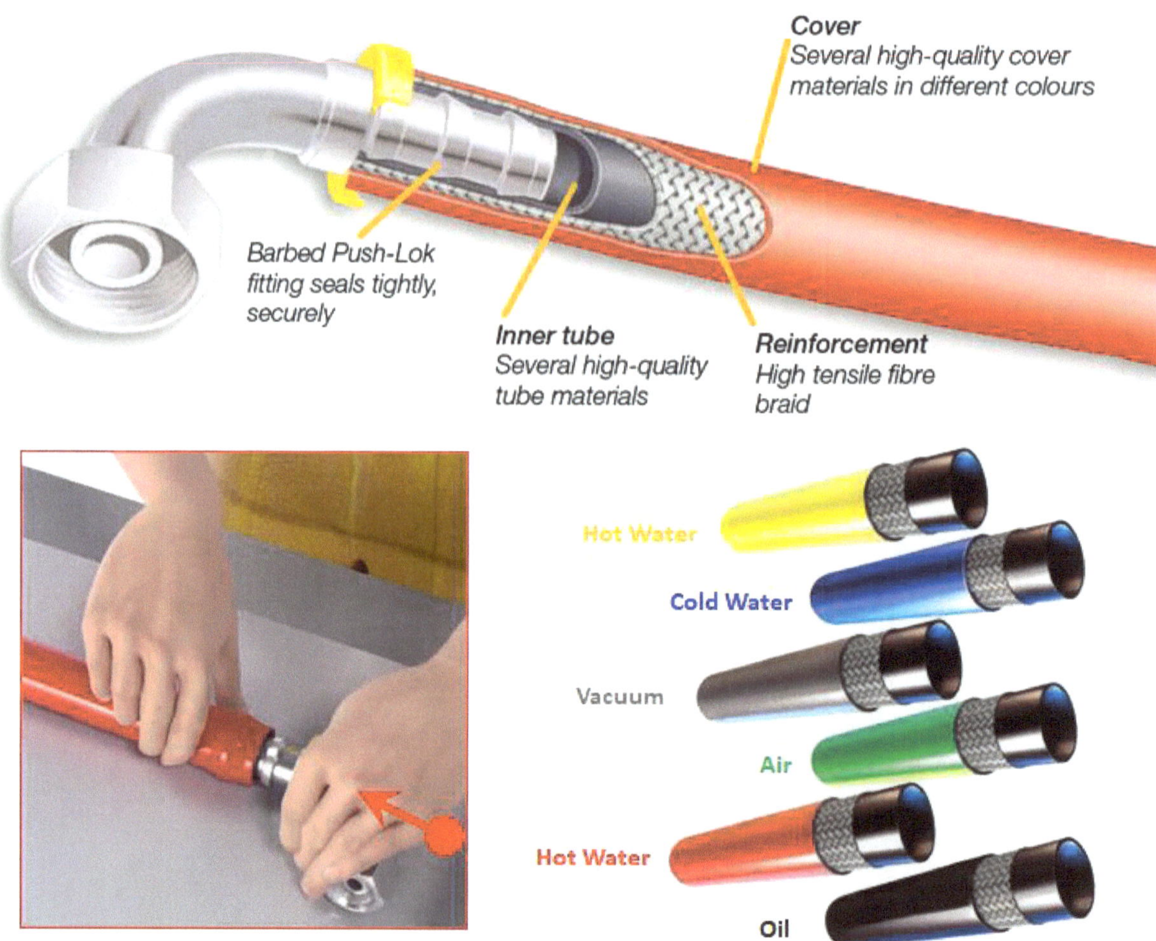

Fig. 1.85 - Push-Lok Hose and Fitting Assembly (Courtesy of Parker)

<u>Assembling and Disassembling (Fig. 1.86)</u>: During assembly, Push-Lok fittings will provide an effective grip only when the Push-Lok hose is pushed fully on the insert, where the cropped end of the hose should be fully concealed by the plastic collar.

Assembly is easy

1. Cut hose cleanly and squarely with a sharp knife or a Parker Push-Lok cut-off tool

2. Lubricate the Push-Lok fitting and/or Hose I.D. with a light oil or soapy water only. Do not use heavy oil or grease.

3. Insert fitting into hose until the barb is in the hose.

4. Place end fitting against a flat object (bench or wall). Grip hose approximately one inch from end and push with steady force until the end of the hose bottoms on the fitting and is covered by the yellow plastic cap.

Disassembles fast

1. Leave fitting in place and cut hose lengthwise from the yellow cap approximately one inch. IMPORTANT: Be careful not to nick barbs when cutting hose.

2. Grip hose and give a sharp downward tug to disengage the fitting.

Caution: Push-Lok fittings will properly grip Push-Lok hose only when pushed all the way in with the cut end of the hose completely concealed by the yellow plastic cap.

Sealing integrity may be damaged by using exterior clamps.

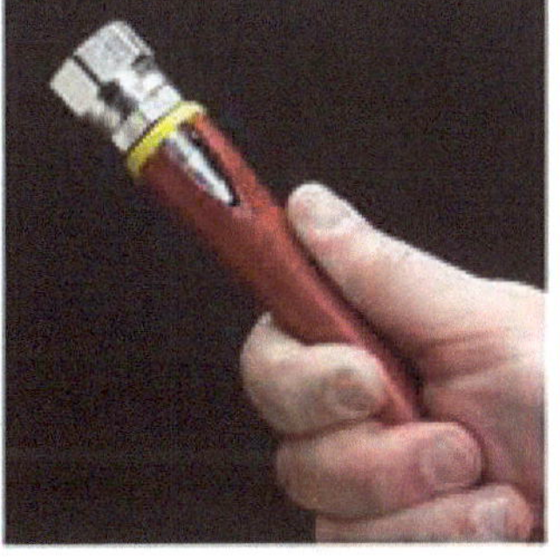

**Fig. 1.86 - Assembling and Disassembling Push-Lok Hose and Fitting Assembly
(Courtesy of Parker)**

Innovative Quick Connect System for Low Pressure (Fig 1.87): The innovative *Quick-Connect* system is an easy to install and cost-effective alternative to conventional hose connection fittings. The patented system is suitable for both <u>suction and return lines</u> and has already been tested more than 100,000 times in many different applications in the field. Tubes that are equipped with these connections can simply be plugged into the appropriate adapters by the user and fixed with the enclosed locking clips. They work in temperature up to 120 ^{0}C (250 ^{0}F) and operating pressure up to 10 bar (150 psi).

Plastics and sealing geometry are selected in such a way that the connections can also be reliably installed to metal adaptors. This makes an attractive product range available which offers the right connection for practically every application.

The locking clip prevents the hose from being pulled off the adaptor. However, the line remains rotatable. In many applications this represents a further advantage since the line can thus move into a position as free of stress as possible.

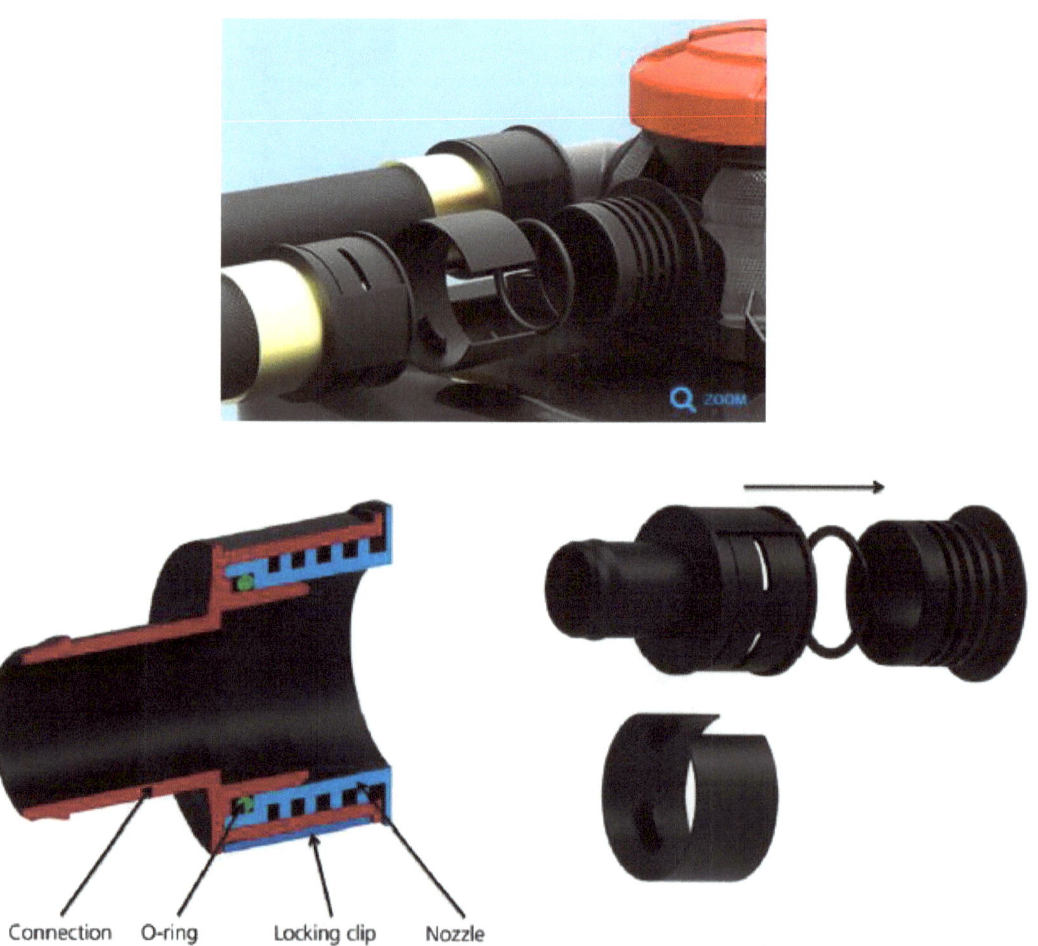

Fig. 1.87 - Innovative Quick Connect Coupling (Courtesy of ArgoHyots)

CEJN Multi-X Quick Connect Couplings (Fig. 1.88): *Multi-X Quick Connect Couplings* are designed to connect multiple lines simultaneously with great flexibility, high performance, easy installation and trouble-free operation. They are designed to meet and exceed the demands of even the most challenging mobile hydraulic application.

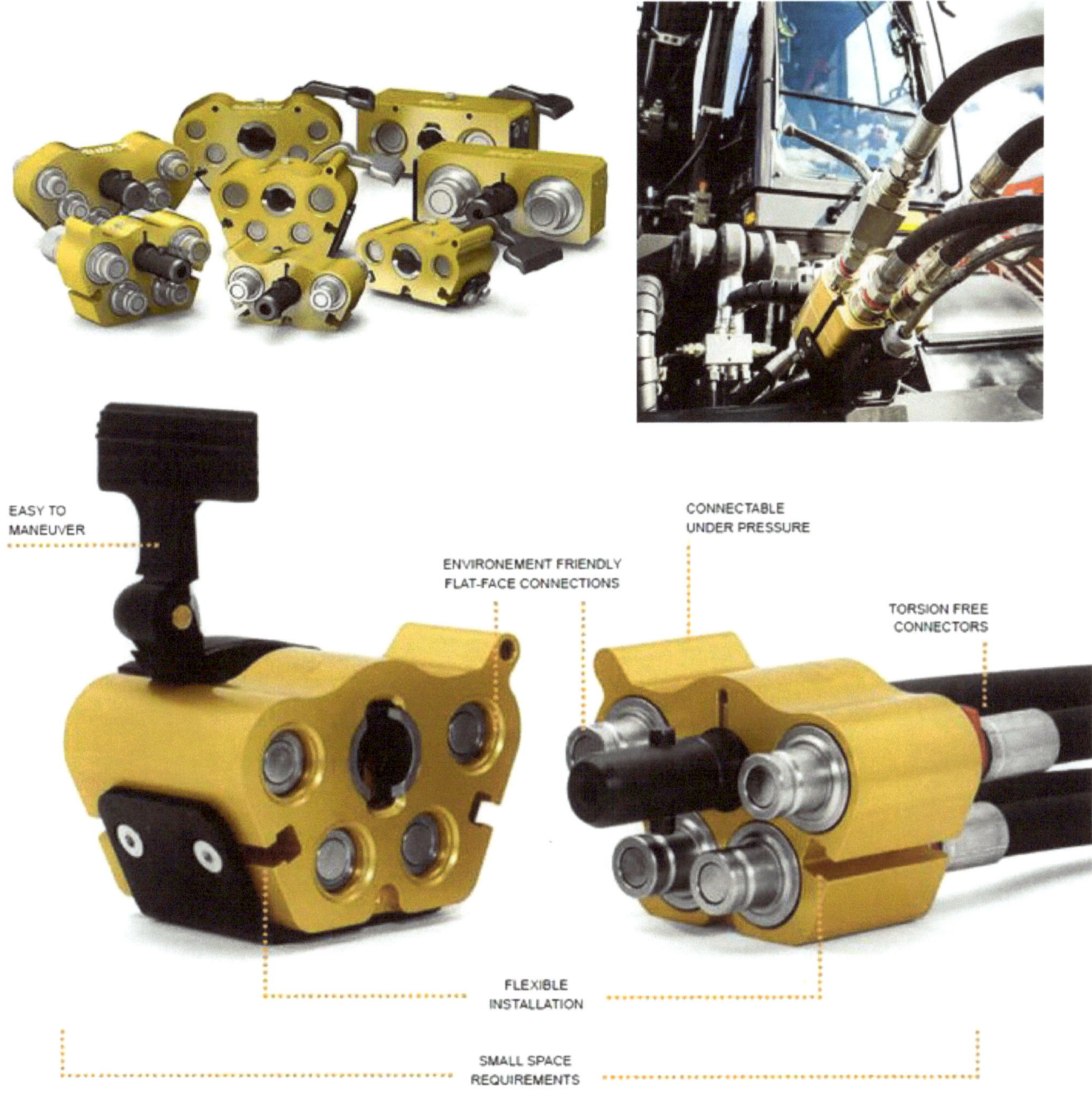

Fig. 1.88 - Multi-X Quick Connect Coupling (Courtesy of CEJN)

Figure 1.89 shows the concept of operation of Multi-X quick connect couplings.

Working pressure of up to 350 bar (5000 psi) can be used on half the port simultaneously with the other half of the port used as return lines with a maximum pressure 50 bar (725 psi).

Both the female and male plate can be used in the fixed part.

Electric connectors can also be easily attached.

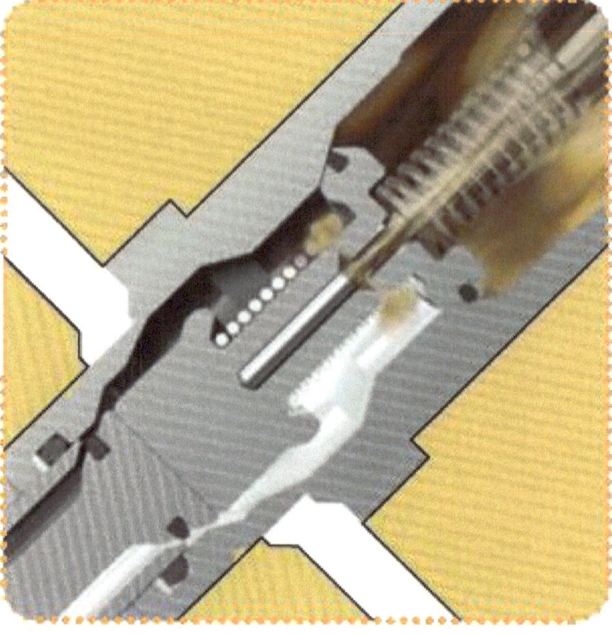

Pressure is connected without spillage making it simple to connect with residual pressure in the system

Fig. 1.89 - Conceptual Operation of Multi-X Quick Connect Coupling (Courtesy of CEJN)

1.6.9- Hose Guards

Using *Hose Guards* is recommended for the following reasons:

- **Hose Protection:** Hose guards prolong the service life of the hose because they offer hose protection against harsh environmental and abrasion conditions.
- **Operator Protection:** Hose guards provide protection for the machine operators against whiplash and oil injection hazards when a hose assembly fails.
- **Machine Protection:** Hose guards prolong the service life of the hose and reduce the costly machine downtime and liabilities.

Hose guards are available in several styles and sizes as follows:

- **Metallic Spring Guards and Armor Guards (Fig. 1.90):** They distribute bending radii to avoid kinking in hose lines and protect hose from abrasion and deep cuts. Guards are constructed of steel wire and plated to resist rust.

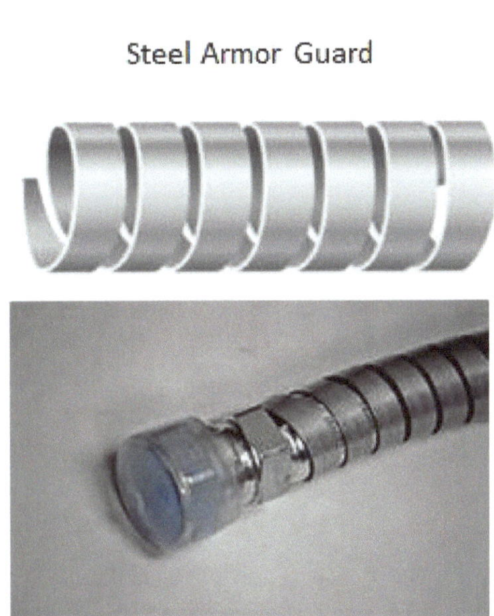

Fig. 1.90 - Metallic Spring and Armor Guards (Courtesy of Parker)

- **Nonmetallic Spiral Wrap Hose Guards (Fig. 1.91):** They protect against abrasion and extreme physical abuse. They are resistant to oil, lubricants, gasoline, most solvents and can withstand ambient temperatures from -40° to +300° F. They are available in different colors to make it easy for hose identifications via color coding, e.g. red color for high pressure hoses and blue color for low pressure hoses.

Fig. 1.91- Nonmetallic Spiral Wrap Guards (www.sapphirehydraulics.com)

- **Hose Shields (Fig. 1.92):** They are durable wear shields made of an abrasion and hydrocarbon resistant material that is impervious to solvents, oils, grease, and gasoline.

- **Hose Protection Sleeves (Fig. 1.93):** They protect hoses from contamination, direct sunlight, and the working environment. They are made of heat and chemical resistant materials.

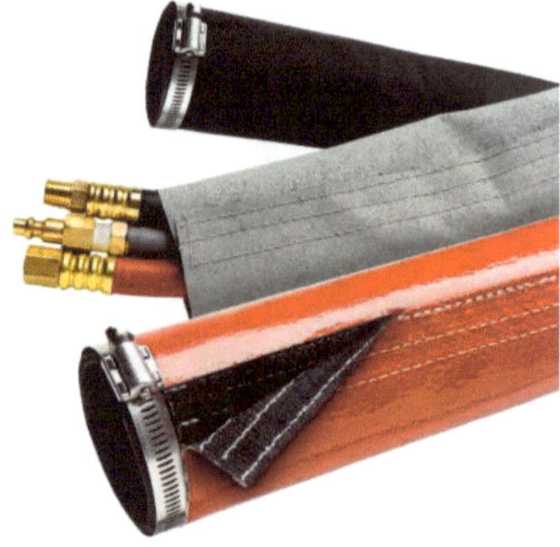

**Fig. 1.92- Hose Shields
(www.epha.com)**

**Fig. 1.93- Hose Protection Sleeves
(www.sealsaver.com)**

Hose Oil-Injection Protection Sleeves (Fig. 1.94):

A pinhole leak in a pressurized hydraulic hose can inject fluid through human body skin into the muscle, nerve, and tissue that can cause loss of life if not treated immediately. Ordinary nylon protection sleeves do not offer the level of safety needed when lives are at risk. As shown in the figure, a special sleeving system from Gates Corporation (LifeGuard™) protects workers within a three-feet circle around the line. This specially designed sleeve has the following characteristics:

- Working pressure 400 bar (6,000) psi and working temperature up to 100 °C (212°F).
- Compatible with a wide range of fluids, including environmentally friendly.
- Its inner layer is made of tightly woven, extruded filament nylon designed to absorb the energy of a hydraulic hose burst or pinhole leak by stretching up to 20 percent.
- The outer sleeve is resistant to abrasion but not specifically designed for abrasive environments. Nylon material layers contains escaped fluids and redirects them to the clamped ends of the hose.
- The sleeve is secured at either end of the hose with special "channel" clamps. The clamps allow leaking fluid to safely escape, so it will not collect behind the sleeve and cause a burst. Plus, leaked fluid allows for fast hose failure detection.

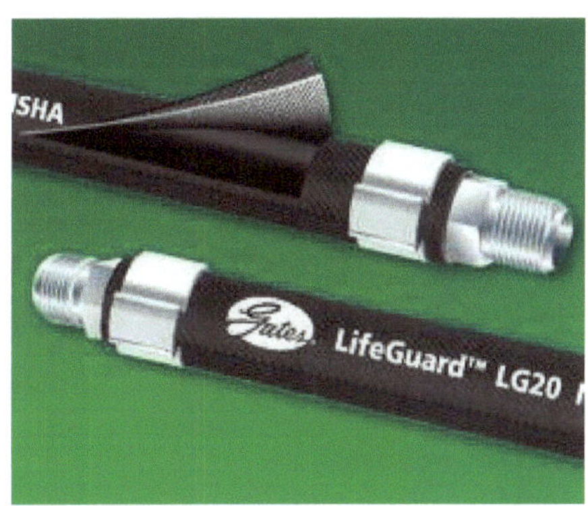

Fig. 1.94- Hose Oil-Injection Protection Sleeve "Lifeguard" (Courtesy of Gates)

1.6.10- Hose Whip Restraints

Major damage or injury can occur from whipping hoses, especially at higher pressures. As shown in Fig. 1.95, *Hose Whip Restraints* are designed to prevent whipping of a pressurized hose in the event of the hose separating from its fitting. So, the *Hose Whip Restraint* provides an additional level of safety by preventing damage to nearby equipment or injury to operators near the failed hose. The system consists of a hose collar and a cable assembly. The hose collar is selected based on the outside diameter of the hose, and the cable assembly is selected based on the type of hose connection.

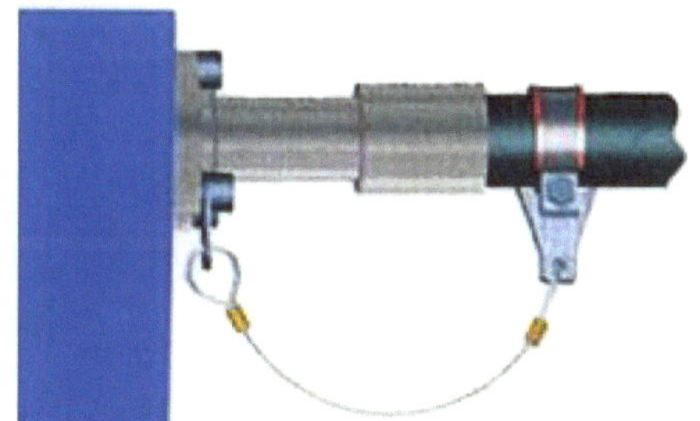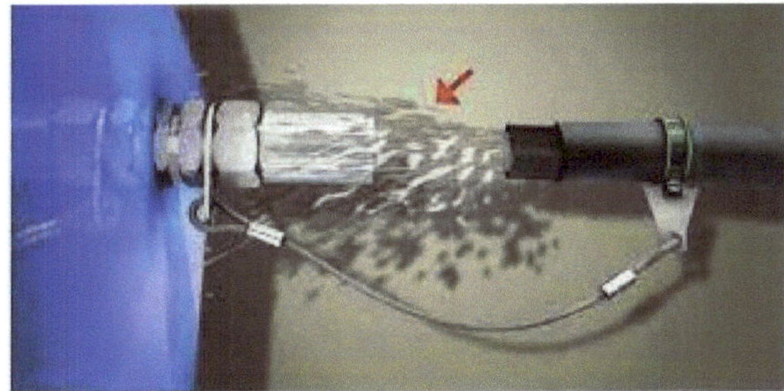

Fig. 1.95- Hose Wipe Restraint (Courtesy of Parker)

1.6.11- Hose Condition Monitoring

As shown in Fig. 1.96, the *LifeSense* from Eaton is a patented technology that intelligently monitors the condition of hose assemblies in real time. LifeSense can detect internal fatigue as well as external abrasion. LifeSense notifies users when a hose assembly is approaching the end of its useful life. Notification can be received on base units and mobile communication devices. The system requires DC power supply.

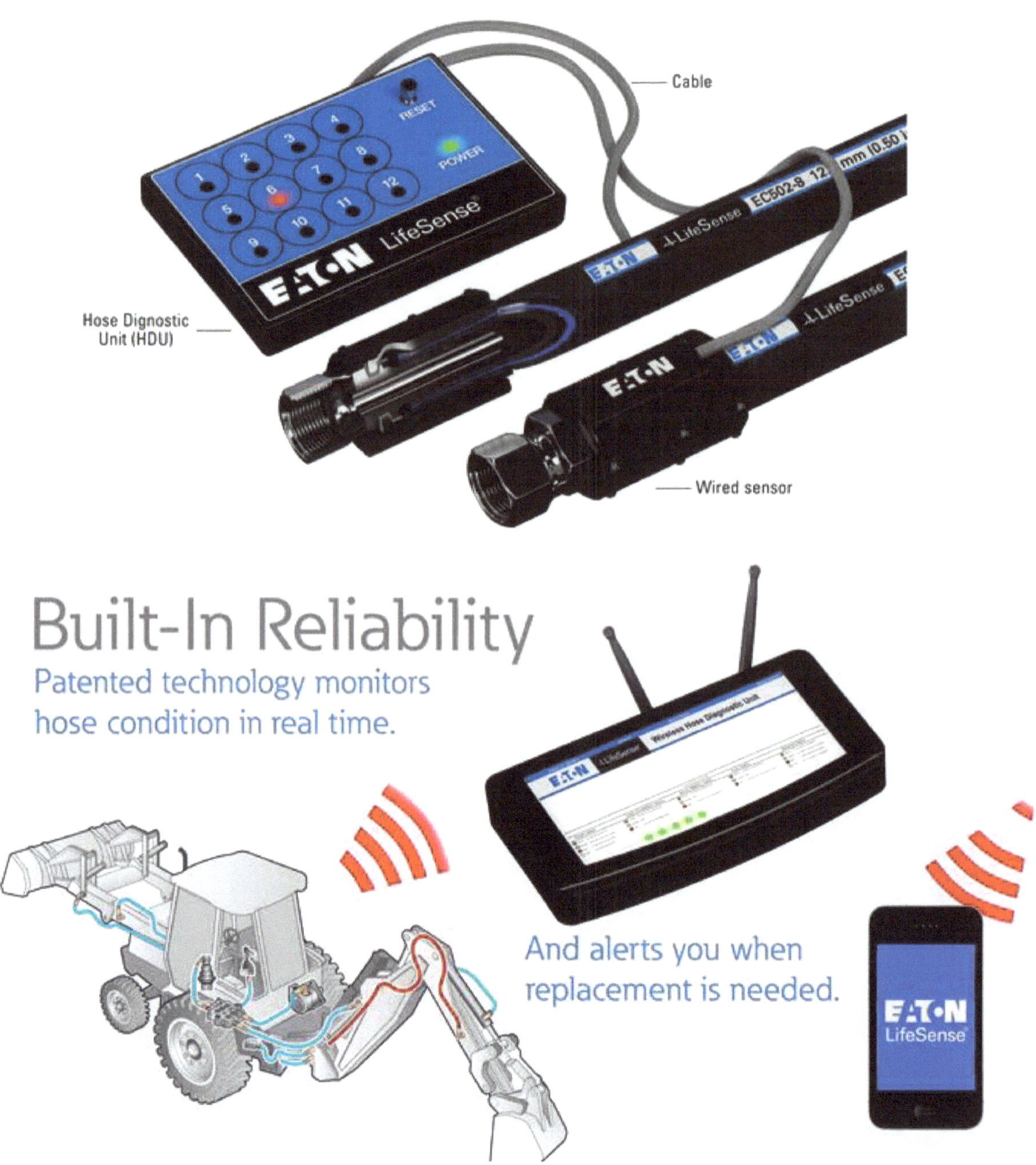

Fig. 1.96 - Real Time Hoses Condition Monitoring (www.hydrotechnik.com)

It was proved that, as shown in Fig. 1.97, using such a method of real time hose condition monitoring extends the service life of hoses as compared to time-bases traditional replacement method.

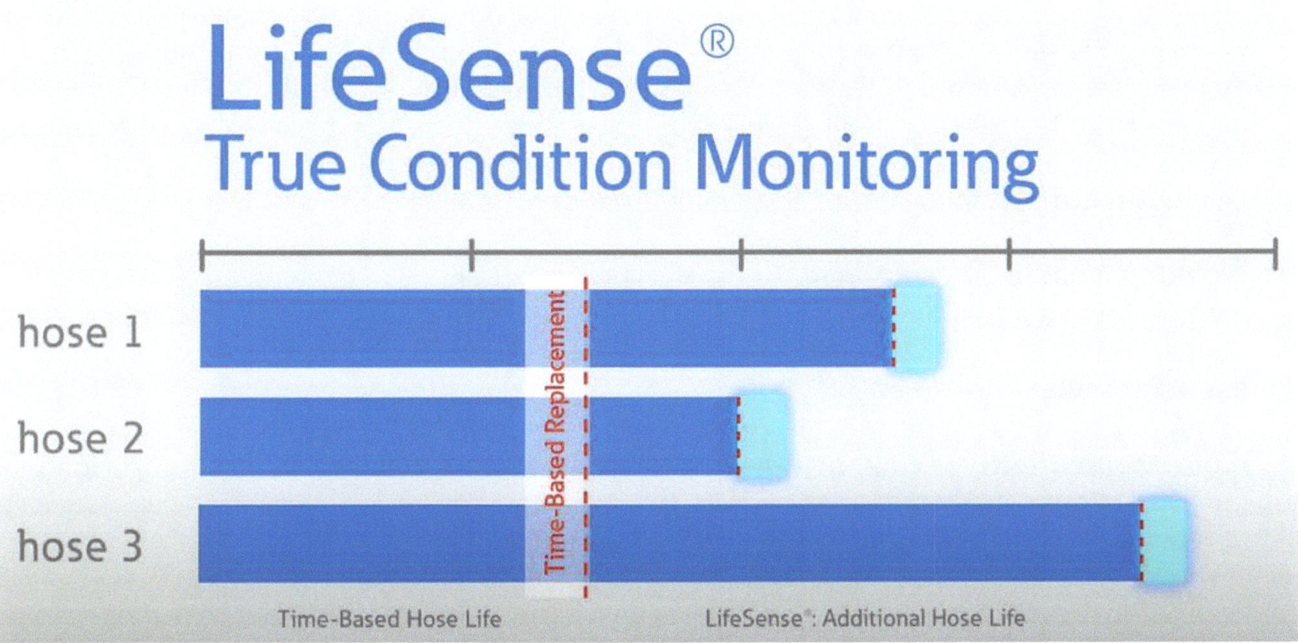

Fig. 1.97- Real Time Hoses Condition Monitoring Maximizes Hose Life (www.hydrotechnik.com)

1.6.12- Selection of Hose Couplings

Size: Make sure coupling size is properly selected in accordance with the size and type of the hose and the port of hydraulic component that the hose is connected to.

Working Pressure: Make sure that maximum allowable pressure for the couplings meets or exceeds the maximum allowable pressure for the hose. Some fittings don't seal well at high pressures and can develop oil leakage. O-ring-type fittings as well as solid port connectors work well at high pressures. Avoid the use of swivel staked nut couplings at extremely high pressures. Also, avoid the use of aluminum ferrules on high impulse applications.

Working Temperature: Mated surfaces type of fittings are recommended under extreme temperature fluctuations. Use of O-rings sealing fittings limits the maximum working temperature. It may be necessary to use O-ring materials that are suitable for high temperatures.

Fluid Compatibility: While hydraulic hose is commonly selected by its compatibility with fluid, couplings usually are not. However, it is advisable to check the chemical resistance charts for compatibility with coupling materials and O-rings.

Corrosion Resistance: In case of hose exposure to corrosive environment, hose couplings should be marked and selected as corrosion resistant and tested under SAE J516 and ASTMB I 17 *Salt-Spray Tests*.

Vibration: Coupling selection should consider if the end connection is exposed to motion and/or vibration, which can potentially weaken or loosen a connection. Use of split flange, or other couplings that use an O-ring for sealing, perform better under vibration. Avoid use of couplings that seal on the threads in such case.

Use of Quick Connect Coupling: check if the application requires the ability to connect and disconnect under pressure.

Use of Adaptors: You may want to select a coupling based on the need of adapters. Some couplings connect directly to a port, while others connect to adapters. Connecting directly to the port eliminates the need for an additional connection but can make installation more difficult. As shown in Fig. 1.98, Adapters can make installation easier; eliminate the need for coupling orientation but introduce an additional connection or possible leak point.

Fig. 1.98 - Use of Adaptors (Courtesy of Gates)

1.6.13- Hydraulic Hoses Selection Criteria

When selecting a hydraulic hose for an application, Fig. 1.99, provides a key word "STAMP" as a reminder for what to check. Interpretation of the key word is as follows:

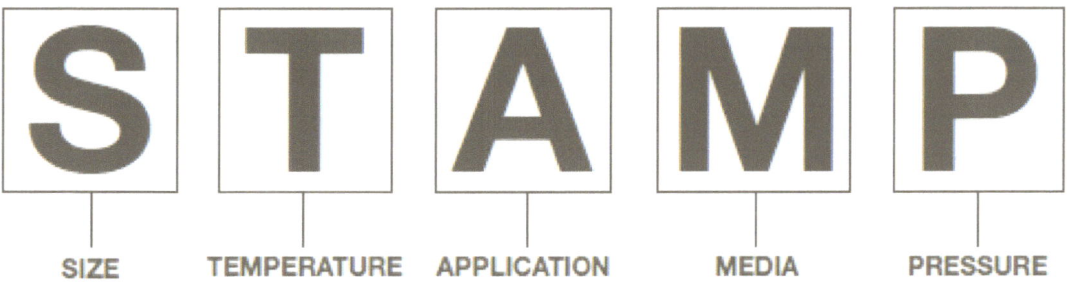

Fig. 1.99 - Keyword for Selecting Hydraulic Hoses (Courtesy of Parker)

Size: Line size has a direct effect on laminar flow condition and pressure losses. Therefore, a hose should be properly sized as per the rules presented in section 1.2.

Temperature: Working temperature of a hose must at least meet or exceed the application working temperature.

Application: Application in which the hose is going to be used should also be considered. For example, applications where hoses will encounter rubbing or abrasive surfaces, would be best handled by a family of abrasion resistant hoses with special cover.

Media (Fluid Compatibility) Material of a hose must be compatible the working hydraulic fluid. Hose manufacturers should provide hose-fluid compatibility charts.

Pressure: Maximum allowable static and dynamic pressure must be considered when specifying a hose as follows:

- Static Pressure: It is the maximum steady state pressure in the part of the system where this hose will be used. A common misconception that is the maximum pressure in a hydraulic system is the pressure setting of the relief valve. This isn't true in all cases. An example of that is pressure intensification at the rod side of a double acting differential cylinder when a system is designed with meter out speed control. In such a case, depending on the system design, configuration, and operating conditions, rod side pressure could be 3-4 times over the relief valve setting.

- Dynamic Pressure: Pressure pulses, spikes, and shocks are different forms of dynamic pressure. Pressure pulsation results from a cyclic load. Pressure spikes likely happened in transient time, e.g., when changing a direction of an actuator. Pressure shocks occur due to rapid increase in the load. Dynamic pressure causes fatigue of the transmission lines. Dynamic pressure likely isn't captured by pressure gauges, it rather requires special instrumentation, e.g., pressure transducer with appropriate readout device.

- Inside Negative Pressure: Suction hoses are structured with reinforcement layers to work under vacuum. If a pressure hose is used as a suction hose, it may collapse because pressure hoses are structured for inside positive pressure only.

- Outside Positive Pressure: If the environmental conditions aren't atmospheric, such as if a hose is used under water, make sure the hose structure can withstand the outside positive pressure. In this case, consult the manufacturer if necessary.

- Low-Pressure Hoses: *Low-Pressure* hydraulic hoses are designed for applications with operating pressures less than 20 bar (300 psi) such as lubrication systems or drain lines using mineral fluids. Their reinforcement layer is usually textile.

- Medium-Pressure Hose: *Medium-Pressure* hoses are designed for applications with operating pressures 20-200 bar (300 – 3000 psi) such as heavy-duty trucks, fleet vehicle, and aircrafts.

- High-Pressure Hoses: *High-Pressure* hoses are designed for applications with operating pressures up to 400 bar (6000 psi) such as construction equipment. These hoses are often called "two-wire" braid hose because they generally have a reinforcement of two-wire braids of high-tensile strength steel.

- Extremely High-Pressure Hoses: *Extremely High-Pressure* hoses are designed for applications with operating pressures above 400 bar (6000 psi) such as off-highway equipment and heavy-duty machinery where extremely high impulse or pressure surges are encountered. These hoses are often called "four-wire" or "six-wire" hoses.

In addition, the following conditions are also important to be reviewed:

Electrical Conductivity: Certain applications require that the hose be *nonconductive* to prevent electrical current flow or to maintain electrical isolation. For applications near high voltage electric lines, only special nonconductive hose can be used. Other applications require the hose and the fittings and the hose/fitting interface to be sufficiently conductive to drain off static electricity.

Length: Hose length must be wisely specified. Neither short nor long hoses are recommended. Short hoses can be overstretched when pressure is applied resulting in detaching end joints. Unnecessary long hoses result in vibration, rubbing, pressure losses, etc.

Hoses Pictogram: Figure 1.100 shows typical manufacturer's pictogram that are printed on the hose cover to indicate main hose specifications.

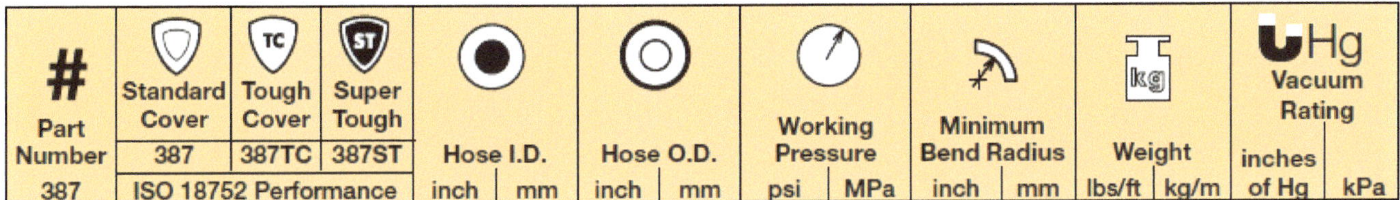

# Part Number 387	Standard Cover 387 ISO 18752 Performance	Tough Cover 387TC	Super Tough 387ST	Hose I.D.		Hose O.D.		Working Pressure		Minimum Bend Radius		Weight		Vacuum Rating inches of Hg	kPa
				inch	mm	inch	mm	psi	MPa	inch	mm	lbs/ft	kg/m	of Hg	kPa

Fig. 1.100 - Hydraulic Hoses Pictogram (Courtesy of Parker)

Figure 1.101 shows examples of hoses with pictogram marking.

Fig. 1.101 - Example of Hydraulic Hoses with Pictogram Marked
(Courtesy of Parker)

Example 1 – High Pressure Hoses: Figure 1.102 shows an example of hoses that are developed for high and ultra-high-pressure applications. They are used in high pressure service tube cleaning applications, such as, heat exchanger tube cleaning in the chemical and refining industries and as waterblast hoses. They can handle pressures above 13,000 psi. Available with a Tough Cover, for improved abrasion resistance in many applications.

PERFORMANCE CHARACTERISTICS	
SPECIFICATIONS MET	DIN EN1829-2 compliant
DESIGN FACTOR	2.5
HOSE INSIDE DIAMETER	1/8, 5/32, 3/16, 1/4, 5/16 inch 3.2, 4, 4.8, 6.4, 7.9 mm
HOSE OUTSIDE DIAMETER	.280, .300, .310, .370, .460, .530 inch 7.1, 7.7, 7.9, 9.5, 11.6, 13.5 mm
MAXIMUM OPERATING PRESSURE	15000, 15950, 17400, 20300, 21750 psi 1034, 1100, 1200, 1400, 1500 bar
MAXIMUM OPERATING TEMPERATURE	158 °F 70 °C
MINIMUM OPERATING TEMPERATURE	14 °F -10 °C
MINIMUM BEND RADIUS	2-3/8, 3, 3-3/4, 4-3/8, 4-3/4 inch 60, 75, 95, 110, 125 mm
COMPATIBLE FITTING	AX, TX (dependent on hose)
MEDIA	Water, Chemicals
APPLICATION	Industrial Cleaning - Waterblast, Shipbuilding, Construction
COVER COLOR	Green, Blue (dependent on hose)
WEIGHT	0.05 - 0.07 lb/ft 0.07 - 0.10 kg/m
HOSE INNER TUBE MATERIAL	Polyoxymethylene
COVER MATERIAL	Polyamide
HOSE REINFORCEMENT MATERIAL	Two spiral layers of maximum tensile steel wire

Fig. 1.102 - Examples of Hoses for High Pressure (Courtesy of Parker)

Example 2 – High Pressure Collapse Resistant Hoses: Figure 1.103 shows an example of *High Pressure Collapse Resistant* (HCR) hose is specifically designed to serve the oil & gas market in subsea hydraulics. Its state-of-the-art construction provides a minimum crush resistance of 6,600 psi external pressure.

PERFORMANCE CHARACTERISTICS	
SPECIFICATIONS MET	ISO 13628-5, API 17E (1.5 design factor)
DESIGN FACTOR	4
HOSE INSIDE DIAMETER	1/2, 1 inch 13, 25 mm
HOSE OUTSIDE DIAMETER	1.04, 1.83 inch 26.4, 46.4 mm
MAXIMUM OPERATING PRESSURE	5000 psi 34.5 MPa 345 bar
MAXIMUM OPERATING TEMPERATURE	131 °F 55 °C
MINIMUM OPERATING TEMPERATURE	-40 °F -40 °C
MINIMUM BEND RADIUS	4, 11.8 inch 102, 300 mm
COMPATIBLE FITTING	HV
MEDIA	Hydraulic Fluids
APPLICATION	Oil & Gas, BOP Stack, Subsea
COVER COLOR	Blue, Yellow
WEIGHT	.45 - 1.44 lb/ft .67 - 2.15 kg/m
HOSE INNER TUBE MATERIAL	Polyamide 11 with 316L SS Carcass
COVER MATERIAL	Polyurethane
HOSE REINFORCEMENT MATERIAL	Aramid Fiber Braid
HOSE I.D. (SIZE)	-8, -16
HOSE TYPE	Subsea, BOP Stack, Oil & Gas
NOTES	Parflex offers an unlimited number of hose assembly configurations. Please contact the division for part numbers and/or ordering assistance.
COLLAPSE PRESSURE	6600 psi 45.6 MPa

Fig. 1.103 - Examples of Hoses for High Collapse External Pressure (Courtesy of Parker)

Example 3 - Low Temperature Hoses: Figure 1.104 shows an example of low temperature thermoplastic hydraulic hoses feature optimum performance in cold temperatures, especially for applications requiring flexibility in cold climates with temperatures as low as -70°F. They are designed for construction and agriculture equipment that work in cold climates.

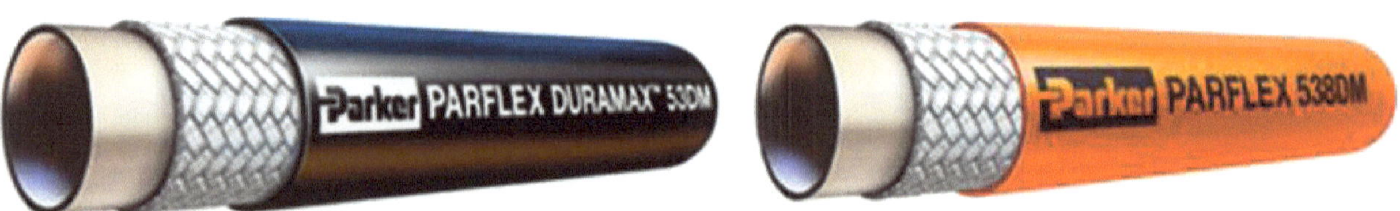

PERFORMANCE CHARACTERISTICS	
SPECIFICATIONS MET	100R17, SAE J517 100R18,SAE 100R18, SAE J517 for less than 50 micro-amps leakage under 75,000 volts per ft.
DESIGN FACTOR	4
HOSE INSIDE DIAMETER	1/4, 3/8, 1/2, 5/8, 3/4 inch 6, 10, 13, 16, 19 mm
HOSE OUTSIDE DIAMETER	.490, .640, .660, .770, .840, .970, 1.080, 1.420 inch 11.9, 16.3, 16.7, 19.5, 21.3, 24.6, 26.2, 28.7 mm
MAXIMUM OPERATING PRESSURE	3000 psi 20.7 MPa 207 bar
MAXIMUM OPERATING TEMPERATURE	250 °F 121 °C
MINIMUM OPERATING TEMPERATURE	-70 °F -57 °C
MINIMUM BEND RADIUS	1-1/4, 2, 2-3/4, 3-1/2, 4, 6-1/2 inch 32, 50.8, 51, 69.9, 88.9, 89, 102, 165 mm
COMPATIBLE FITTING	56 Series, CG Series, 43 Series
MEDIA	Air, Hydraulic Fluid, Oil Water
APPLICATION	Cold Applications, Construction Equipment, Hydraulics, Industrial Equipment, Over-the-Sheave, Pneumatics - dependent on size
COVER COLOR	Black, Orange
WEIGHT	0.07 - 0.40 lb/ft 0.10 - 0.60 kg/m
HOSE INNER TUBE MATERIAL	Copolyester, Non conductive Copolyester
COVER MATERIAL	Copolyester
HOSE REINFORCEMENT MATERIAL	Fiber, High Tensile Steel Wire Braid
HOSE I.D. (SIZE)	-4, -6, -8, -10, -12

Fig. 1.104 - Example of Hoses for Cold Temperature (Courtesy of Parker)

Example 4 - High Temperature Hoses: Figure 1.105 shows an example of high temperature hoses. PTFE hoses feature superior chemical compatibility, resist moisture, perform over a wide temperature range (up to 230 ^{0}C = 450°F) and offer a low coefficient of friction to reduce pressure drop.

PERFORMANCE CHARACTERISTICS	
SPECIFICATIONS MET	SAE 100R14A, SAE 100R14B, FDA CFR 177.1550 (natural)
DESIGN FACTOR	4
HOSE INSIDE DIAMETER	3/16, 1/4, 5/16, 13/32, 1/2, 5/8, 7/8, 1-1/8 inch 5, 6, 8, 10, 13, 16, 19, 22, 29 mm
HOSE OUTSIDE DIAMETER	.330, .400, .460, .560, .580, .660, .790, 1.060 inch 8.2, 10.1, 11.6, 14.3, 16.8, 20.1, 26.9 mm
MAXIMUM OPERATING PRESSURE	1000, 1200, 1500, 2000, 2500, 3000 psi 6.9, 8.3, 10.3, 13.8, 17.2, 20.7 MPa 69, 83, 103, 138, 172, 207 bar
MAXIMUM OPERATING TEMPERATURE	275, 450 °F 135, 232 °C
MINIMUM OPERATING TEMPERATURE	-40 to -73 °C -40 to -100 °F
MINIMUM BEND RADIUS	2, 3, 4, 5, 6-1/2, 7-1/2, 9 inch 50, 75, 100, 127, 165, 191, 229 mm
COMPATIBLE FITTING	90 Series, 91/91N Series
MEDIA	Adhesive, Air, Chemicals, Cooking Oil, Fluid, Water
APPLICATION	Fluid Handling, Chemical Transfer, Manufacturing / Industrial, Medical/Pharmaceutical, Packaging, Instrumentation, Transportation
COVER COLOR	Red, Brown
WEIGHT	0.06 - 0.39 lb/ft 0.09 - 0.58 kg/m
HOSE INNER TUBE MATERIAL	Natural, Static Dissipative PTFE
COVER MATERIAL	Silicone, Polyurethane
HOSE REINFORCEMENT MATERIAL	Stainless Steel Braid
HOSE I.D. (SIZE)	-4, -5, -6, -8, -10, -12, -16, -20
HOSE TYPE	Natural, Static-Dissipative
VACUUM PRESSURE RATING	10, 12, 14, 28 Inch-Hg 254, 305, 355, 711 mm-Hg

Fig. 1.105 - Examples of Hoses for High Temperature (Courtesy of Parker)

Example 5 – Nonconductive Hoses: Figure 1.106 shows an example of nonconductive hydraulic hose. It is the optimal solution for medium pressure hydraulic service where hydraulic circuit exposure and contact with high voltage may be encountered.

PERFORMANCE CHARACTERISTICS	
SPECIFICATIONS MET	SAE J517 electrical standards (non-conductivity), SAE J517 100R7, DNV-GL Type Approved, ANSI A92.2, ISO 3949-R7-2
DESIGN FACTOR	4
HOSE INSIDE DIAMETER	1/4 inch 6 mm
HOSE OUTSIDE DIAMETER	.470 inch 11.9 mm
MAXIMUM OPERATING PRESSURE	3,000 psi 20.7 MPa 207 bar
MAXIMUM OPERATING TEMPERATURE	212 °F 100 °C
MINIMUM OPERATING TEMPERATURE	-40 °F -40 °C
MINIMUM BEND RADIUS	1-1/2 inch 38 mm
COMPATIBLE FITTING	56 Series, 51R Series
MEDIA	Hydraulic Fluids, Lubricants
APPLICATION	Aerial Lift, Hydraulics, Mobile Equipment
COVER COLOR	Orange
WEIGHT	0.06 lb/ft 0.09 kg/m
HOSE INNER TUBE MATERIAL	Nylon
COVER MATERIAL	Polyurethane
HOSE REINFORCEMENT MATERIAL	Fiber
HOSE I.D. (SIZE)	-4

Fig. 1.106 - Examples of Nonconductive Hydraulic Hoses (Courtesy of Parker)

Example 6 – Fluid Compatibility: Table 1.15. shows an example of fluid compatibility chart provided by a manufacturer.

Chemical Name	Neoprene (Poly-Chloroprene)	Nitrile (Acrylonitrile and Butadiene)	Butyl (Isobutylene and Isoprene)	Hypalon (Chlorosulfonated Polyethylene)	EPDM (Ethylene Propylene Diene)	CPE (Chlorinated Polyethylene)
ASTM-SAE Designation SAE J14 & SAE J200	SC BC	SB BG	R AA	TB CE	R AA	None None
Flame Resistance	Very Good	Poor	Poor	Good	Poor	Good
Petroleum Base Oils	Good	Excellent	Poor	Good	Poor	Very Good
Diesel Fuel	Good to Excellent	Excellent	Poor	Poor	Poor	Very Good
Resistance to Gas Permeation	Good	Good	Outstanding	Good to Excellent	Fair to Good	Good
Weather	Good to Excellent	Poor	Excellent	Very Good	Excellent	Good
Ozone	Good to Excellent	Poor for Tube Good for Cover	Excellent	Very Good	Outstanding	Good
Heat	Good	Good	Excellent	Very Good	Excellent	Excellent
Low Temperature	Fair to Good	Poor to Fair	Very Good	Poor	Good to Excellent	Good
Water-Oil Emulsions	Excellent	Excellent	Good	Good	Poor	Excellent
Water-/Glycol Emulsions	Excellent	Excellent	Excellent	Excellent	Excellent	Excellent
Diesters	Poor	Poor	Excellent	Fair	Excellent	Very Good
Phosphate Esters	Fair (For Cover)	Poor	Good	Fair	Very Good	Very Good
Phosphate Ester Base Emulsions	Fair (For Cover)	Poor	Good	Fair	Very Good	Very Good

Table 1.15 - Hose-Fluid Compatibility Chart (Courtesy of Gates)

Example 7- Hoses for Extreme Conditions: Figure 1.107 shows an example of extreme high-pressure hoses. In such hoses, the central tube is made from synthetic combatable rubber and surrounded by 4-6 reinforcement layers of spiraled high-tensile steel wire. These reinforcement layers are separated by thin adhesive layer of rubber to keep the wires in place. The reinforcement layers are spiraled in opposite direction to balance the pressure force and increase the ability to withstand pressure impulses. As shown in the figure, compared with conventional spiral hose, the *Compact Spiral* design 797 hose offers measurably greater advantages in routing and installation, product size and weight

797
Hydraulic – Constant Working Pressure
ISO 18752 - AC/BC/CC/DC

Part Number	Standard Cover 797	Tough Cover 797TC	Super Tough Cover 797ST	Hose I.D.		Hose O.D.		Working Pressure		Minimum Bend Radius		Weight		Parkrimp 43 Series	Parkrimp 77 Series
	ISO 18752 Performance			inch	mm	inch	mm	psi	MPa	inch	mm	lbs/ft	kg/m		
797-4	AC	AC	AC	1/4	6,3	0.51	13,0	6000	42,0	2	50	0.21	0,31	●	
797-6	BC	CC	CC	3/8	10	0.66	17,0	6000	42,0	2-1/2	63	0.31	0,46	●	
797-8	BC	DC	DC	1/2	12,5	0.83	21,1	6000	42,0	4	100	0.45	0,67		●
797-10	BC	DC	DC	5/8	16	0.94	23,9	6000	42,0	4-1/2	115	0.54	0,80		●
797-12	BC	DC	DC	3/4	19	1.10	27,9	6000	42,0	5-1/4	135	0.78	1,16		●
797-16	BC	DC	DC	1	25	1.40	35,7	6000	42,0	6-1/2	165	1.17	1,74		●
797-20	BC	DC	DC	1-1/4	31,5	1.77	44,9	6000	42,0	8-3/4	225	1.95	2,89		●
797-24	BC	CC	CC	1-1/2	38	2.08	52,8	6000	42,0	12	305	2.66	3,96		●
797-32	BC	CC	CC	2	51	2.66	67,6	6000	42,0	15	380	4.37	6,50		●

Fig. 1.107 - Hydraulic Hoses for Extreme Conditions (Courtesy of Parker)

1.7- Flanges for Transmission Line Connections

Flanges versus Fittings:
- <u>Connection and Sealing:</u> The major difference between them is how they achieve an effective seal. In fittings, the connection and seal are achieved by tightening threads between the mating halves. In flanges the connection and seal are achieved by bolting together the flanges on two mating halves that have an O-ring in between.

- <u>Line Size:</u> As shown in Fig. 1.108, fittings are most effective and easy to install for small hose diameters (< 1" OD). Large size hoses (> 1" OD) are exposed to severe forces, high pressures, vibration, and shock loads. Therefore, flanges are required to achieve a secure connection. Flanges are also ideal for installation in areas where there is not enough swing clearance for a wrench.

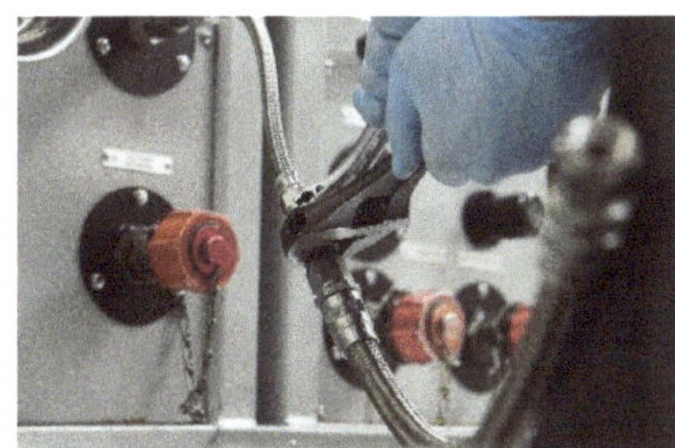

Fig. 1.108 - Use of fittings in Hoses (mac-hyd.com)

Flanges According to SAE-J518 Standard: There are two 4-Bolts flanges known worldwide as "*SAE Flanges*" per the codes shown below. As shown in Fig. 1.109, the SAE standard specifies the dimensions of the flange based on the line size. However, some OEM have their own flange dimensions.

- **SAE Standard Code 61** is for Pmax = (3000 – 5000) psi, depending on the size.
- **SAE Standard Code 62** is for Pmax = 6000 psi regardless of size.

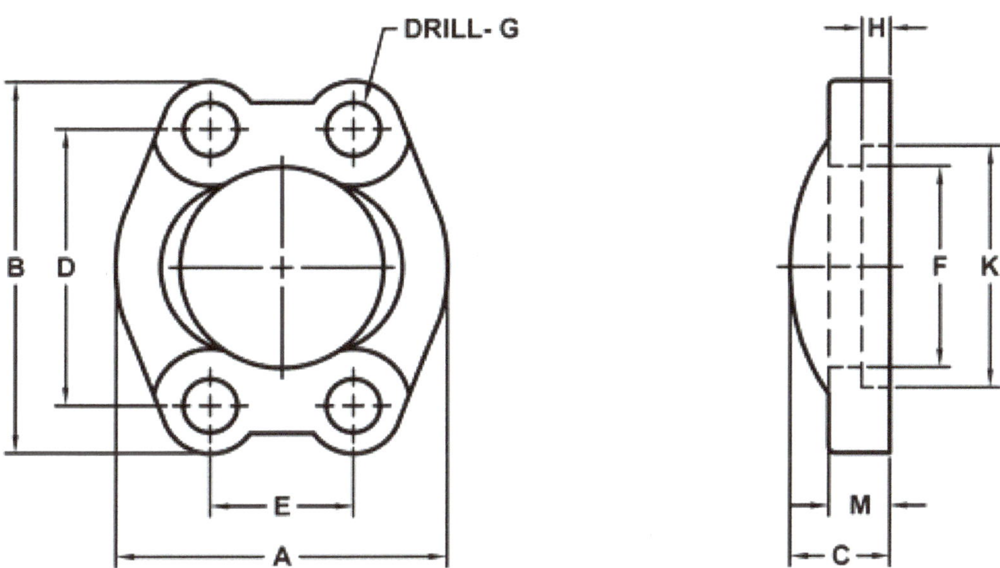

Fig. 1.109 - Standard Flange Dimensions

Components of Flange Assembly: As shown in Fig. 1.110, flanges could be one-piece or Split Flange, i.e. two pieces. Flange connections use a static face seal (a high durometer O-ring), and clamps/bolts for holding power as shown in the figure. The O-ring seal is compressed between the bottom of the groove in the flange head and the flat surface of the port or flange pad, providing a reliable soft seal. The flange, the O-ring, and four bolts are sold as a kit. Having an O-Ring limits using flange connections in high-temperature applications.

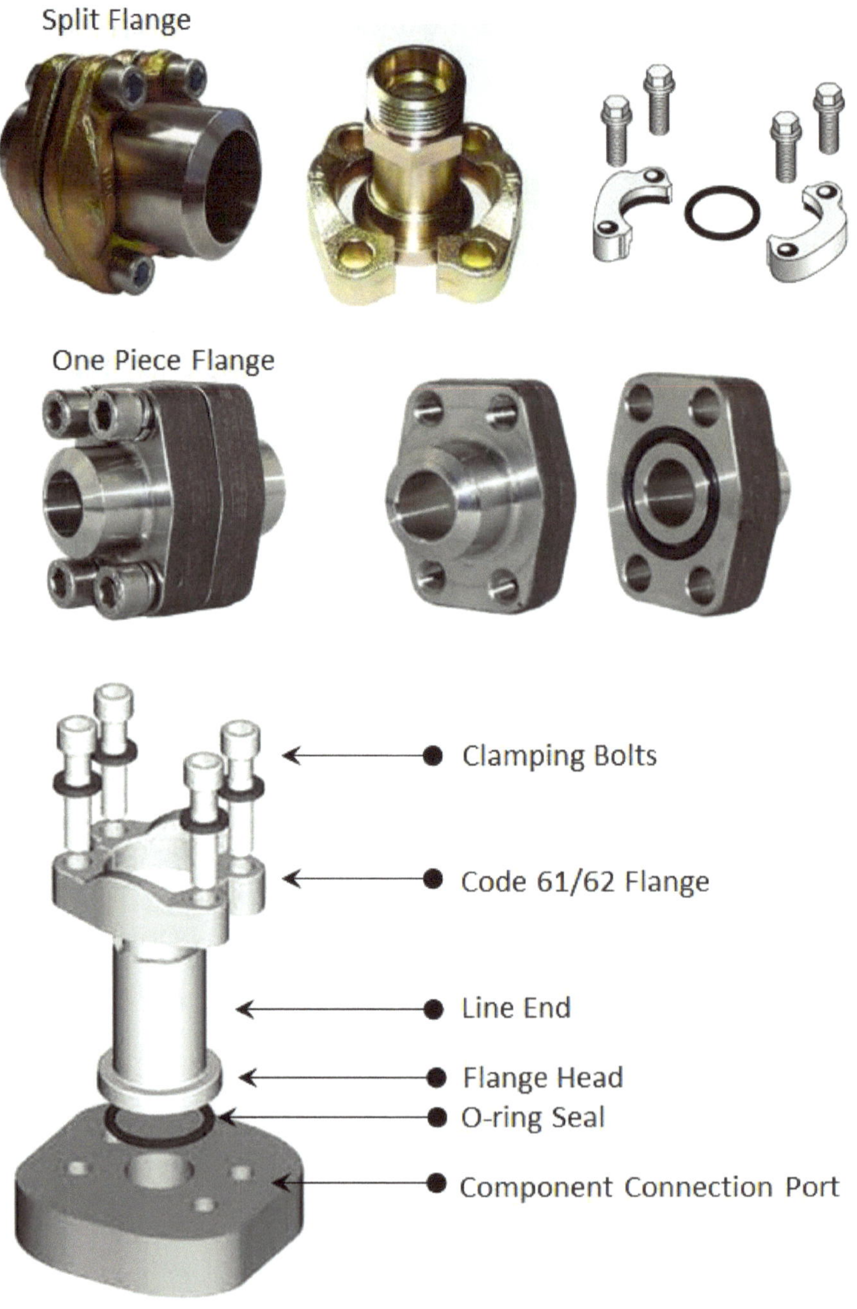

Fig. 1.110- Components of Flange Assembly

One-Piece versus Split Flanges: Flanges are used with all kinds of lines (Pipes, tubes, and hoses) usually as a connection on pumps and motors. As shown in Fig. 1.111, 4-bolt one-piece welded-end flanges could connect pipe sizes up to 5 inches. Figure 1.112 shows that use of split flanges is more adequate for connecting tubes and hoses.

Fig. 1.111 - Hydraulic Pipe Flange Assembly

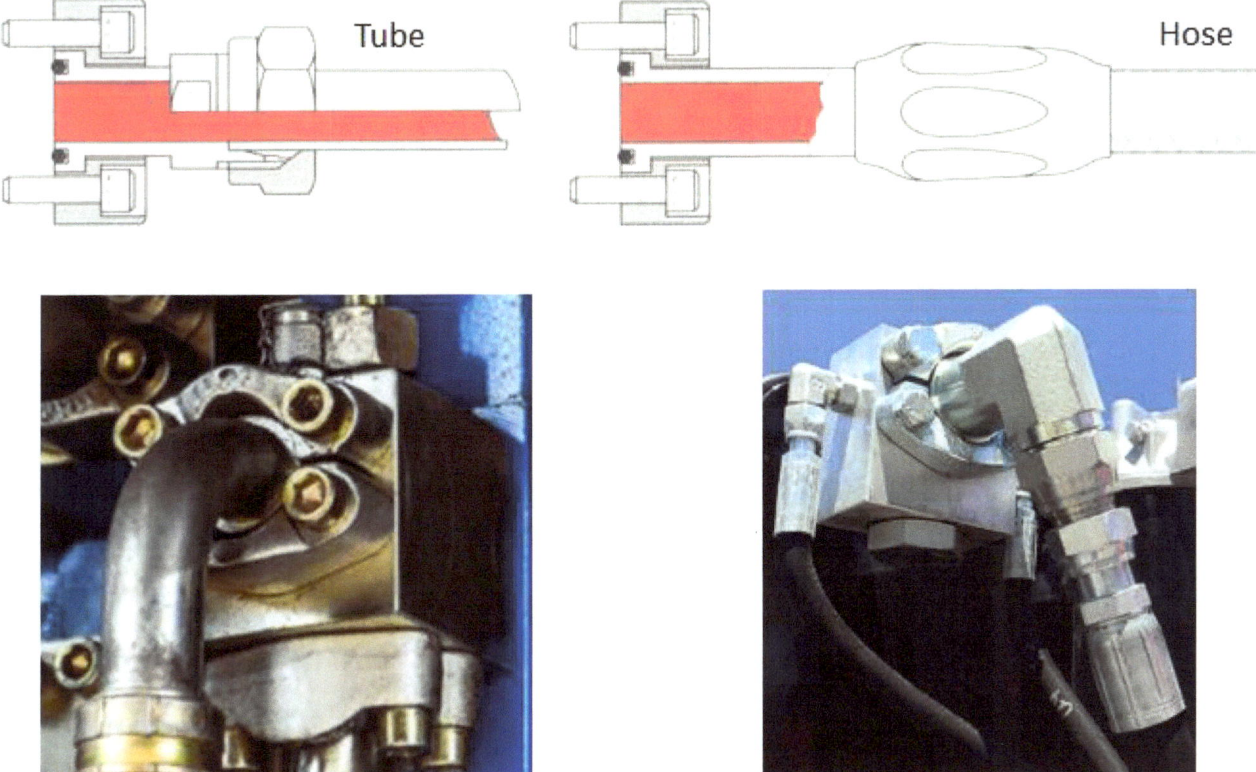

Fig. 1.112 - Hydraulic Tube and Hose Flange Assembly

Square Flanges: As shown in Fig. 1.113, *Square Flanges* are widely used to connect between two tubes or two pipes in hydraulic systems. Square flanges are suitable to connect a tube or a pipe to some hydraulic component, such as a hydraulic valve, hydraulic block, or other component. They are available in full range of dimensions and materials, mainly carbon steel.

Square flange usually is designed and manufactured in different standards in different countries as follows:

- ANSI/ISO 6164/ASME Square flanges.
- JIS B2291, JIS F7806, DIN 3901, BS 16.5.
- Customized square flanges are also available according to customers' drawing and specifications.

Fig. 1.113 - Square Flanges (https://flanges-pipe.com)

1.8- Rubber Expansion Fittings

Reasons to use Rubber Expansion Fittings: They are used in a pipework system to:
- Absorb expansion and contraction caused by thermal stresses due to temperature fluctuation.
- Absorb expansion and contraction caused by mechanical stresses due to movement of foundations, fluctuating loads or vibration during startup and flow surges.
- Isolate structure-borne noise.
- Accommodating inaccuracy and misalignment in installation.

Material for Rubber Expansion Fittings:
- They are made of natural rubber or synthetic rubber depending on the pressure.

Working Conditions for Rubber Expansion Fittings:
- The pressure strength of rubber expansion fittings depends on the size and design of the fittings. Usually, they are good for pressures below 20 bar (300 psi). The preferred point for locating rubber expansion fittings in hydraulic systems is in the suction and drain lines. It can also connect to isolation valve.
- Usually, expansion fittings with synthetic fiber reinforcement can be used up to +10°C and with steel wire reinforcement up to + 130°C.

Figure 1.114 shows various styles of rubber expansion fitting.

Fig. 1.114 - Rubber Expansion Fittings (www.grainger.com)

1.9- Test Points

Function of Test Points: *Test points* were developed and widely used in the design of hydraulic systems for the following purposes:
- Pressure monitoring and venting.
- Sampling of hydraulic fluids.
- Connection with instruments and sensors.

Features of Test Points (Fig. 1.115): Test points have the following features:
- Ensures 100% leakage free coupling while under system pressure.
- Available sealing materials are NRB, FKM and EPDM.
- Available in various styles and sizes.
- Available standards are ISO, SAE, and international threads.
- Available in steel and stainless steel.

Fig. 1.115 - Series 1620 Test Points (www.hydrotechnik.com)

Innovative Test Point (Fig. 1.116): An innovative test points (ICON™) were developed to allow easy and tight seal connection with various fluid power components. As shown in the figure, they can connect to various tube ends configurations. Moreover, they add sensing capabilities, include instant LED visual feedback, and data output to verify sealing connection accuracy. Hence, they maximize production efficiency, and optimize maintenance.

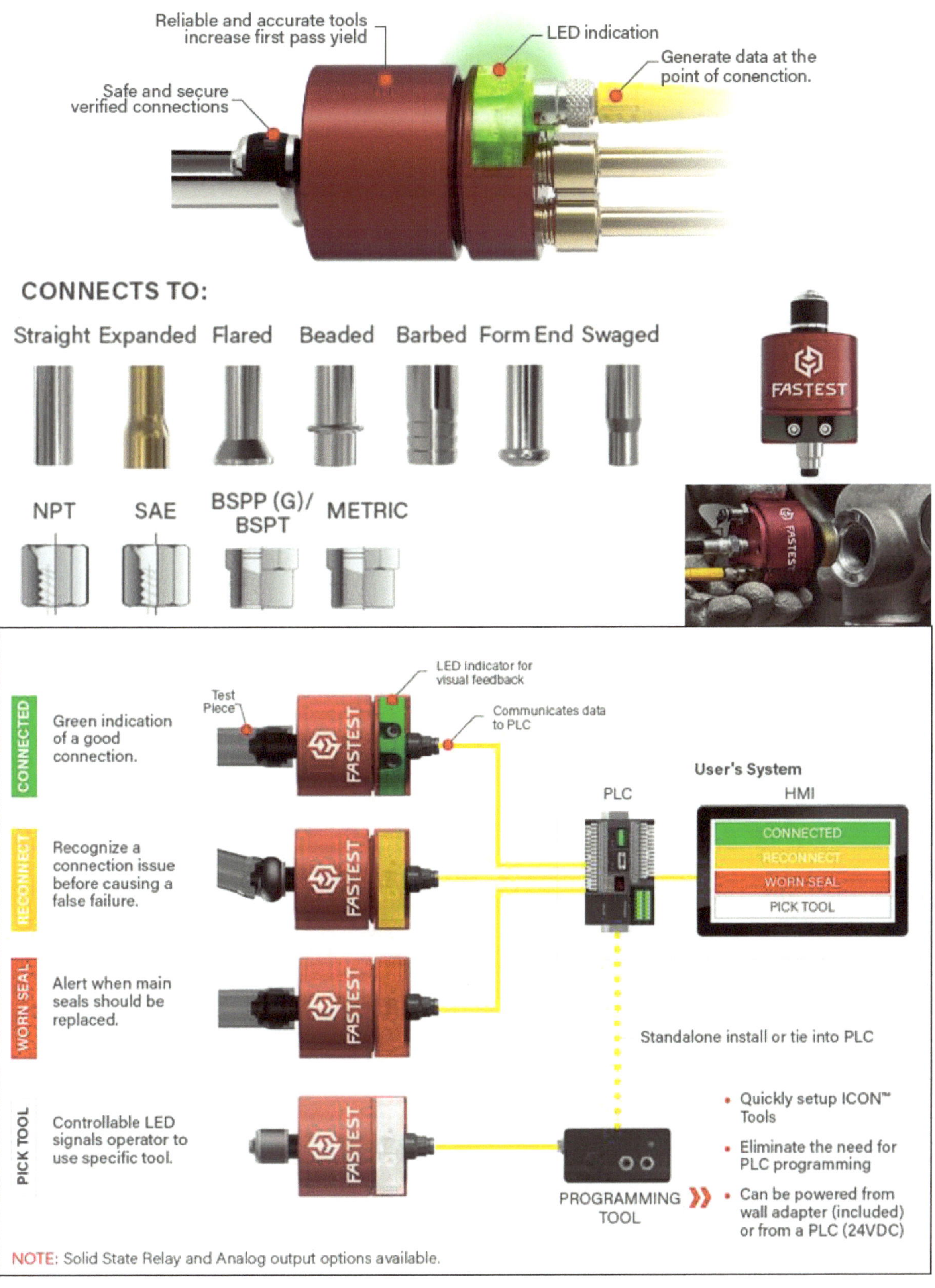

Fig. 1.116 - Innovative Test Points (Courtesy of CEJN)

1.10- Pressure Measurement Hoses

Function of Measurement Hoses: *Measurements Hose Lines* are used for purposes of pressure measuring, transmitting control pressure, and system diagnostic.

Features of Measurement Hoses: Measurement hoses have the following features:
- They have been developed for applications with pressures of up to 600 bar (9000 psi).
- They are characterized by their resistance to aggressive fluids.
- As shown in Fig. 1.117, they are constructed from a core, reinforcement layer, and protective cover.
- As shown in Fig. 1.118, they are supported by anti-kinking spirals or Aluminum protective covers.
- As shown in Fig. 1.119, they can be combined with test points.
- As shown in Fig. 1.120, use of these hoses is associated with pressure drop that should be reported by the hose manufacturer.

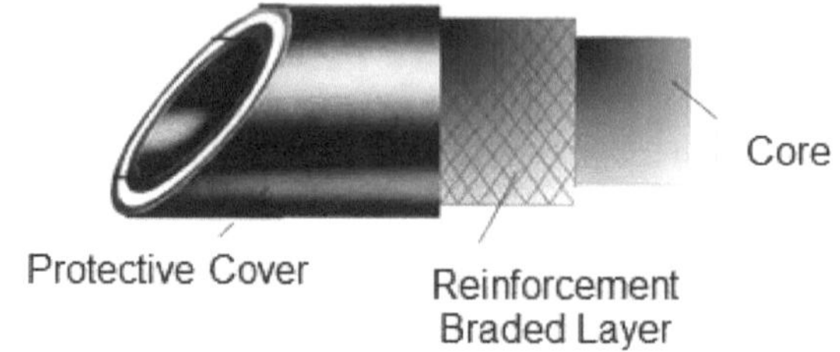

Fig. 1.117 - Construction of Measurement Hoses (www.hydrotechnik.com)

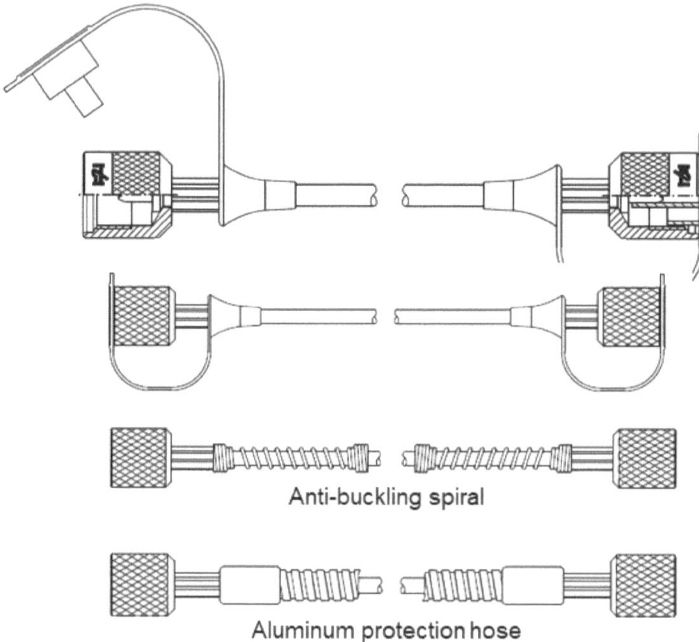

Fig. 1.118 - Measurement Hoses Supported by Anti-Buckling and Protection Covers (www.hydrotechnik.com)

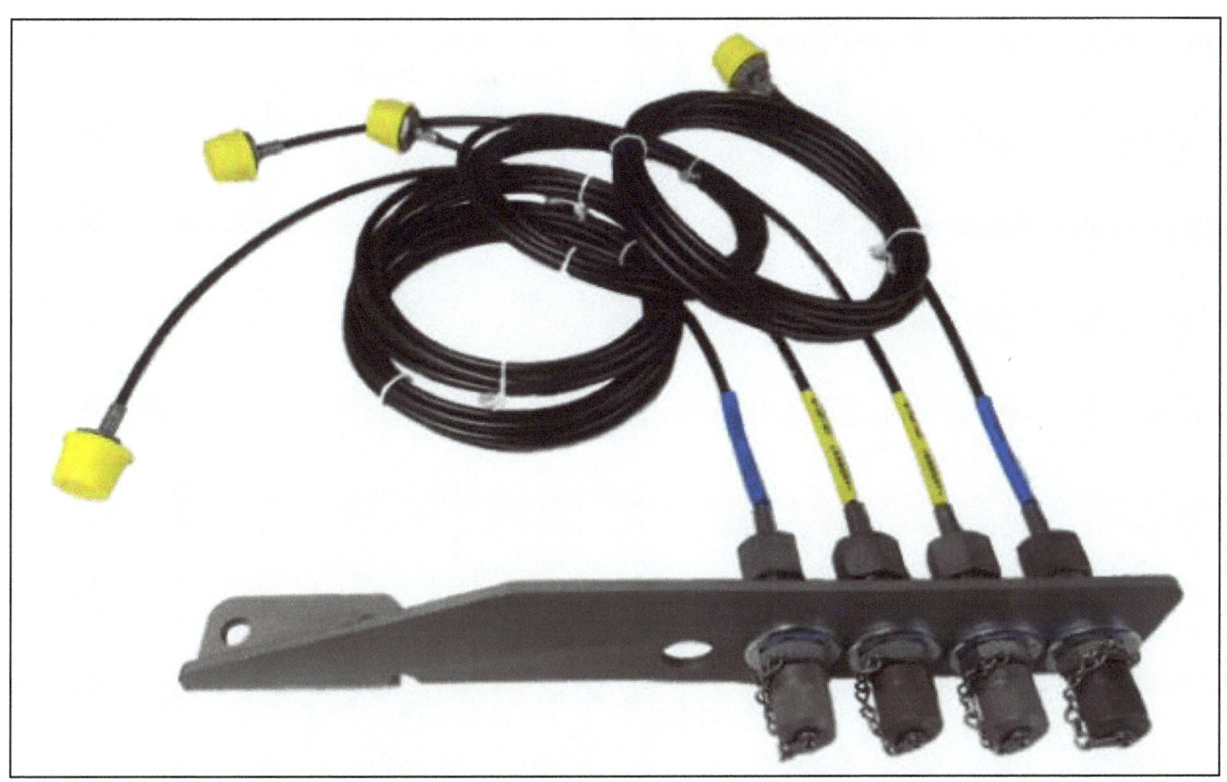

Fig. 1.119 - Measurement Hoses Combined with Test Points (www.hydrotechnik.com)

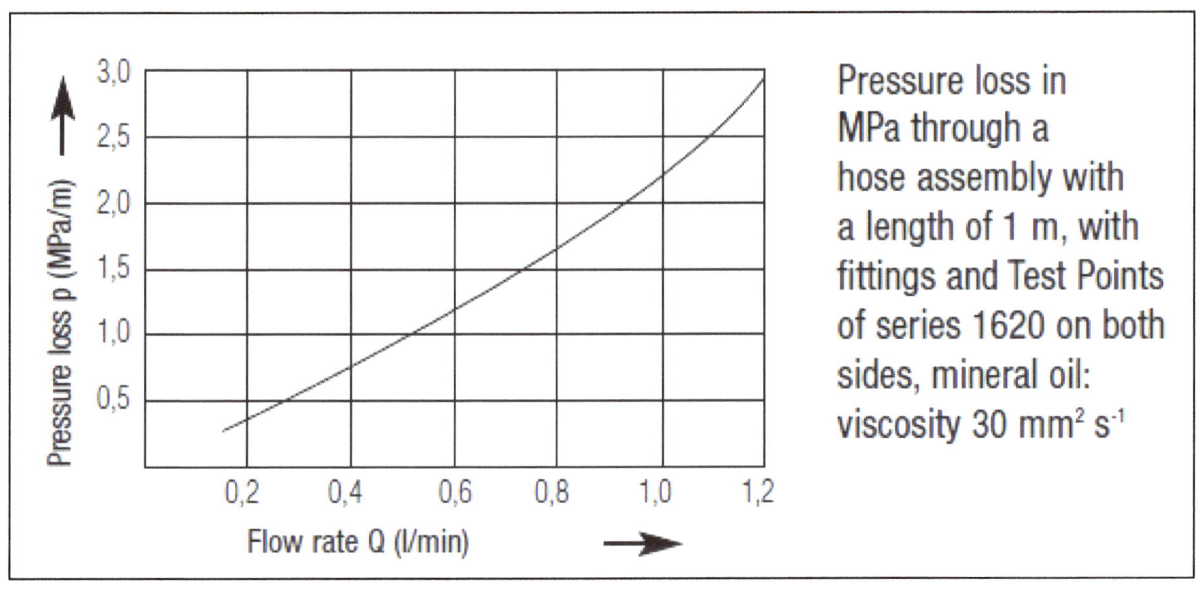

Fig. 1.120 - Pressure Drop in Measurement Hoses (www.hydrotechnik.com)

Routing of Measurement Hoses: Reliability of the system and lifetime of the hose assembly are dependent on the correct installation. Measurement hoses can easily be routed to make maintenance and diagnostic activities easier. However, as shown in Fig. 1.121, routing them should follow the best practices to avoid hose failure.

1. Under load, the length of a hose pipe can change. shortening causes an additional tensile stress of the hose and the connections. Therefore, the hose pipe should be in an unpressurised state. Also, union nuts must only be tightened for the torque specified by manufacturer.

2. With curved assemblies, attention must be paid to the bending radius. Sharp bends must be avoided wherever possible.

3. 90° hose fittings are also available to aid in the fitting of hose assemblies to maximize life and operation of the assembly.

4. 90° hose fittings can also aid in installing hose assembly in tight positions.

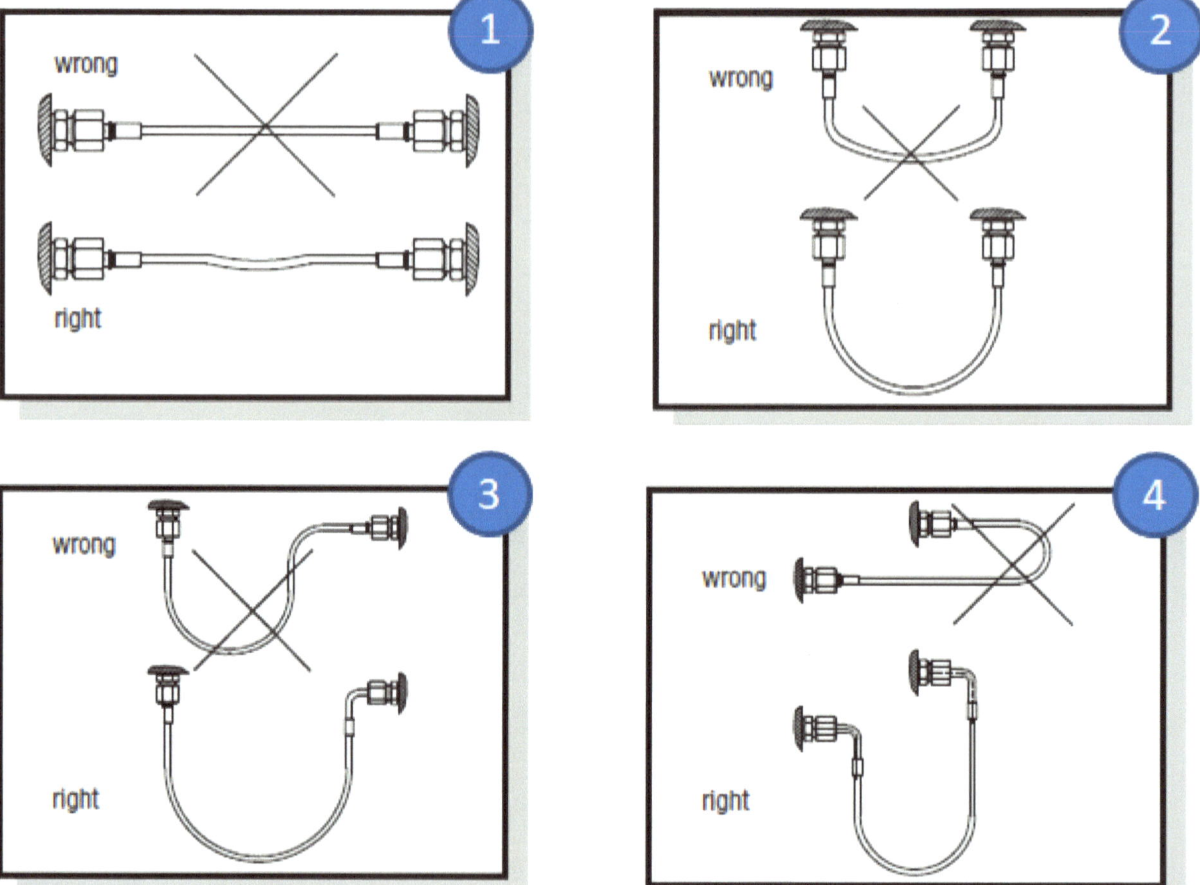

Fig. 1.121 - Best Practices of Routing Measurement Hoses (www.hydrotechnik.com)

1.11 Manifolds

Manifolds are machined blocks with internal passages to connect hydraulic components. The main purpose of using such manifolds is to minimize use of transmission lines for the sake of developing compact, fast response, and energy efficient systems. Subplates are manufactured according to ANSI/NFPA standard for various types and sizes of hydraulic valves. They are also customized for group of components. There are several software platforms, such as AutoCAD and Automation Studio, used to customize a manifold based on hydraulic circuit diagram and components of specific brands. Figure 1.122 shows common type of manifolds.

Subplates (1): *Subplates* allow a simple solution to mount a single valve in a hydraulic system. Several different valve sizes, port sizes, and mounting configurations are available.

Cover Plates (2): Also known as "blanking plates," "blocking plates," or "cap-off plates". A *Cover Plates* is used to cover a subplates after a valve is removed or to cover a valve pattern intended for future use.

Bar Manifolds (3): *Bar Manifolds* reduce plumbing and the number of leak points. Hence, improve maintenance of hydraulic systems. Multi-station manifolds allow several valves to be mounted to a single manifold which has common pressure and tank connections.

Tapping Plates (4): Also referred to as *Sandwich Plates* or *Module Plates*. They provide threaded port access to the A, B, P, T, X and Y passages of an existing valve manifold to add a test port or gauge ports.

Valve Adaptors (5): This *Stack Module* mounts between the existing manifold and a valve. Used to convert the existing manifold valve pattern to an alternative size valve without disrupting the existing manifold configuration.

Din Bodies (6): These are single valve bodies to house DIN slip-in logic valves used in applications that require higher flow than a typical subplate mounted or threaded cartridge valve.

Header & Junction Blocks (7): Also known as *Distribution Manifolds*. Used to connect several threaded connections into one common block to eliminate leaks and reduce plumbing labor.

Custom Engineered Manifolds (8): They are designed for specific applications after the circuit is designed and components are selected. These designs optimize application with a single integrated manifold solution, eliminating leak points and labor costs associated with standard off-the-shelf discrete manifolds.

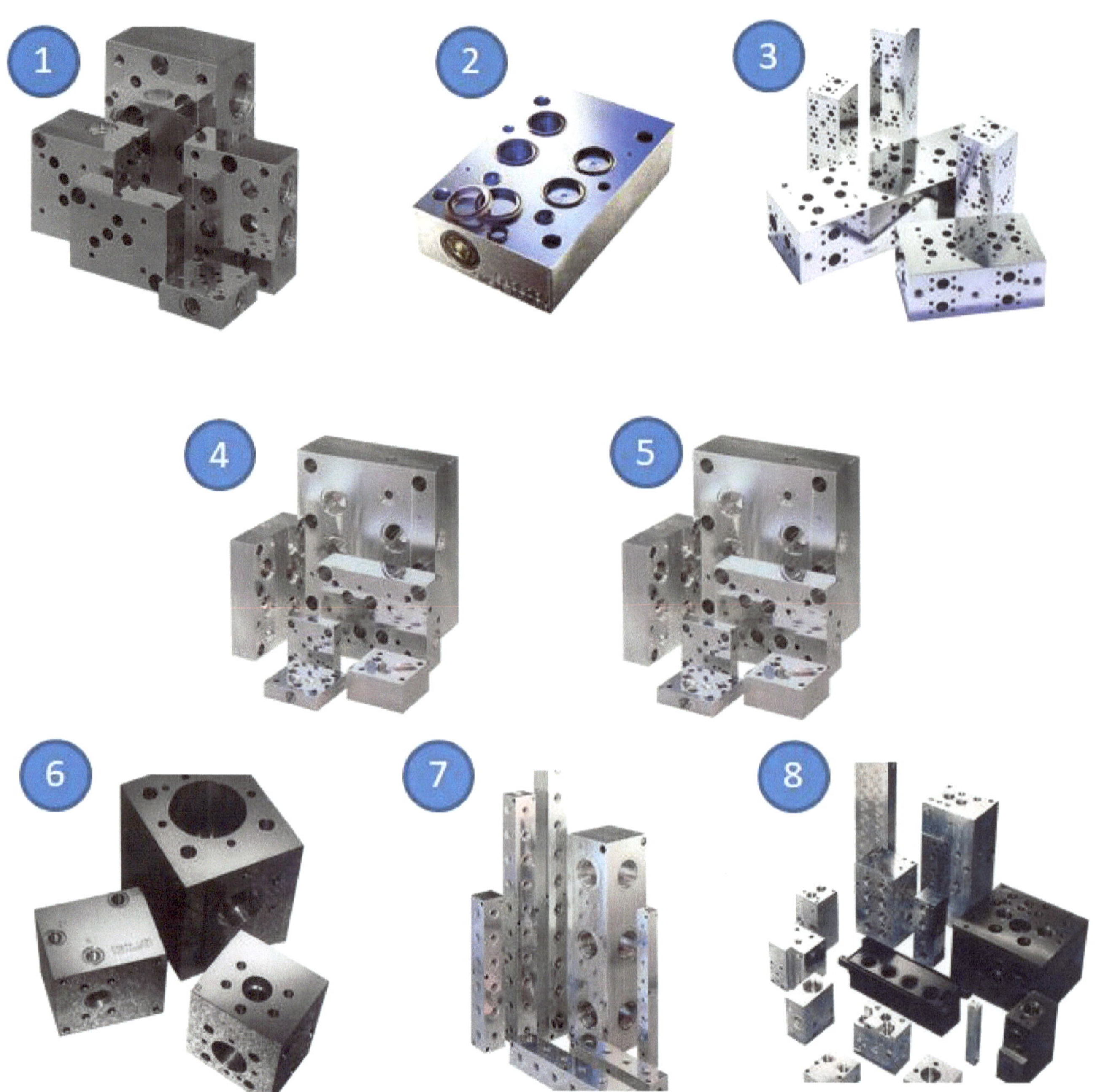

Fig. 1.122 – Hydraulic Manifolds (www.daman.com)

Chapter 2

Contamination Control in Hydraulic Transmission Lines

Objectives

This chapter discusses best practices for controlling contamination in hydraulic transmission lines including projectile cleaning and hydraulic system flushing.

Brief Contents

2.1- Contamination in Hydraulic Transmission Lines
2.2- Projectile Cleaning
2.3- Pickling of Hydraulic Transmission Lines
2.4- Flushing of Hydraulic Transmission Lines

Chapter 2 – Contamination Control in Hydraulic Transmission Lines

2.1- Contamination in Hydraulic Transmission Lines

In hydraulic systems, power is transmitted through pressurized fluid within transmission lines. The initial cleanliness level of a hydraulic system can affect its performance and service life. Unless removed, contaminants present after manufacturing and assembling a system can circulate through the system and cause damage. To limit such damage, the fluid and internal surfaces of the hydraulic fluid power system must be cleaned to an acceptable level.

ISO 1643 describes a clean-up procedure that uses filters after final assembly of the system. This practice is not a substitute for the use of good practices to achieve and maintain cleanliness prior to final assembly.

2.1.1- Sources of Contamination in Hydraulic Transmission Lines

Particulate contamination in hydraulic transmission lines can be:

- **Built-in:** during manufacturing and/or assembly (cutting, crimping, bending, flaring, etc.). This process generates significant amount of contamination that must be removed before pulling the system in service.
- **Introduced:** Ingested from surrounding (e.g., during long storage).
- **Introduced:** Induced during system repair (e.g., during untightening & retightening).
- **Generated:** Generated and settled inside transmission lines during system operation (e.g., due to bearing wear).

2.1.2- Methods for Cleaning Hydraulic Transmission Lines

The common basic methods used to assure the cleanliness of transmission lines are:

- Projectile Cleaning.
- Pickling using Chemicals.
- Hydraulic System Flushing.

In order to minimize generated contaminants during transmission line fabrication, the following best practices should be considered:

Cut the Line to the Correct Length: Cutting the line to the correct length the first time will eliminate additional processes for adjusting the line length that produce more contaminants.

Use the Right Saw Blade: Abrasive wheeled chop saws are the worst tool for hose cutting. As shown in Fig. 2.1, switching to a metal blade whenever possible reduces the amount of debris generated by the cutting process. Use a *"Clean Cut"* saw or blade that is specifically designed to minimize the ingress of contamination.

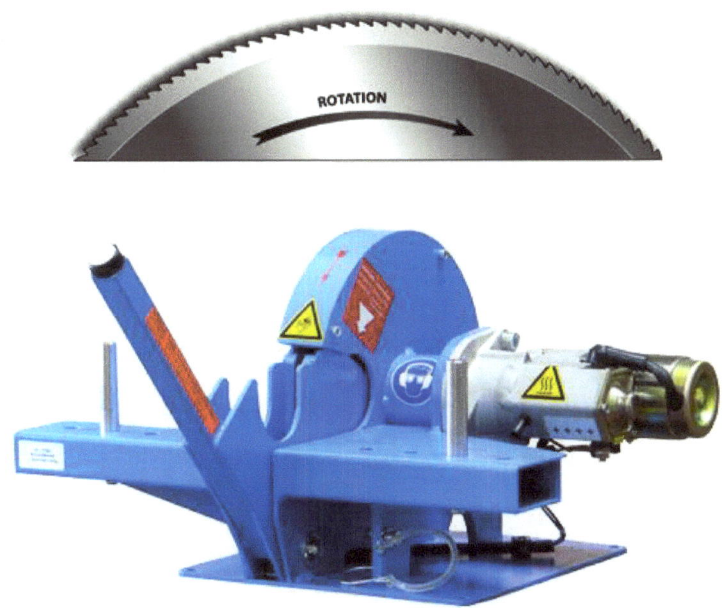

Fig. 2.1- Transmission Line Chop Saw

Use Clean Air: Using compressed air is a common method of cleaning a hose after cutting. However, using dirty air is like mopping a floor with muddy water. Moist air is even worse. Water particles will adhere to the walls on the inside of hose or tubing, capturing, and retaining airborne dust particles and contaminating the hydraulic oil on the equipment. Using clean, dry air is one of the best ways to avoid recontamination.

Clean Work Area: Transmission line fabrication should be done in a clean area of the facility isolated from manufacturing processes, that generates airborne contaminants, such as welding, grinding, sand blasting, painting, etc. Keeping workshop floor clean prevents dust and airborne particles from circulating around the shop and settling all over the place. It is recommended to use a vacuum instead of a broom to avoid stirring up the dirt.

2.2- Projectile Cleaning

2.2.1- Projectile Cleaning Overview

Distributors of transmission lines today are feeling more and more pressure from their customer base and OEM's to meet or maintain cleanliness levels. Transmission lines suppliers are judged based on their conformance to cleanliness specs. Hydraulic transmission lines manufacturers may not be as concerned about cleanliness as distributors need to be, but their actions can still affect distributors' cleanliness results. However, as an end user for transmission lines, should never assume that transmission lines are clean before assembling the transmission line in the system.

Projectile Cleaning of transmission lines is a relatively new cleaning technique. As shown in Fig. 2.2, projectile cleaning process is a sponge-like projectile which is shot through hose and tubing assemblies by a blast of compressed air. The Projectile strips out the internal contamination as it travels through line, forcing the contamination out in front of it.

Research conducted by Fluid Power Institute at Milwaukee School of Engineering (FPI-MSOE) several years ago indicated that projectile cleaning could reduce the contamination level by up to four ISO Codes in hydraulic hoses with diameters ranging between ½ to 1-inch I.D. Multiple projectiles were shot through the hose until they exited the transmission line visibly clean.

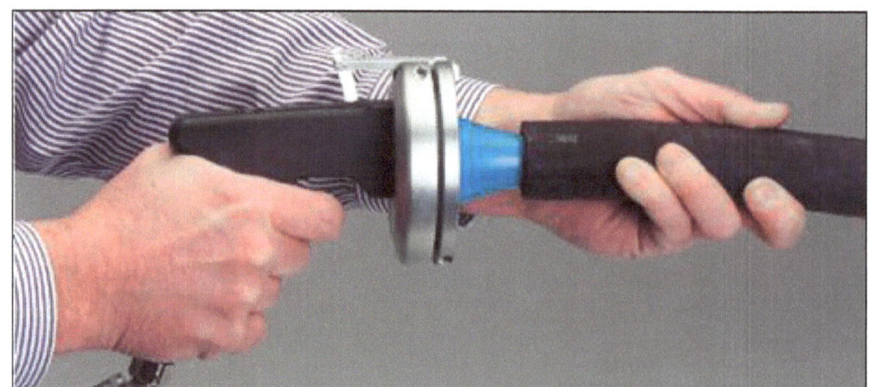

Fig. 2.2- Transmission Lines Projectile Cleaning (Courtesy of Ultra Clean Technologies)

2.2.2- Projectile Cleaning Equipment

The equipment for this procedure is much more affordable than that for high-velocity flushing. It cleans hydraulic lines much more effectively than compressed air alone and in a fraction of the time it takes for high-velocity flushing. The main equipment used are as follows:

Compressed Air: As shown in Fig. 2.3, there should be a source of clean-dry compressed air with the following requirements:
- 80 PSI (5.5 Bar) minimum to 110 PSI (7.5 Bar) maximum.
- 1/2" ID air hose to ensure 110 SCFM (3.1 m³/min) air flow.
- 5-micron filter and regulator with gauge are strongly suggested!

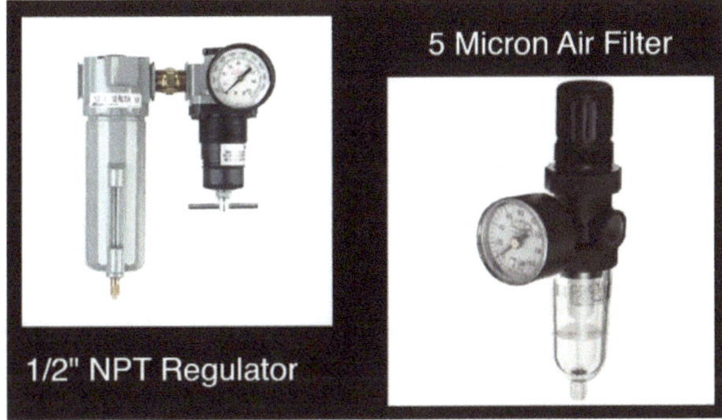

Fig. 2.3- Compressed Air Requirements (Courtesy of Ultra Clean Technologies)

Launcher: A *Launcher* is required to hold and fire the projectile in the line. As shown in Fig. 2.4, various sizes and styles of launchers are available to meet the different types of hydraulic lines. A one launcher can be used for various types of nozzles.

Fig. 2.4- Launchers (Courtesy of Gates)

Projectiles: As shown in Fig. 2.5, various types of projectiles are available with different sizes. Usually, projectiles are sized 20% to 30% larger than the I.D. of the cleaned line. *Regular Projectiles* (1) are used for all lines, particularly hoses. These types of projectiles are non-reusable. If tubes or pipes appear to contain rust, weld slag or other corrosion particles on the inside surface, then *Abrasive projectiles* (2) should be used first, as many times as is necessary to remove the corrosion. Grinding Projectiles (3) are recommended for all types of carbon steel piping products. Abrasive and grinding projectiles may be used more than once, if necessary, or until they wear out.

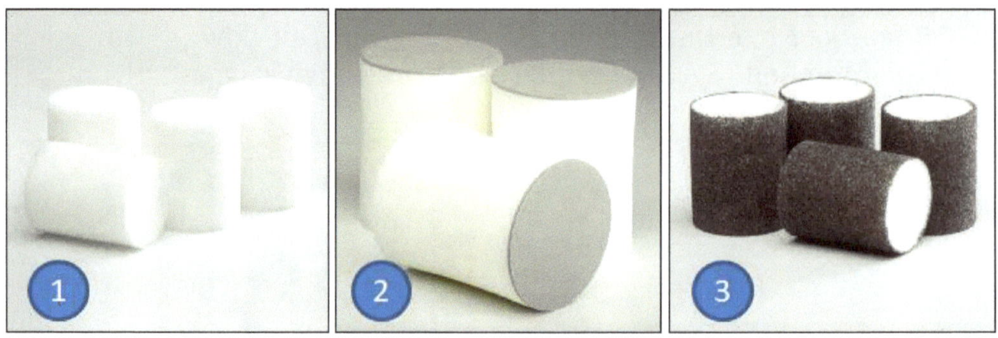

Fig. 2.5- Projectiles (Courtesy of Ultra Clean Technologies)

Cap Seals: As shown in Fig. 2.6, once the transmission line assembly is cleaned, protective caps or plugs are used at both ends to prevent contamination from entering it. These should not be removed until the transmission line is being installed on the equipment.

A Complete Kit: As shown in Fig. 2.7, a complete kit is available that includes a launcher, set of nozzles, set of projectiles, and a bucket for used projectiles.

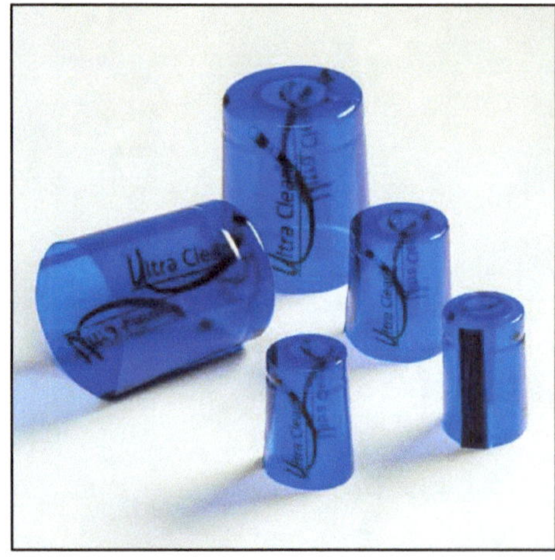

**Fig. 2.6- Cap Seals
(Courtesy of Ultra Clean Technologies)**

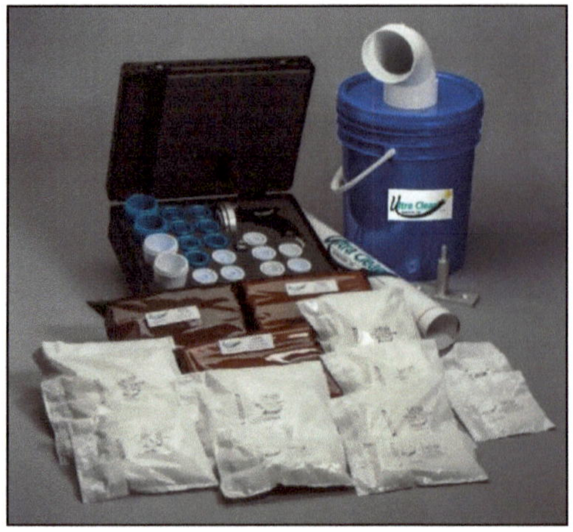

**Fig. 2.7- Complete Kit
(Courtesy of Ultra Clean Technologies)**

2.2.3- Hydraulic Hose Projectile Cleaning

As shown in Fig. 2.8, looking inside a hose to check if it is clean or not is not a good practice. As it has been mentioned earlier, naked eyes can't see particles below 40 μm.

Fig. 2.8- Hydraulic Hose Cleanliness (Courtesy of Ultra Clean Technologies)

As shown in Fig. 2.9, a hydraulic hose is highly recommended to be cleaned right after cutting for the following reasons:

- Heat from the cutting process will cause the rubber & metal dust to stick or adhere to the hose tube as it cools. Contamination from freshly cut hose is much easier to remove.
- The Hose end joints are difficult to insert over the contamination at each end of the hose, removing the contaminants makes stem insertion easier.
- Contamination that is trapped between the hose stem and rubber tube could form a leak path for the hydraulic fluid.

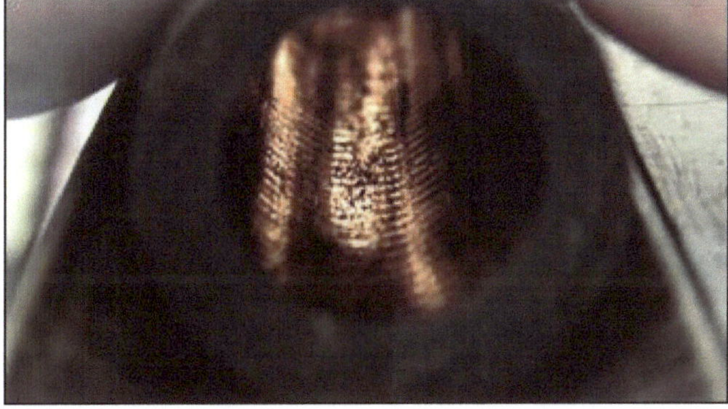

Fig. 2.9- Hydraulic Hose Cutting Causes Contamination (Courtesy of Ultra Clean Technologies)

Figure 2.10 shows the recommended hose cleaning procedure after hose cutting. These steps should be repeated for each end of the hose.

1. Connect your launcher to a dry filtered and regulated air source.
2. Load the recommended nozzle and projectile into the launcher's face plate.
3. Now close the face plate of the launcher. The safety release bar will lock it into position.
4. Insert the nozzle into the hose. Secure the other end of the hose into the containment barrel or catcher bucket.
5. Press the trigger until the projectile has exited the opposite end of the hose. The projectile strips out the internal contamination as it travels through the hose and around bends forcing the contamination out in front of it.
6. Wipe the end of the nozzle.
7. Repeat this process now through the other end of the hose. Continue launching projectiles until they exit the transmission line visibly clean. At this point, launching additional projectiles typically will not remove any more contaminants.

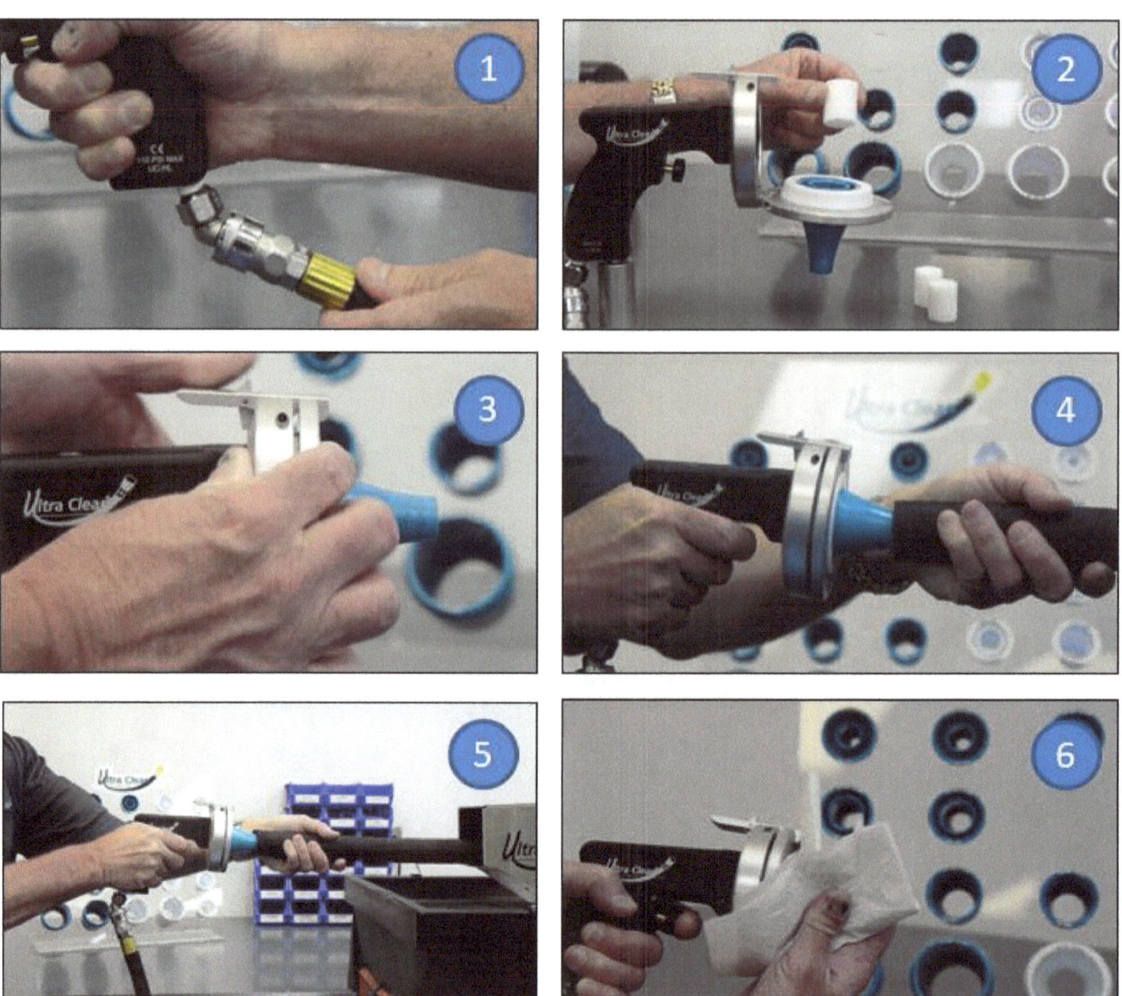

Fig. 2.10- Hydraulic Hose Cleaning Procedure (Courtesy of Ultra Clean Technologies)

Figure 2.11 shows the recommended hose cleaning after hose ends have been attached. The attachment process causes stem deformation to achieve the proper coupling retention. Internal metal flash contamination will occur, and it must be removed. Projectile must continue to be fired through the entire assembly until they exit visibly clean, then hose ends are wiped and covered by cap seals.

**Fig. 2.11- Hydraulic Hose Cleaning after Attaching Hose Ends
(Courtesy of Ultra Clean Technologies)**

2.2.4- Hydraulic Tubes and Pipes Projectile Cleaning

As shown in Fig. 2.12, tube cutting, bending, flaring, and de-burring all leave both visible and microscopic particles inside the line that can effectively destroy precision of hydraulic components in the system.

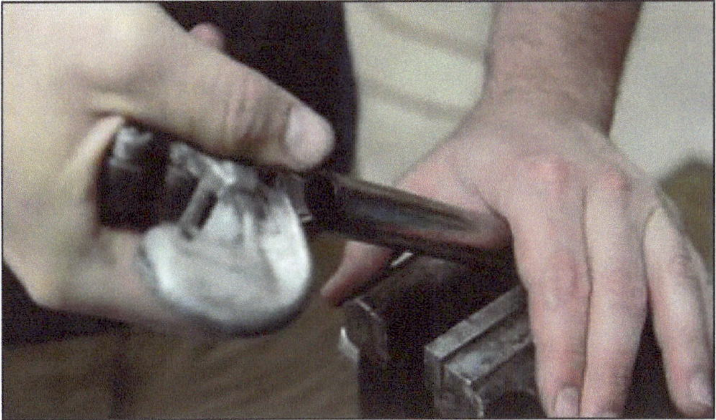

Fig. 2.12- Tube Bending and Flaring (Courtesy of Ultra Clean Technologies)

After processing a tube or pipe, check first if the inside surface contains rust, weld slag or other corrosion particles, then abrasive or grinding projectiles are used until these contaminants removed. After that, as shown in Fig. 2.13, regular projectiles may then be implemented to clean the tube from both sides. When the projectiles exit the tube/pipe visibly clean, tube or pipe ends are covered by cap seals.

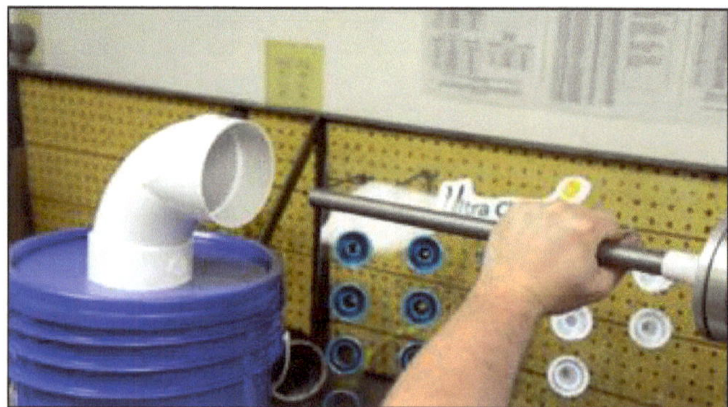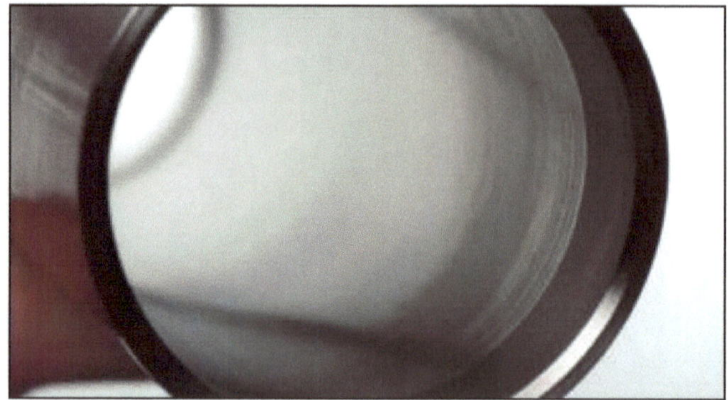

Fig. 2.13- Tube Projectile Cleaning (Courtesy of Ultra Clean Technologies)

2.2.5- Clean Seal Capsules

Ultra Clean Technologies developed a *Clean Seal* system as an advanced capping method to raise the level of transmission line cleanliness. This innovative technology uses heat to shrink *Clean Seal Capsules* onto the ends of cleaned hose and tube assemblies. This eliminates possible re-contamination that is problematic with traditional capping and plugging methods. Re-contamination occurs when traditional caps and plugs are forced onto assemblies, causing plastic particles to shear off into the hose or tube. As shown in Fig. 2.14, The clean seal capsule comes pre-stacked in sticks forms, so the inside is always clean and contamination free. Injection molded caps and plugs are loose in a box and will become contaminated when exposed to airborne contamination in a hose or tube fabrication area. Clean seal capsules are available in 16 sizes to fit most hose and tube assemblies.

Fig. 2.14- Pre-Stacked Clean Seal Capsules (Courtesy of Ultra Clean Technologies)

As shown in Fig. 2.15, the heat source for shrinking the clean seal capsule is controlled by a timer that can be set up to 60 minutes. The temperature setting is preset to the temperature dial which is approximately 165 degrees Celsius (325 degrees Fahrenheit). The machine will need few minutes to reach the correct temperature from a cold start.

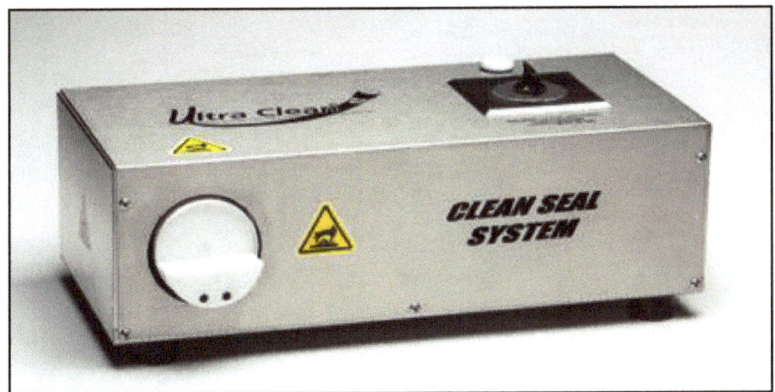

**Fig. 2.15- Heat Source for Shrinking the Clean Seal Capsules
(Courtesy of Ultra Clean Technologies)**

Figure 2.16 shows the simple process of applying clean seal capsules:

1. Choose the closest fit capsule and place it over the coupling. Slide the correct size clean seal capsule over the end fitting.
2. Place the clean seal capsule against the white plunger and push in. A complete seal takes place in less than two seconds.

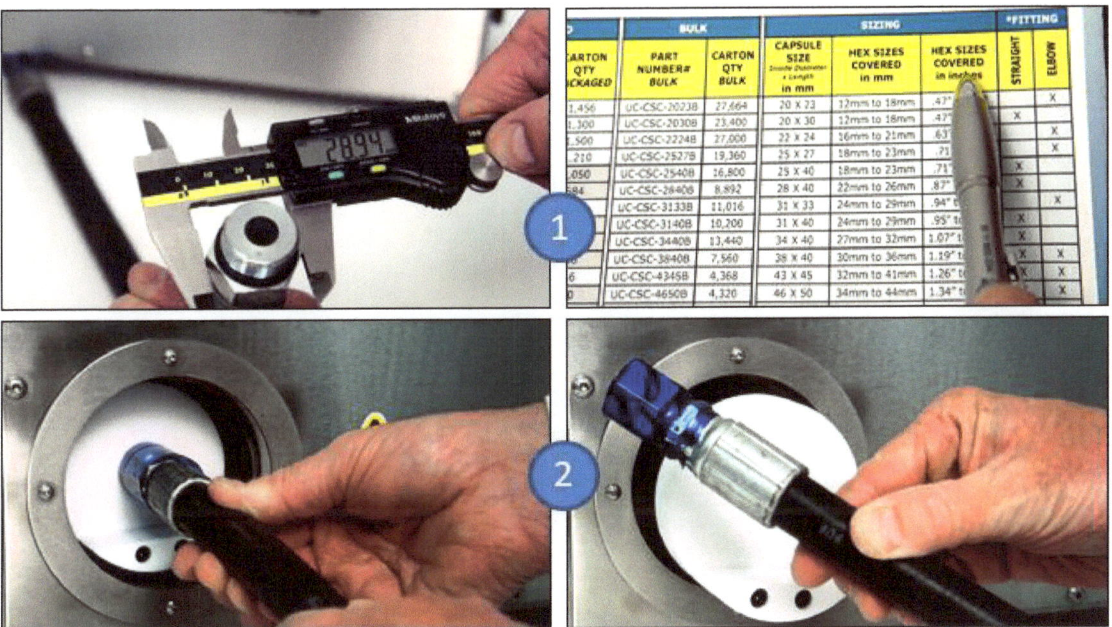

Fig. 2.16- Process of Applying Clean Seal Capsules (Courtesy of Ultra Clean Technologies)

At the time of using the transmission line, follow these three easy steps shown in Fig. 2.17.

1. **Grip** the black pull tab.
2. **Rip** the pull tab upwards.
3. **Slip** the Clean Seal Capsule off of the assembly.

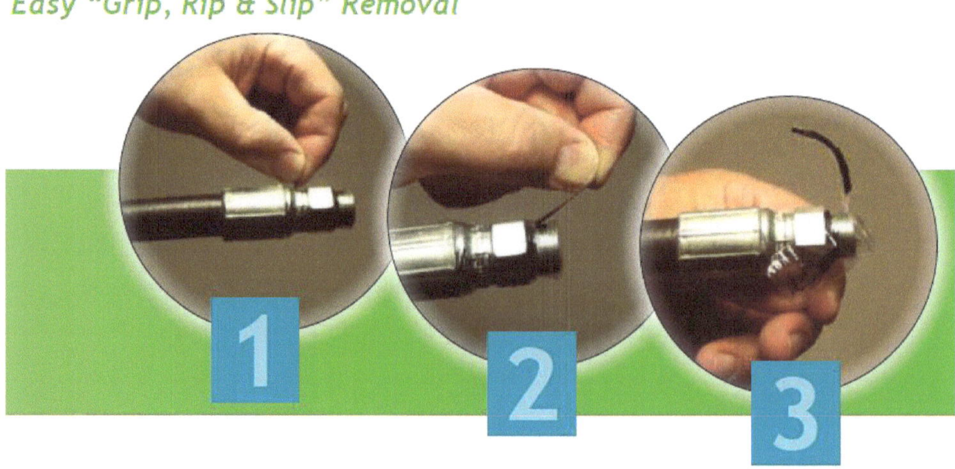

Fig. 2.17- Process of Removing Clean Seal Capsules (Courtesy of Ultra Clean Technologies)

2.2.6- Clean Seal Flange

As shown in Fig. 2.18, the *Clean Seal Flange* is a tool that easily attaches to SAE flanges. It is used to prevent dirt and other contaminants from entering hydraulic transmission lines when switching out or removing components such as pumps, cylinders or valves from heavy duty equipment, particularly in the field. Once the change is complete, the Clean Seal Flange is removed, and normal operation is resumed. No tools are needed to connect or remove the Clean Seal Flange!

Designed with the environment in mind, Clean Seal Flange also prevents oil from spilling out of hydraulic transmission lines and contaminating surrounding soil and water.

Fig. 2.18- Clean Seal Flanges (Courtesy of Ultra Clean Technologies)

2.3- Pickling of Hydraulic Transmission Lines

2.3.1- What is Pickling?

Hydraulic lines *Pickling* simply means "cleaning by acids". Hydraulic pipes and tubes are contaminated by mandrel lubricants and other greasy products during fabrication. Additionally, particles of rust or oxide are produced from long storage. These products are the first contaminants introduced to the new line and one of the most difficult to be removed after running the machine. Therefore, removing these products is necessary before the transmission line can be used in the hydraulic system.

2.3.2- Pickling Process

For small individual lines, spray an alcohol-based solvent inside the line. The solvent should stay inside the line for certain time to break up the greasy products residue. The line is then cleaned using dry projectiles. For multiple long hydraulic pipes and tubes, a more in-depth process is required. Normally this process is done by a vendor who specializes in this process. However, if no instructions were found, the following sequence of steps can be considered as a guide:

- **Degreasing:** Pipes/tubes are closed from both ends and filled with the degreasing liquid (typically 5% soda) to a pressure slightly higher than atmospheric pressure.
- Degreasing fluid stays inside the pipe for certain time (typically 1 hour).
- Pipes/tubes are emptied, rinsed by soft water for 15-20 minutes.
- **De-rusting** by dipping the pipes/tubes in Hydrochloric acid (18-20% concentration) tank for 20-30 minutes.
- Rinsing by soft water for 20-30 minutes.
- **Drying** by compressed air or nitrogen.
- **Projectile Cleaning:** Apply Projectile cleaning.
- **Flushing** with compatible hydraulic fluid, for making a layer of protective oil inside.

2.4- Flushing of Hydraulic Transmission Lines

2.4.1- What is Flushing?

Hydraulic lines *Flushing* simply means "cleaning by oil". Hydraulic system flushing is a kind of advanced *Kidney Wash* or *Offline Filtration* that is performed under certain conditions following a certain procedure. In the process of flushing, flushing oil is forced through the system at high velocity. In theory, this leaves the inside walls of the fluid transmission lines at the same cleanliness level as the new fluid to be installed. Then, during normal operation, the system will experience only externally and internally generated contamination that can be controlled with filtration.

2.4.2- Reasons to Flush a Hydraulic System

The process of flushing a hydraulic system is required in one or more of the following cases:

Newly Built Hydraulic Systems: Even brand-new components shouldn't be assumed clean. They came with built-in contaminants due to various manufacturing processes such as drilling, honing, grinding, casting, sand blasting, cutting, welding, etc. Additionally, during shipment and storage they may receive contaminants. Therefore, a new or rebuilt hydraulic system should be flushed before it becomes operational.

Majorly Repaired Hydraulic Systems: A hydraulic system requires major repair when a catastrophic failure occurs for one or more components. This is most likely due to contamination. Major repair processes include disassembling, machining, reassembling, plumbing work, etc. Hence, there is a good chance for contamination to get into the system.

Hydraulic Fluid Degradation: If the working hydraulic fluid degrades for any reason, such as thermal, chemical, hydrolysis, oxidation. Flushing is required to remove sludges, acids, varnishes, chemicals, etc.

Contamination Level was Exceeded: The fluid was tested and found that the cleanliness class recommended by the system manufacturer was exceeded. This may occur due to filter collapse that results in migrating the debris back into the system. Or it occurs when a hydraulic system works in harshly contaminated environments beyond the ability of the filter.

Failure of an Oil Water-Cooler: When an oil water-cooler fails, oil mixes with water, antifreeze, other contaminants carried by cooling water.

Mixing of Incompatible Hydraulic Fluids: It is always advisable to use the hydraulic fluid that is predefined by the system manufacturer. When incompatible fluids are mixed, something bad will occur right away. An example of that is when the base fluids are incompatible, such as *Polyglycols* and *Mineral Oils*, once they are mixed, they form thick sludge. Another example is when additives work against each other; such as Demulsifiers that improve surface tension and Foam suppressant that reduce surface tension.

Bacteriological Contamination: When water-based hydraulic fluids or mineral oils have been invaded by bacteria that grows fast in the presence of water and heat. That results in green sludge, varnish, acids, etc.

After Pickling Process: Flushing by an appropriate fluid is required to leave a protective oil layer after pickling process for transmission lines.

2.4.3- Flushing System Requirements (ISO 23309)

Hydraulic system flushing must follow a thorough analysis of system constraints and a clear definition of the task objectives. In this regard, all transmission lines shall be flushed to accepted standard such as **ISO 23309 and ISO 16431**. The following sections present the flushing system requirements:

Flushing Flow: In order to remove the particulate contaminants out of the system, hydraulic fluid should be forced at a flow rate higher than the flow rate of the flushed system. Flushing flow rate should be determined so that turbulent flow is developed throughout the whole system. The *Reynolds Number* must be higher than 3,000.

Flushing Fluid: It is recommended to use a fluid with low viscosity (ISO 15 is recommended) to be able to easily reach into the sharp corners and pass through flushing filters with reduced differential pressure. The flushing fluid shall be specified by the equipment manufacturer and shall not contain suspended solids that may plug small lines. Flushing fluid must be compatible with components and seals in the flushed system.

Flushing Filters:

- Flushing Filters Efficiency: Efficiency of flushing filters is based on the targeted cleanliness level.

- Flushing Filters DHC: Dirt holding capacity of flushing filters are typically higher than normal for the purpose of retaining the contaminants out of the flushing process. Flushing DHC is proportional to the difference between the cleanliness level at the initial state and the targeted cleanliness level. It is also proportional to the flushing oil volume.

- Flushing Filter Medium: Depends on the reason for flushing, flushing filters mediums can be selected to remove water contents, varnish, chemicals, fluid degradation products, or combination of them.

- Flushing Filters Size: Size of flushing filters must have surface area large enough to pass the high flushing flow rate at an acceptable differential pressure. If one filter doesn't have the required surface area, several filters can be arranged in parallel to increase the surface area.

Flushing Temperature: The flushing process should be performed at a higher than normal working temperature, typically not less than 50 $^{\circ}$C (122 $^{\circ}$F). Temperature control system is recommended to maintain an approximate constant flushing temperature.

Flushing Oil Volume: Flushing fluid volume should be able to fill the volume of the hydraulic lines plus volume of fluid required in the reservoir to maintain satisfactory and safe suction conditions for the flushing pumps.

Flushing Duration: As shown in Eq. 2.1, flushing time is calculated based on the number of fluid circulations (typically 200 times) in the flushed system.

$$\text{Minimum Flushing Time (minutes)} = \frac{200 \times \text{Flushing Fluid Volume (liters)}}{\text{Flow Rate}\left(\frac{\text{liters}}{\text{min}}\right)} \qquad 2.1$$

For example: flushing fluid volume of 100-liter, flushing pump flow rate of 10 liter/min, and a requirement of circulating the flushing fluid 200 times, minimum flushing time equal (200 x 100)/ 10 = 2000 min = 33 hours and 20 minutes.

However, flushing may continue for 30 minutes after the desired overall system cleanliness level is achieved just for assurance of system cleaning.

Flushing Evaluation: Fluid samples should be taken at various intervals during the process. More efficient, is to have an online electronic particle counter installed for continuous monitoring during the flushing process. The system is considered clean when samples from the system indicate that the specified cleanliness level has been reached for three consecutive test intervals. The process should continue until cleanliness level is one code below the system's target cleanliness level. For example, if the target is ISO 15/13/11, continue to flush the system until ISO 14/12/10 is reached.

Flushing Power Units: As shown in Fig. 2.19, a number of commercially available units can be used for this. However, flushing power units can be customized to meet the flushing requirements.

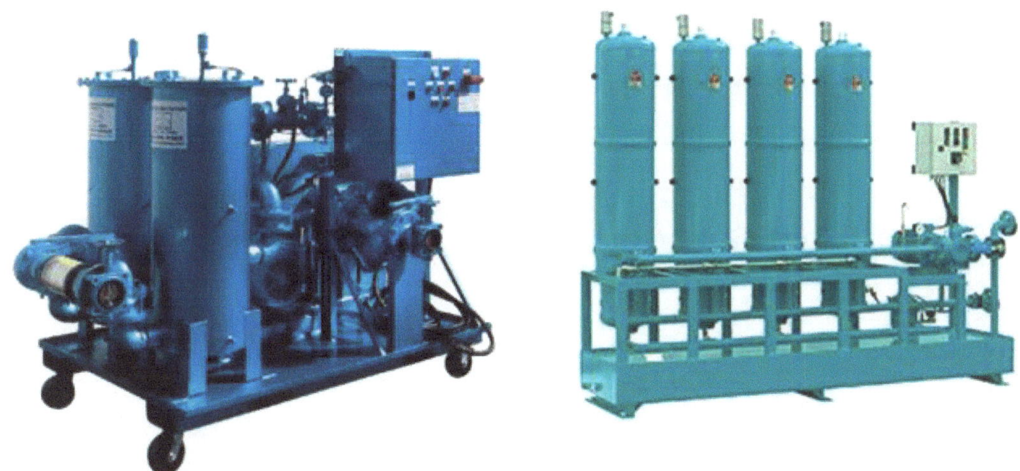

Fig. 2.19- Examples of Hydraulic System Flushing Power Units

2.4.4- Flushing Process

If no instructions are provided by the equipment manufacturer, the following set of bullets (in sequence) provides guideline for system flushing assuming the system reservoir is used:

1. Operate the system at minimum load/pressure for 15 minutes after reaching the regular operating temperature.
2. Completely drain the system. Make sure all components that contain fluids are completely drained including the reservoir, all lines, cylinders, accumulators, filter housings.
3. With a lint-free rag, clean the reservoir of all sludge and deposits. Make sure the entire reservoir is free of any soft or loosened paint.
4. Prepare the system for flushing with the following considerations.
   - Dead ends without circulation shall be avoided.
   - Fluid must circulate through all points in the circuit being flushed.
   - Circuits being flushed are recommended to be connected in series not in parallel to guarantee that same flow rate will pass through all lines.
   - As shown in Fig. 2.20. components that can restrict or damaged by a high flow velocity shall be bypassed using *Jumpers*. For multiple circuits/functions connected in parallel as shown in the figure, individual circuits can be flushed one at a time by shifting the appropriate directional control valve. Start first with the circuit with the longest lines (largest fluid volume), and then finish with the circuit with the smallest fluid volume.
5. Flushing unit shall be located as close as possible to the flushed system to minimize pressure losses.
6. Hook the flushing unit to the system.
7. Every effort should be made to avoid spilling oil and to prevent spilled oil from returning directly to the reservoir.
8. Run the flushing unit to force the flushing fluid through the system.
9. Flushing temperature and flow should be measured near the return line of the circuit being flushed.
10. If directional valves are included in the circuit being flushed, stroke the valves frequently to ensure they are thoroughly flushed, and the direction of flushing fluid flow is reversed through the circuit being flushed.
11. The fluid should be filtered, and flushing should continue until reaching one cleanliness level below the system's target cleanliness level.
12. Drain the flushing fluid when hot and as quickly as possible.
13. Inspect and clean the reservoir again. Chemical cleaning techniques will not only clean the hydraulic oil reservoir but also provide a protective oxide layer that will further inhibit build up in the future.
14. Replace the system filters if used in the flushing process.
15. Fill the system with the fluid to be used.
16. Return the system to its original circuit design.
17. Bleed and vent the pump.
18. Run the system's original pump at no load for about 10 minutes.

19. Cycle the actuators to return oil to the reservoir and bleed air from the system.
20. Keep an eye on the fluid level in the reservoir, and refill if needed.
21. Run the system for 30 minutes to bring it to normal operating temperature.

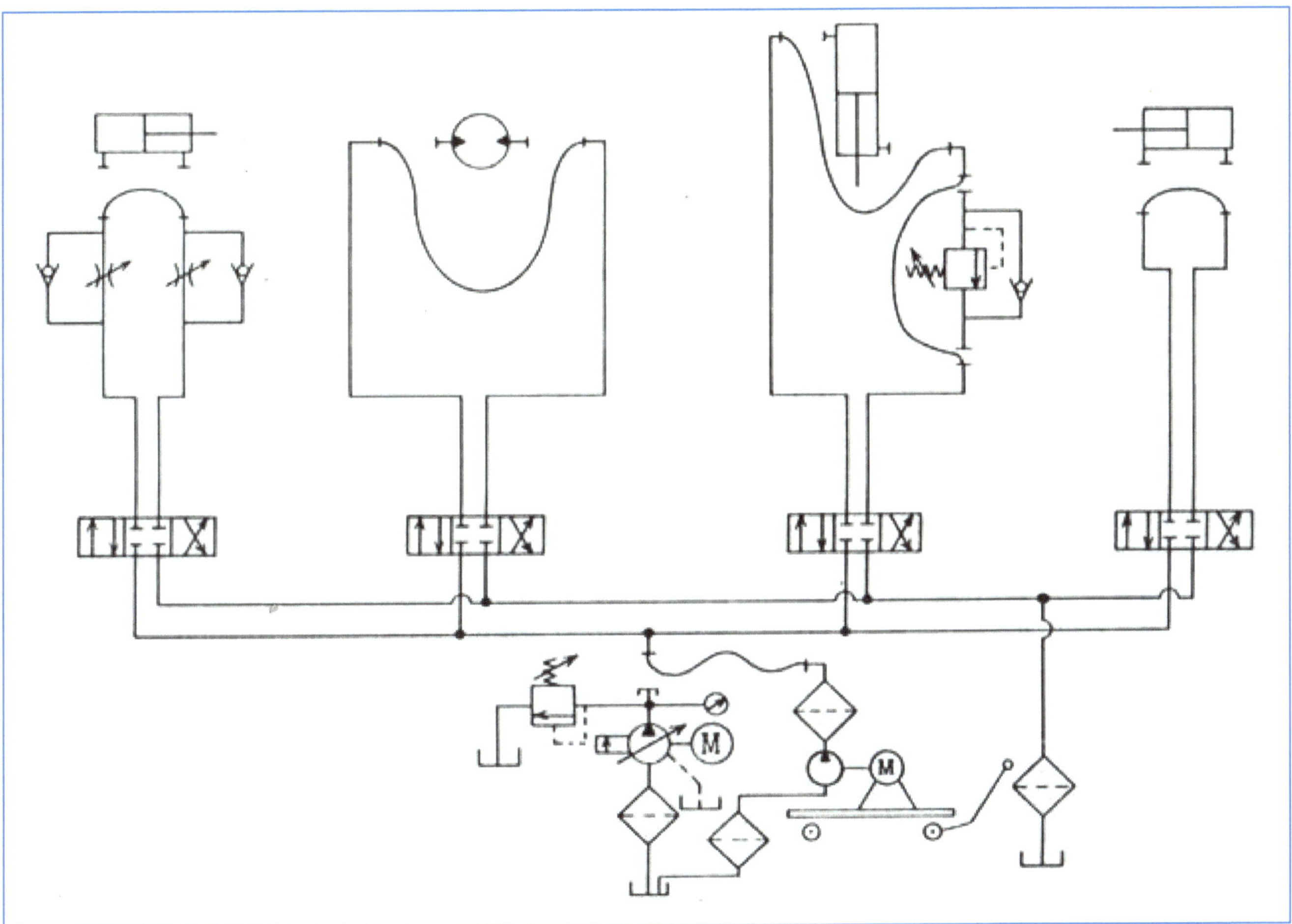

Fig. 2.20- Examples of Flushing Industrial Machine Function/Circuit

Chapter 3
Maintenance and Safety of Transmission Lines

Objectives

This chapter provides guidelines for **transmission lines** selection, replacement, maintenance scheduling, installation, testing, storage and transportation. This chapter is supported by examples and figures granted by leading fluid power manufacturers.

Brief Contents

3.1-BP-Transmission Lines-01-Selection and Replacement
3.2-BP-Transmission Lines-02-Maintenance Scheduling
3.3-BP-Transmission Lines-03-Installation and Maintenance
3.4-BP-Transmission Lines-04-Standard Tests and Calibration
3.5-BP-Transmission Lines-05-Transportation and Storage
3.6- Oil Injection Avoidance and Treatment

Chapter 3: Maintenance and Safety of Transmission Lines

The following set of best practices provide general guidelines and may not be applicable for all cases. They are not intended to replace the instructions given by the component manufacturer. It is strongly advisable to adhere to instructions provided by the manufacturer.

3.1-BP-Transmission Lines-01-Selection and Replacement

During system design, *Transmission Lines* were originally specified to consider leak-free connections and laminar flow conditions. During system maintenance, when replacing an existed transmission line, the following specifications shall not be changed:

Size: Line size has a direct effect on laminar flow condition and pressure losses. Therefore, when a line is replaced, the new line must have the same size. It is to be reminded that, if the line is a *flexible hose*, the line size is based on the inner diameter. If the line is a *tube* or *pipe*, then the size is based on the outer diameter.

Temperature: Particularly for hoses, minimum and maximum permissible temperatures must be checked. If the new hose is not within the same temperature rating, there is a good chance it will prematurely fail. If the original hose is a thermoplastic type, new hose must be the same.

Application: If a pressure hose is used to replace a suction hose, there will be a chance for the inner layers to collapse due to vacuum. As a result, partial line blockage and pump cavitation could occur. If the original hose has abrasion resistant outer cover, the new hose should be the same.

Media (Fluid Compatibility): Particularly for hoses, the new hose must be compatible with the hydraulic fluids. Otherwise, hose may be deteriorated over the time and become a source of contamination. It also can fail due to losing the attachment force between the hose and the end connections.

Pressure: Two lines that have the same size and look like each other DOES NOT mean they can replace each other. Rated pressure is directly related to the safety of the machine operation. Maximum allowable static and dynamic (spikes and pulses) pressure must be considered when replacing a hose.

Electrically Nonconductivity: Hydraulic hoses that are used nearby high voltage lines, must be electrically nonconductive. When any of these hoses are replaced, the new ones must also be electrically nonconductive. Otherwise, there is a good chance to accumulate static charges resulting in a potential explosion and possible fire.

Length: Generally speaking, pipes, tubes, and hoses are preferred for long, medium, and short line lengths; respectively. However, original line length should be respected. DO NOT reduce the hose length because this may result in stretching the hose when it is pressurized and may result in detaching the end connections. DO NOT increase the line length because this increases the line losses and may result in pressure hammering.

Hard Line (Pipes and Tubes) Material: Same inner diameter and wall thickness are not enough for tubes/pipes to replace each other. The steel grade, from which the tubes/pipes are made, should also be the same. Otherwise, maximum permissible pressure might be different.

End Joints: Size, type and pressure rating of *End Joints* must be kept the same. Avoid using adaptors because it causes more pressure drop and affects the overall length of the line.

3.2-BP-Transmission Lines-02-Maintenance Scheduling

Unless otherwise is stated by components and systems manufacturer, Table 3.1 provides guidelines for *scheduling* preventive maintenance actions for hydraulic transmission lines.

#	Preventive Maintenance Actions	Daily	Weekly	Monthly	Biannually	Annually
1	Line visual inspection **(Note 1)**	✔	✔	✔	✔	✔
2	Inspect for leakage **(Note 2)**		✔	✔	✔	✔
3	Clean around and the outer surface including the end joints **(Note 3)**		✔	✔	✔	✔

Table 3.1- BP-Transmission Lines-02-Maintenance Scheduling

Note 1: Like a vehicle tire, a hose should be replaced based on a given service life no matter how it looks like. When it comes to visual inspection, a hose service life is shorter than hard tubing. Therefore, they should be visually inspected more frequently than hard tubing. Any of the following conditions require immediate shut down and replacement of the *Hose Assembly* no matter what its lifetime is:
- Kinked, crushed, flattened or twisted Hose.
- Cracks from minimum bend radius exceeded and around fittings.
- Brittleness or loss of flexibility.
- Frayed protective layers or broken reinforcement layers.
- Outer cover pulled back from the end of the coupling.
- Fitting slippage on Hose.
- Rusted, broken, cracked, damaged, leaking, or badly corroded fittings.

Note 2(Refer to Fig. 3.1): All types of transmission lines must be inspected for leakage. As best practices for *leakage* inspection, consider the following actions:

- Never tighten a fitting while the system is under pressure.
- Avoid possible oil *injection*. DO NOT use your hand to check leakage from a line.
- Wear proper industrial gloves that prevent oil injection.
- Use the technique of *fluorescent* dye and *UV light* for leakage detection. For details on leakage detection kit, refer to Volume 6 (Troubleshooting and Failure Analysis, Chapter 2).
- If special leakage detection equipment isn't available, use a piece of wood or cardboard that is 1-2 feet long. Hold one end of the wood or cardboard, place the other end approximately 1-2 inches away from the inspected part of the line, and move it around the line.

Fig. 3.1- Bet Practices for Hydraulic Transmission Line Leakage Inspection (www.bondfluidaire.com)

Note 3 (Refer to Fig. 3.2): Outer surfaces including the end joints of the transmission lines must be frequently cleaned to allow better heat dissipation and to maintain overall acceptable cleanliness level of the hydraulic fluid.

Fig. 3.2- Keep Outer Surfaces including End Joints (mac-hyd.com)

3.3-BP-Transmission Lines-03-Installation and Maintenance

Proper installation and maintenance of hydraulic transmission lines are important processes for trouble-free operation of the system. The following best practices provide guidelines for *installation and maintenance* of transmission lines.

BP-Transmission Lines-03-Installation and Maintenance:
1. Proper Line Cleaning before Assembling.
2. Proper Hose Crimping.
3. Proper Hose Routing.
4. Proper Hose Assembling.
5. Assemble for Leakage Prevention.

The following sections provide detailed interpretations, with examples, for the actions listed in the previous best practices list.

3.3.1- Proper Line Cleaning before Assembly

Contamination settled inside new and used hydraulic lines is very dangerous for hydraulic system operation. Therefore, it is necessary to clean inside the transmission lines before assembly. One or combination the following cleaning methods are used to clean inside the transmission lines, *Pickling*, *Flushing*, and *Projectile Cleaning.* Review Chapter 2 for more details.

3.3.2- Proper Hose Crimping

Crimping process should be explained in the operating manual of the crimping machine. Referring to Fig. 10.3, the following is an example of hose crimping:

1. Setup the crimping machine referring to the manufacturer' instructions identifying the crimp specifications shown in the figure.
2. Select the proper die series. The dies are color coded and stamped on the top.
3. Before loading the die, brush the inside surface by a lubricant.
4. Load the selected die into the crimper.
5. Place the die ring above the die.
6. Determine the hose insertion depth.
7. Insert the hose into the crimpable fitting until the insertion mark aligned with the end of the fitting.
8. Insert the hose into the die and properly align the coupling with the die fingers.
9. Finish the crimping, remove the hose, check the crimping diameter using a caliper.

Fig. 3.3- Hose Crimping Process (Courtesy from Parker)

Hose Insertion Depth: Chapter 1 provides information about hose *Cut Length* and the hose *Overall Assembly Length*. However, when crimping a hose in a permanent fitting, *Hose Insertion Depth* must be checked and marked. Estimating the depth of the hose by eyes can be extremely dangerous. If the hose isn't inserted into the shell of the fitting per the recommended length, the fitting can blow off. Therefore, as shown in Fig 3.4, it is recommended to check the hose insertion depth using standard gauge blocks.

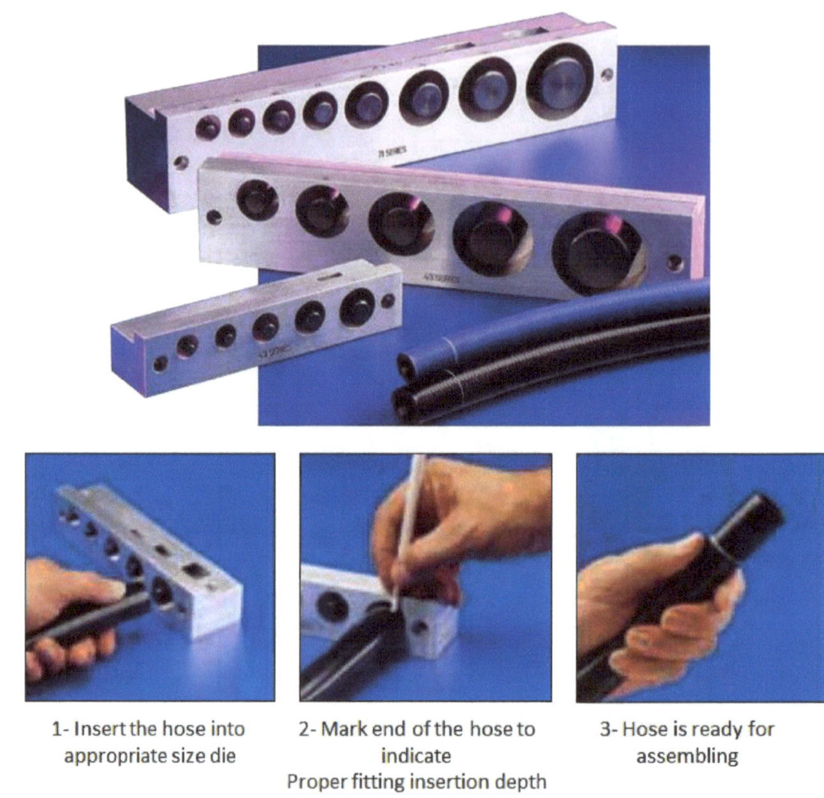

Fig. 3.4- Hose Crimping Machines (Courtesy from Gates)

3.3.3- Proper Hose Routing

Referring to Fig. 3.5, the following guidelines shall be considered when routing hydraulic hoses lines:

Route for Better Appearance and Serviceability (1): Hoses must be routed so that the final assembly looks ordered and not messy. Also, hose ends shouldn't be close to each other for better serviceability. Use 45-degree or 90-degree adapters in order to route hoses directly and to provide clearance for wrenches.

Avoid Abrasion and Sharp Edges (2): Line must be routed to avoid abrasion with moving or sharp elements. Otherwise, the outer surface will continuously be rubbing causing line failure.

Respect Minimum Bend Radius (3): Hose and tubes manufacturers specify minimum bend radius depends on the line size, length, and pressure ratings. Lines must be routed with full respect to the minim bend radius and to avoid tight bend. Otherwise, the outer surface will crack, and the line prematurely fails. Avoid using hoses that are unnecessarily long.

Shield from Heat Sources (4): Lines, particularly hoses, must be protected against heat sources such as exhaust or hot steam pipes. Protection includes isolating using heat resistant shields of baffles. Otherwise, hoses may be burned.

Do Not Twist Hydraulic Hoses (5): Hose ends must be assembled to the last turn without twisting the hose. If a hose is twisted, the hose length is shortened, its pressure carrying capacity is reduced, and hose end connection will loosen over the time.

Allow Clearance for Hose Length Change (6): When a hose is pressurized, it loses (2-4) % of its length. If a hose is stretched during the first assembly, the hose ends likely will detach under pressure pulsation. Therefore, when installing hoses, leave a little extra length to allow for changing in length when pressure is supplied. The other best practices in this regard is not to clamp a low pressure and high-pressure hose together. Otherwise, they will rub against each other.

Allow Clearance for Flexing Component Movement (7): If a hose is connected to a moving component, it must be routed to allow clearance so that the hose doesn't twist and/or exceed its minimum bend radius throughout the range of component movement. Use of swivel joints could be a solution for such a case.

Use Clamps (8): Use clamps or other devices to secure hoses so they do not rub against machine parts, or they do not vibrate during machine operation.

Reduce Connections (9): Reduce number of pipe threads joints by using hydraulic adapters instead of pipe fittings.

Separate Hydraulic Lines from Electric Lines: Hoses must be routed to avoid intersection or contact with electrical lines. That is to avoid possibility of explosion or ignition in case of developing spark and oil spray.

Shield Hoses from Heat Sources: Use guards and shields to keep Hoses away from extreme temperatures or heat sources.

Protect the Hoses from Hazardous Material and Conditions: Use armures, guards, or sleeves to protect the hoses as needed. Also, keep hoses away from chemicals where possible

Quick-Disconnect Couplings: When assembling quick connect couplings, make sure they are matched. Ensure they are connected/disconnected properly. If in doubt, disconnect and re-connect the couplings again.

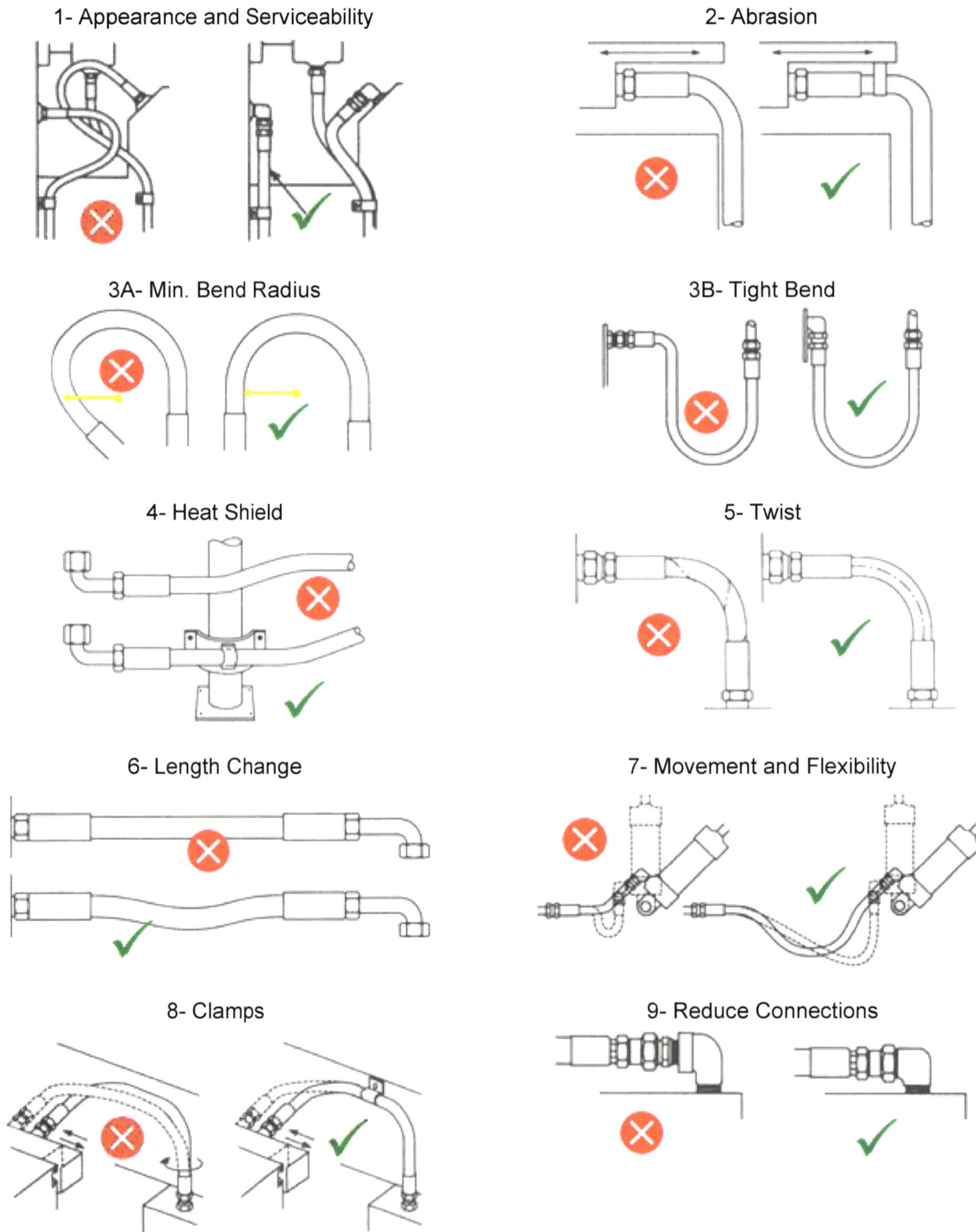

Fig. 3.5- Best Practices for Hydraulic Hose Routing

3.3.4- Proper Hose Assembling

Best Practices for Hose Assembling: Referring to Fig. 3.6, the following steps provides best practices for installing hose assembly:
1. Clean the surrounding area where connections are to be made. Make sure no dirt or contamination gets into hydraulic openings.
2. Install adapters into ports (if used). Torque to manufacturers specifications.
3. Lay the hose assembly into routing position to verify length and correct routing.
4. Thread one end of hose assembly onto port (or adapter). If the hose assembly uses an angled fitting, always install it first to ensure proper positioning.
5. Thread other end of the assembly without twisting the hose. Use a wrench on the backup hex on the fitting while tightening.
6. Properly torque both ends with manufacturer's torque considered.
7. Run the hydraulic system to circulate oil under low pressure and safely reinspect for leaks and potentially damaging contact.

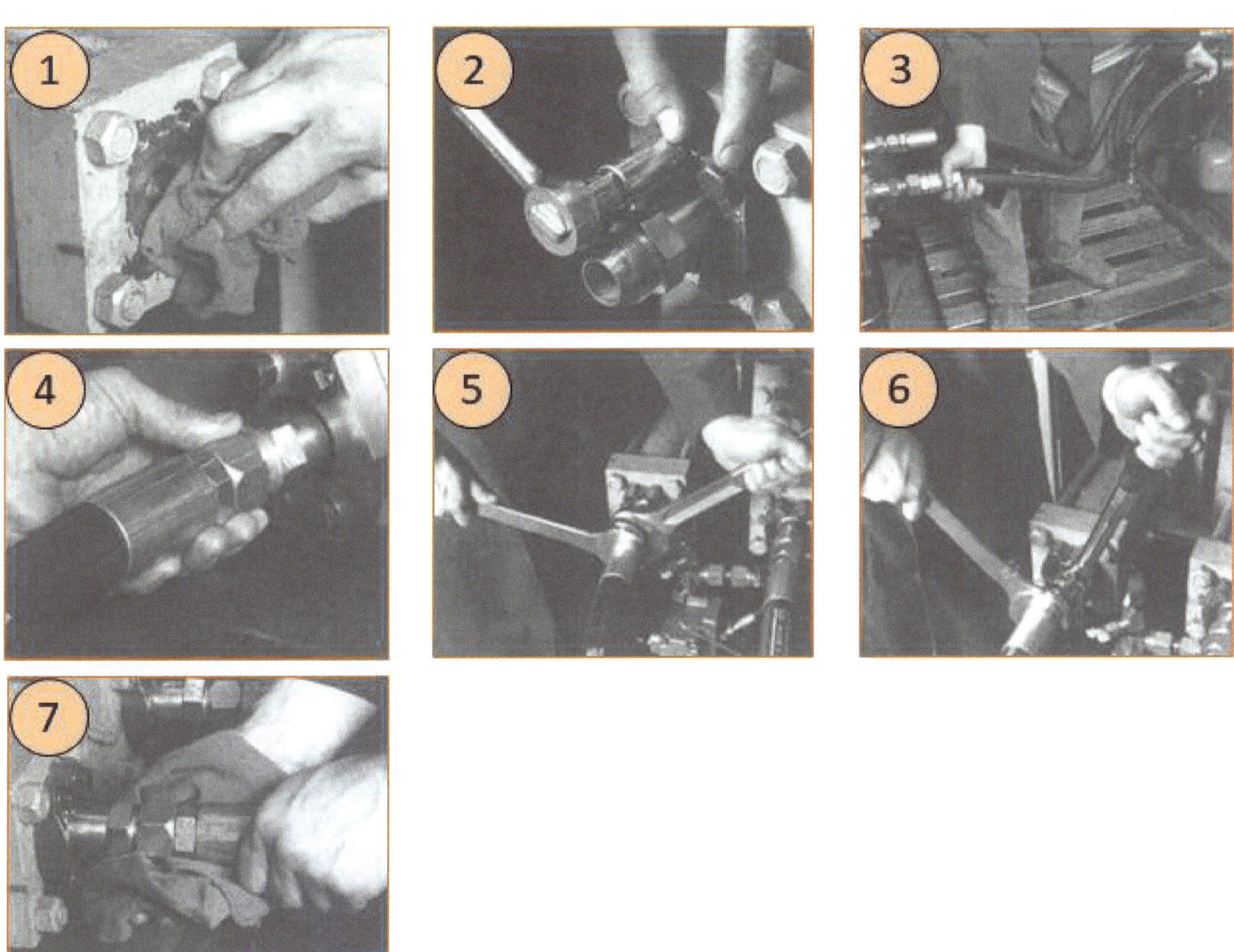

Fig. 3.6- Steps to Install Hose Assembly (Courtesy of Gates)

3.3.5- Assemble for Leakage Prevention

As shown in Fig. 3.7, fluid stains or puddles under hydraulic equipment or transmission lines indicate the presence of a leak in the line.

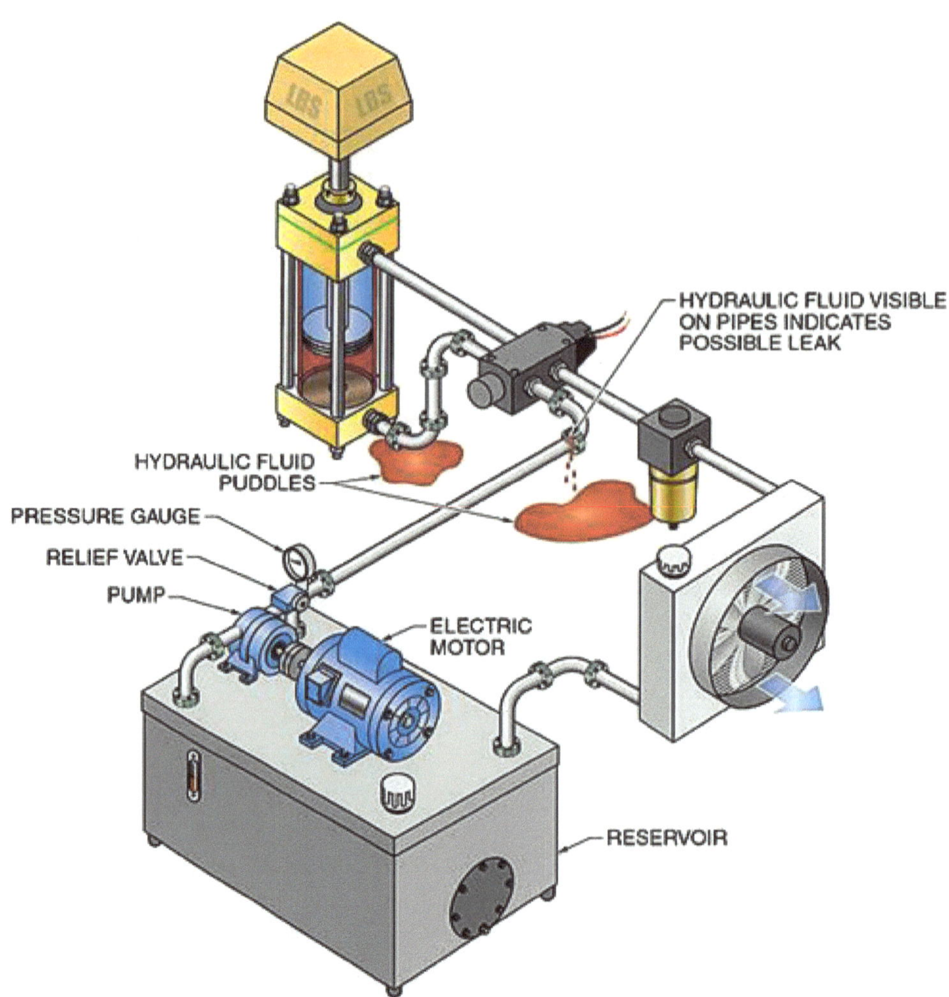

Fig. 3.7- Transmission Line Signs of Leak (Courtesy of American Technical Publisher)

The following best practices help minimize the possibility of line leakage when considered at the time of line selection and assembly:

❖ **Quality of Product:**
- Never use galvanized steel or commercial "from-off-the-shelf" fittings.
- Since fittings, and crimpers are designed to work together as a system, avoid mixing of fittings and hoses form different suppliers even if they follow same standard (SAE, ISO, etc.). This can result in hose assembly leakage, hose separation or other failures which can cause serious bodily injury or property damage from spraying fluids, flying projectiles, or other incidences.

❖ **Determining the Thread Type (Fig. 3.8):** As shown in the figure, threads of various fittings look similar. Mixing fittings of different thread types causes leakage. Therefore, it is advisable to assure correct thread identification before using them. Threads are measured as follows:

▪ **Thread Gauge:** Using a *Thread Gauge*, the number of threads per inch can be determined. Holding the gauge and coupling threads in front of a lighted background helps to obtain an accurate measurement.

▪ **Caliper Measure:** A *Vernier Caliper* should be used to measure the thread OD and ID diameters.

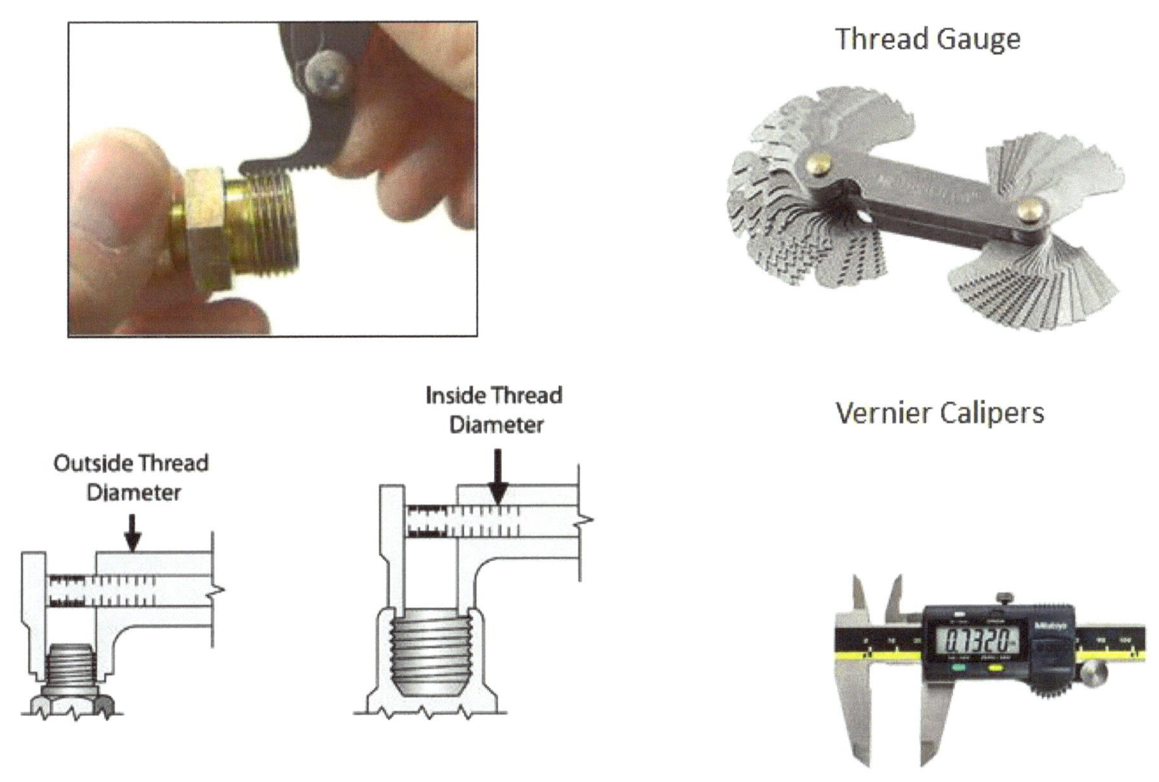

Fig. 3.8- Determining the Thread Type of Hydraulic Fittings

❖ **Fittings Assembly Torque:**
▪ Over torqueing a fitting DOES NOT mean it seals better. Overtightening means overstressing or cracking. Manufacturers specify assembly torque which depends on the type, size and pressure rating. Instructions from the manufacturers should be fully respected.

▪ Never tighten a fitting while the pump is running. For your safety, release the system pressure first in accordance with manufacturer's recommendations.

- Tables 3.2, 3.3, and 3.4 show examples of typical data provided by a manufacturer. The torque values in the following tables give minimum and maximum torque recommendations. The minimum value will create a leakproof seal under most conditions. Applying torque values greater than the maximum recommendation will distort and/or crack the fitting.

Size		Steel			
		Ft. Lbs.		Newton-Meters	
Dash	Inches	Min	Max	Min	Max
-4	1/4	10	11	13	15
-5	5/16	13	15	18	20
-6	3/8	17	19	23	26
-8	1/2	34	38	47	52
-10	5/8	50	56	69	76
-12	3/4	70	78	96	106
-16	1	94	104	127	141
-20	1-1/4	124	138	169	188
-24	1-1/2	156	173	212	235
-32	2	219	243	296	329

Table 3.2- Recommended Tightening Torque for 37° & 45° (Machined or Flared Fittings) (www.new-line.com)

Size		Newton-Meters	
Dash	Inches	Min	Max
-4	1/4	14	16
-6	3/8	24	27
-8	1/2	43	54
-10	5/8	60	75
-12	3/4	90	110
-14	7/8	90	110
-16	1	125	240
-20	1-1/4	170	190
-24	1-1/2	200	245

Table 3.3- Recommended Tightening Torque for Flat-Face O-Ring Seal (Steel) (www.new-line.com)

Size		Ft.Lbs. Working Pressures 4,000 Psi (27.5 Mpa) And Below		Newton-Meters Working Pressures 4,000 Psi (27.5 Mpa) And Below		Ft.Lbs. Working Pressures Above 4,000 Psi (27.5 Mpa)		Newton-Meters Working Pressures Above 4,000 Psi (27.5 Mpa)	
Dash	Inches	Min	Max	Min	Max	Min	Max	Min	Max
-3	3/16	–	–	–	–	8	10	11	13
-4	1/4	14	16	20	22	14	16	20	22
-5	5/16	–	–	–	–	18	20	24	27
-6	3/8	24	26	33	35	24	26	33	35
-8	1/2	37	44	50	60	50	60	68	78
-10	5/8	50	60	68	81	72	80	98	110
-12	3/4	75	83	101-1/2	113	125	135	170	183
-14	7/8	–	–	–	–	160	180	215	245
-16	1	111	125	150	170	200	220	270	300
-20	1-1/4	133	152	180	206	210	280	285	380
-24	1-1/2	156	184	212	250	270	360	370	490

Table 3.4- Recommended Tightening Torque for SAE Steel O-Ring Boss (www.new-line.com)

❖ **Proper Assembling of Flareless Fitting:** Figure 3.9 shows an example of best practices of assembling of *flareless* fitting as follows:

1- Properly burr the tube end, clean the burring products, and visually inspect the tube end to make sure it is burred properly.
2- Check the 90^0 tube end trimming using proper tool.
3- Apply lubricant to the fitting.
4- Assemble the nut and the cutting ring with the tube.
5- Assemble the tube assembly with the fitting.
6- Tight by hand.
7- Tight by a wrench to torque specified by manufacturer.
8- Disassemble the tube assembly and check the proper attachment between the tube and cutting ring. The cutting ring may rotate in place around the tube but must not move axially along the tube. A special light source tool is used to make sure light isn't seen from between the tube and the cutting ring. If so, fluid leak will not occur.

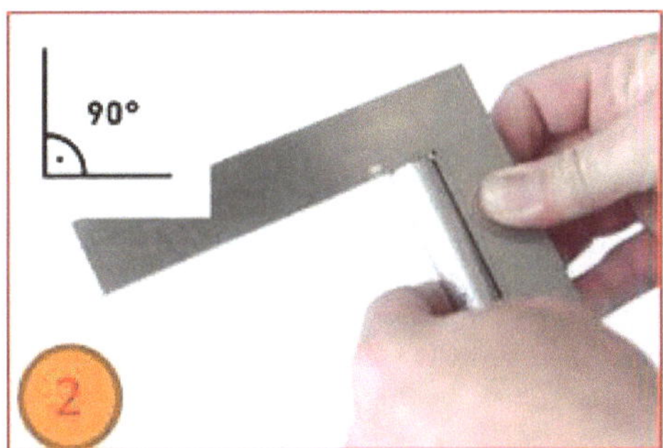

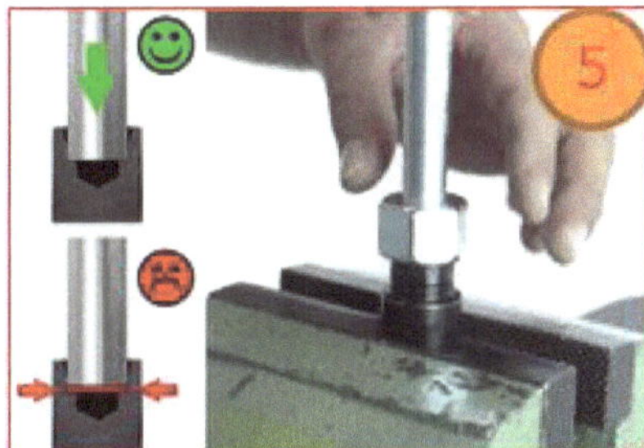

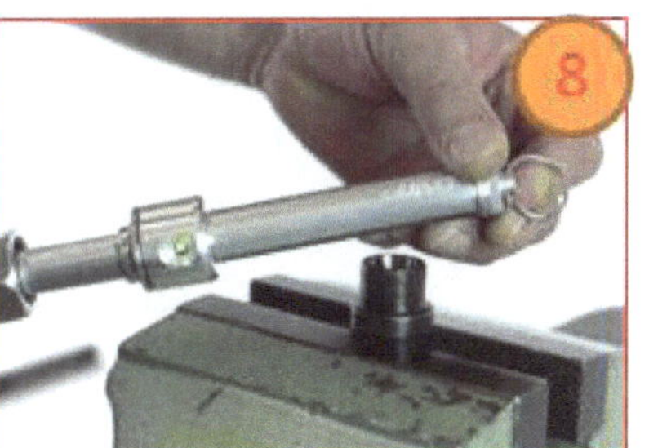

Fig. 3.9- Best Practices for Assembling Flareless Fitting (Courtesy from Parker)

❖ **Proper Tube Flaring (Fig. 3.10):** As shown in the figure, improper tube flaring results in cracked tube nose and fluid leakage. Follow the standard tube flaring procedure and dimensions for every tube size. The following steps provide best practices for tube flaring:

1. Select proper tubing based on (fluid compatibility, operating pressure, etc.).
2. Review the standard flaring dimensions based on single or double flare. For detailed information about the dimensions of tube flaring, refer to Chapter 1.
3. Cut the tube to required length.
4. End should be trimmed square within +/- 2° tolerance.
5. Properly burr O.D. and l.D. of tube.
6. Clean tube to remove all dirt from both O.D. and I.D. of tube.
7. Assemble tube nut and sleeve on tube. The threaded end of nut and flared end of sleeve must point toward the end of the tube to be flared.
8. Flare the tube end using the correct flaring tool for the tube size and desired flare angle.
9. Inspect flare to the dimensions indicated in the standard tables.
10. In addition, flare should be checked for concentricity, thin out, cracks, or other defects which may prevent sealing.

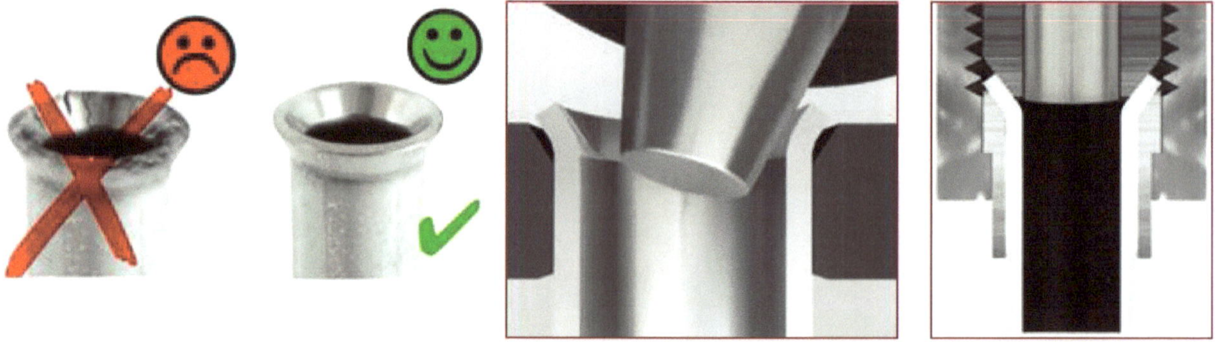

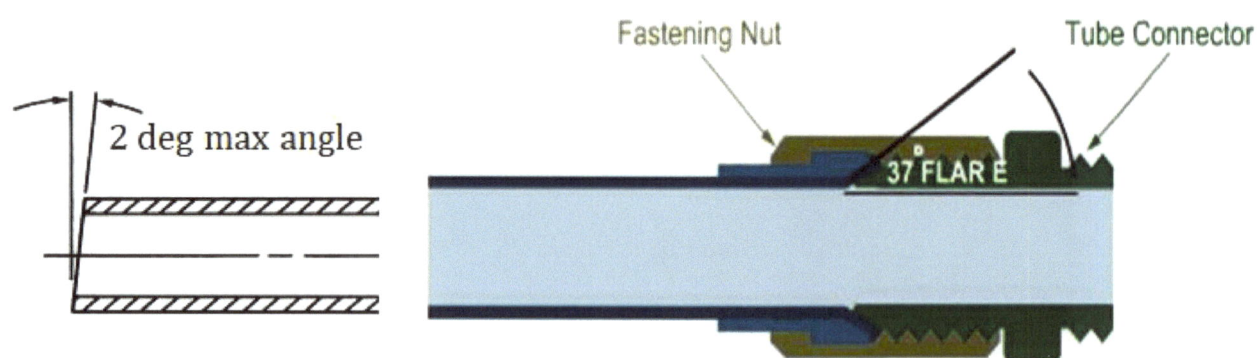

Fig. 3.10- Best Practices for Tube Flaring (Courtesy from Parker)

❖ **Proper Assembling of Flared Fitting:** Figure 3.11 shows an example of simple three steps of assembling a 37º *flared* tub and tightening to torque specified by manufacturer. The catalogue indicates either the tightening torque or the how many faces the nut should rotate after tightening by hand.

1- Clap the body on a Vice

2- Tightened by Hand

3- Tightened by Wrench to Specified Torque

Fig. 3.11- Best Practices for Assembling 37º Flared Fitting (Courtesy from Parker)

❖ **Flared Tube Sealing:** Sealing between a fitting and a transmission line is an additional challenge. Imperfection of tube flaring may cause leakage. Figure 3.12 shows a patentable sealing element named *"Flaretite"* and pronounced "Flare Tight". Such sealing element conceptually compressed to match the misalignment between the flared surfaces.

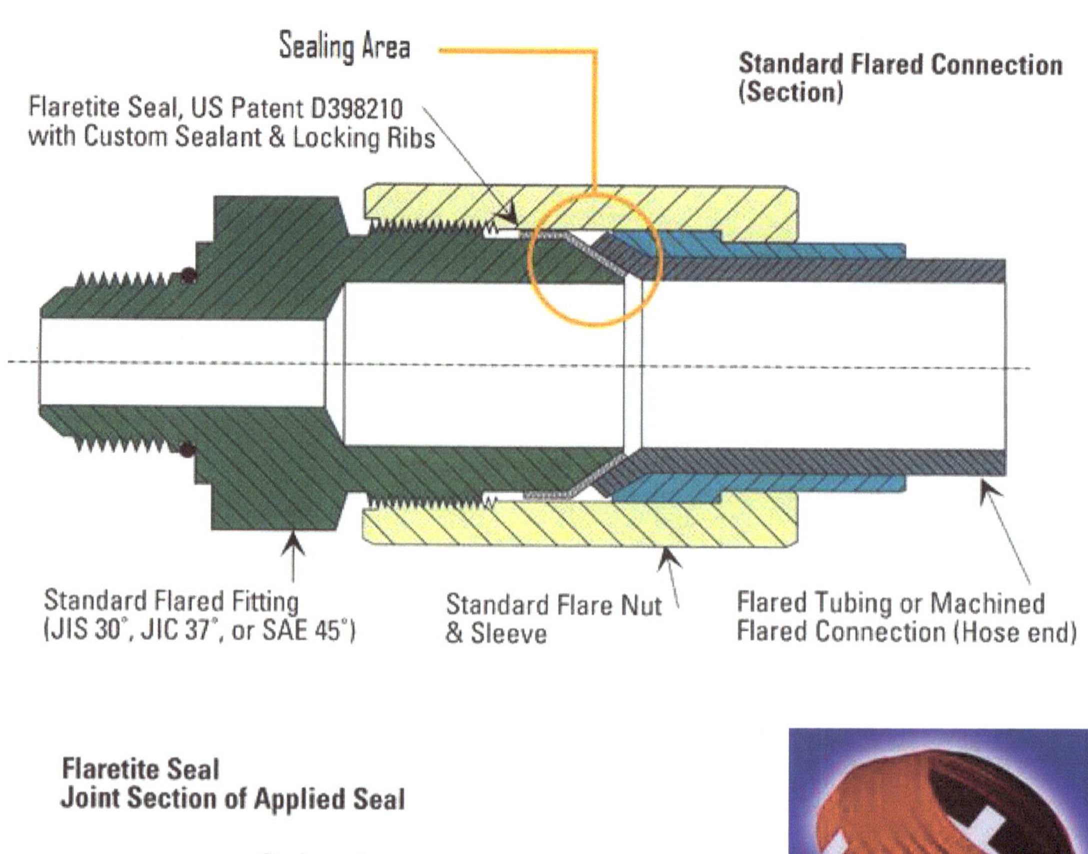

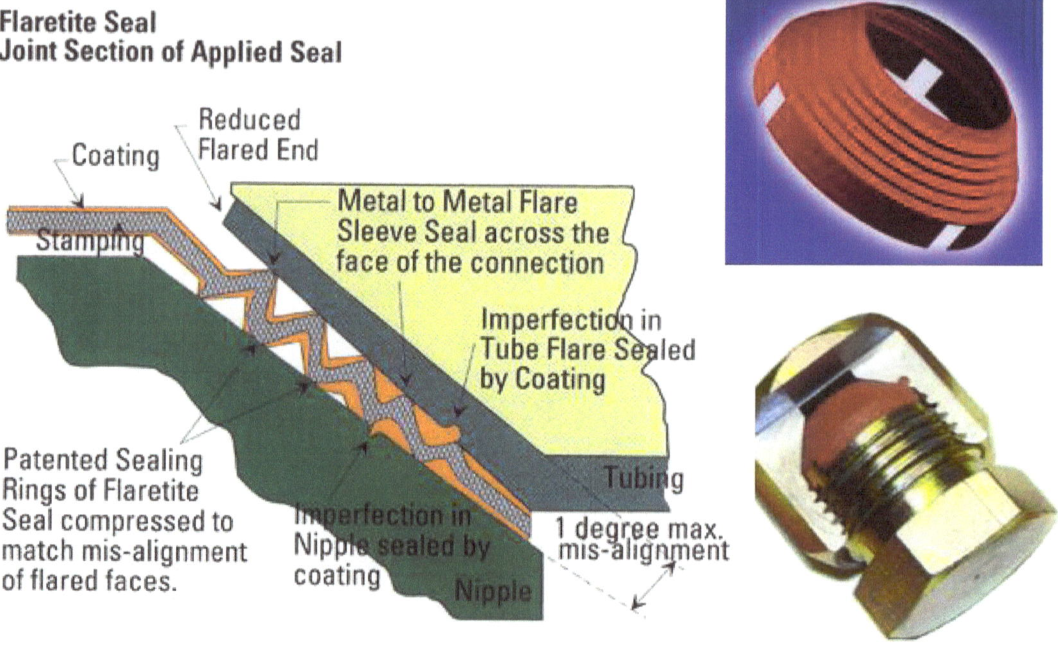

Fig. 3.12- Flare Tight Seal (www.Flaretite.com)

❖ **Proper Tube Bending:** Tubes should be bent in accordance with minimum bend radius instructed by the tube manufacturer based on the tube size, pressure ratings, etc. Tubes can be bent by hand, or by press depending on the tube size. Figure 3.13 shows perfect bend versus other incorrect bend cases.

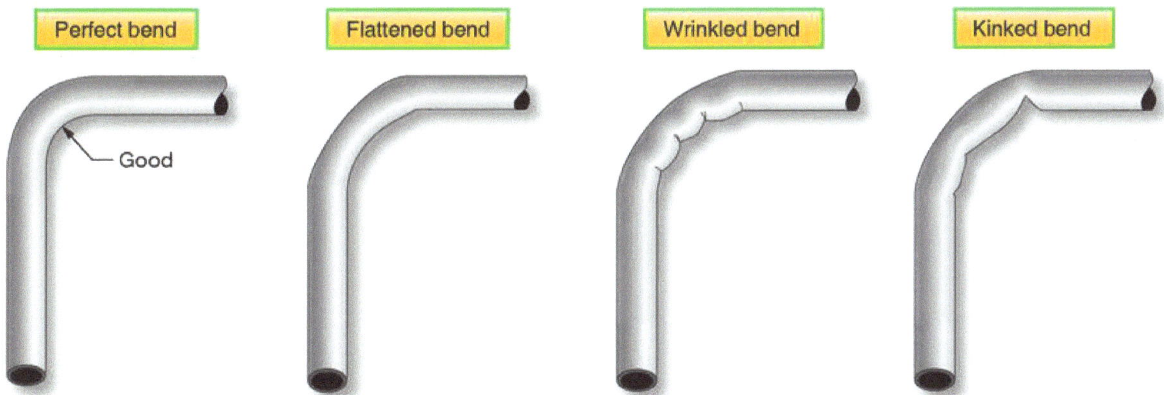

Fig. 3.13- Perfect and Incorrect Tube Bends (www.aircraftsystemstech.com)

❖ **Proper Pipe Welding:** The following shall be meticulously considered when welding pipes, particularly high-pressure ones. Otherwise, pipes may leak.

- Pipe assemblies must be properly prepared for welding to avoid excessive contamination in the line.
- Components which can be affected by the extreme heat must be properly isolated.
- Welding personnel must be certified and well trained.
- Inspect the pipe after welding and before installation, the sealing surfaces and internal surfaces of piping shall be free of any visible detrimental foreign matter such as scale, burrs, swarf, etc.

❖ **Proper Pipe Assembly:** As shown in Fig 3.14, when assembling a pipe, leave the end pipe threads free from pipe thread sealant or Teflon® tape to ensure that it does not contaminate the system.

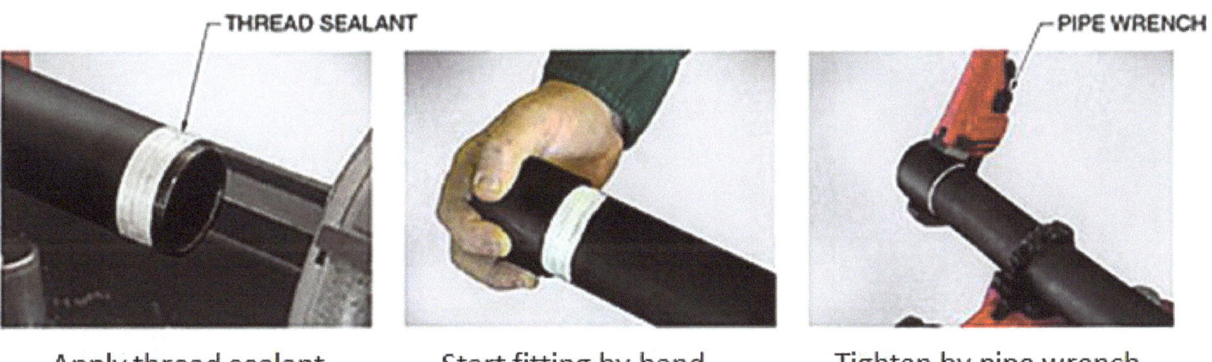

Fig. 3.14- Pipe Assembly (Courtesy from American Technical Publishers)

❖ **Transmission Line Proper Clamping:**
▪ **Why should a hydraulic line be clamped?**
  o Unclamped transmission lines transmit noise and vibration.
  o If a pressurized transmission line assembly blows apart, the fittings can be thrown off at high-speed causing risks of injuries.
  o If an unclamped pressurized hose is accidentally detached from one of its ends, it whips like a snake with great force.

▪ **Best practices for hose clamping:**
  o Do Not temporarily drop a return line hose into the reservoir.
  o Do Not allow hoses to drag in the dirt or lay on the ground.
  o Clamp hoses to the structure; not to components to minimize vibration.
  o Clamp far from moving parts.
  o Clamp to avoid surface abrasion as shown in Fig. 3.15.
  o Use hose restraints, shown in Fig. 3.16, to protect against injury when a hose is blown.

Fig. 3.15- Proper Hose Clamping

Fig. 3.16- Restrains for Hoses Clamping

- **Best practices for tubes/pipes clamping:** As shown in Fig. 3.17, different standard tube clamping collars are available in the market. Collars can be made of polypropylene, self-extinguish polyamide, aluminum or zinc-coated steel with plastic supports and they are fastened via a weld plate, a thread, or a guide. Figure 3.18 provides guidelines for clamping distance. Random clamping distance could lead to generating of noise and vibration.

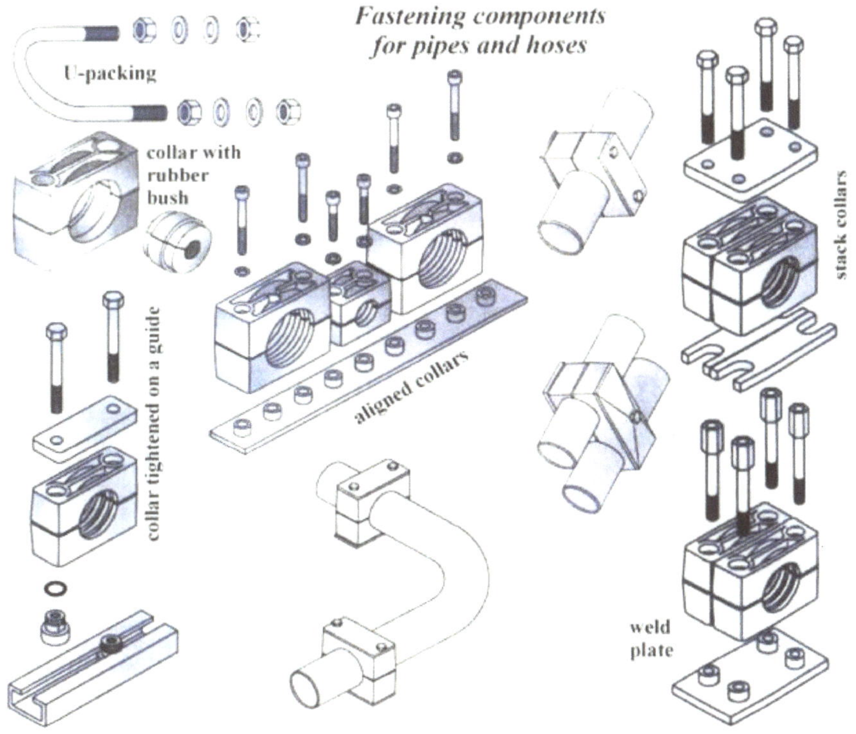

Fig. 3.17- Clamping Collars for Tubes and Pipes (Courtesy of Assofluid)

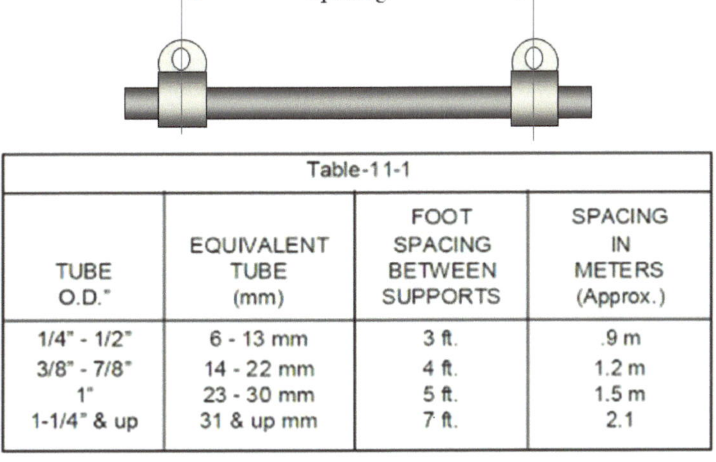

Table-11-1			
TUBE O.D."	EQUIVALENT TUBE (mm)	FOOT SPACING BETWEEN SUPPORTS	SPACING IN METERS (Approx.)
1/4" - 1/2"	6 - 13 mm	3 ft.	.9 m
3/8" - 7/8"	14 - 22 mm	4 ft.	1.2 m
1"	23 - 30 mm	5 ft.	1.5 m
1-1/4" & up	31 & up mm	7 ft.	2.1

Fig. 3.18- Guidelines for Tube Clamping Distances

Avoid Mechanical Stresses on Hard Transmission Lines: Referring to Fig. 3.19, the following set of examples provide guidelines to avoid developing mechanical stresses and consequently reduce the chance of leakage and permanent failure.

Example 1: Dead weight of the components stressing the transmission lines. Avoid excessive stress on joints by selecting appropriate clamping spots for plumbing and components.

Example 2: Hard transmission lines shall not be used for equipment support. If a hard line connects to a component, an additional support is required to reduce the stresses on the line.

Example 3: Avoid poor tube bending that may result in stressing the fittings. DO NOT heat to ease tube bending. The tube may lose its ductility and thereby be subject to failure under high-pressure conditions.

Example 4: Do NOT lay or stand on transmission lines during system servicing.

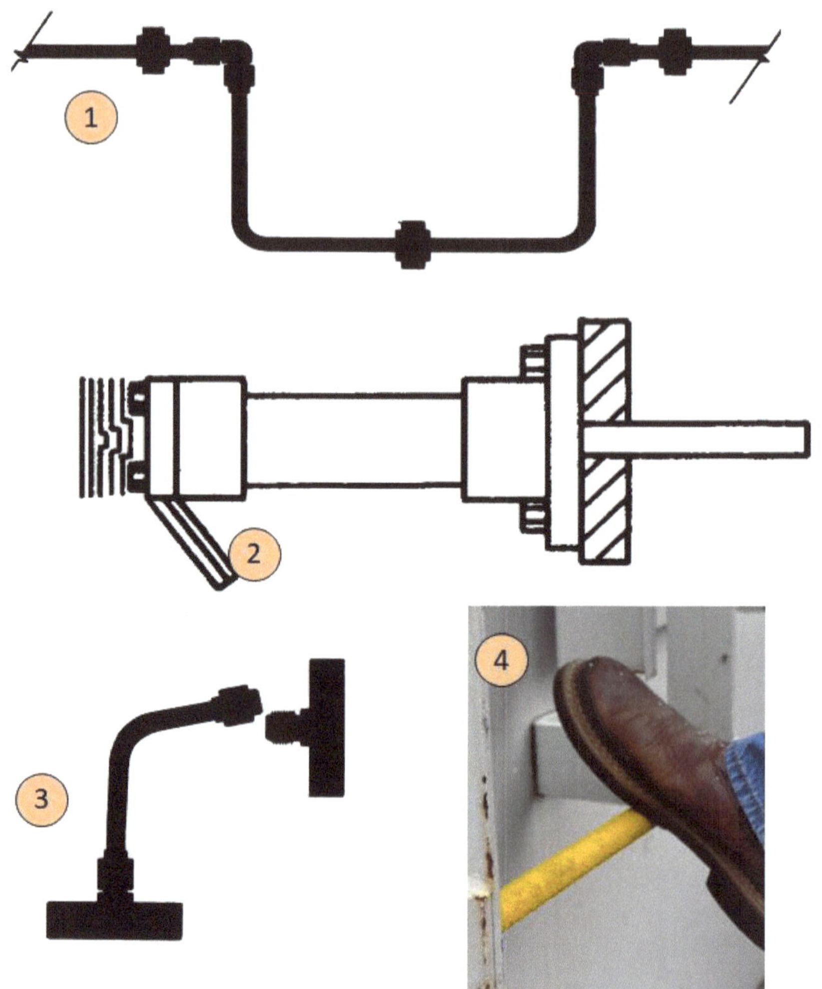

Fig. 3.19- Guidelines to Avoid Mechanical Stresses (Courtesy from Parker Hannifin)

3.4-BP-Transmission Lines-04-Standard Tests and Calibration

Hoses: ISO 6605 2002 Hydraulic Hoses and Hose Assemblies-Test Procedures. This international *standard* specifies test methods for evaluating the performance of hoses and hose assemblies used in hydraulic fluid power systems. The following are common tests:

Pressure Proof Test (ISO 1402): *Pressure Proof Test* is a nondestructive test. The hose assembly is hydrostatically tested to the specified proof pressure in accordance with the relevant product specification using the method specified in ISO 1402 for a period 30-60 seconds. The assembly is considered as passing the test with no signs of leakage or failure.

Burst Pressure Test (blog.parker.com): This is a destructive test. A *Burst Test* is a hydrostatic pressure test of a hose assembly that determines the actual burst strength of the assembly. Any signs of leakage, bulging, coupling ejection or hose burst below the specified minimum rated burst pressure of the assembly are considered a failure. Minimum burst values are used as one factor in the establishment of a reasonable and safe maximum work pressure and hose lifetime. As shown in Fig. 3.20, maximum rated working pressure is based on safety factor of 4.

Fig. 3.20- Hose Burst Test

Cold Flexibility Test (ISO 10619): A *Cold Flexibility Test* is a destructive test specified in ISO 10619. It is used to determine the ability of a hose to operate under cold weather. Condition hose assembly at a temperature equal to the minimum application temperature of the relevant product specification in a straight position for 24 hours. While still at the minimum application temperature, bend the sample on a mandrel over a time of 8-12 seconds. For hose (ID) sizes <= 22 mm bend them through 180^0 mandrel. For hose (ID) sizes > 22 mm bend them through 90^0 mandrel. After bending, allow the sample to warm to room temperature, visually examine it. The hose or cover should not crack; and when warmed to ambient temperature, the hose should not leak or crack when subjected to proof pressure.

Impulse Test: An *Impulse Test* for 100Rxx hydraulic hose is one of the key predictions of hose life. Impulse testing involves pressurizing the hose cyclically at 133% of working pressure, at rates up to 1 cycle per second while the hose is held in either a 90° or 180° configuration. To pass the test, the hose must meet or exceed the minimum number of impulse cycles based on the applicable industry standard.

Salt Spray Test (ASTM B117): *Salt Spray Test* (also called *Salt Fog Test*) is a standardized and popular corrosion test method for hydraulic fittings used to check corrosion resistance of materials and surface coatings. Salt spray testing is a destructive test. It is an accelerated corrosion test that produces a corrosive attack to coated samples in order to evaluate and compare the suitability of the coating for use as a protective finish.

Most commonly, the time taken to oxides the sample under test is considered to determine whether the test is passed or failed. Figure 3.21 shows SAE J516 standard hydraulic fittings after the 144-hour salt spray test in accordance with ASTM B117. The appearance of red rust within this period indicates failure.

Fig. 3.21- Salt Spray Test (Courtesy from Gates)

Gravimetric Cleanliness Measurement (ISO 4405): Gravimetric measurement is a reporting method that references the total mass of contaminant found inside a hydraulic component as normalized by the total internal component surface area of a hydraulic component. A fluid is used to dislodge contamination in a hose assembly and is then poured through a membrane catch filter. The filtered particles are dried and weighed, and results are typically stated in milligrams of contaminant per square meter of internal surface area. Figure 3.22 shows an example determining the cleanliness level when 52 mg of contaminant is found in a hose assembly:

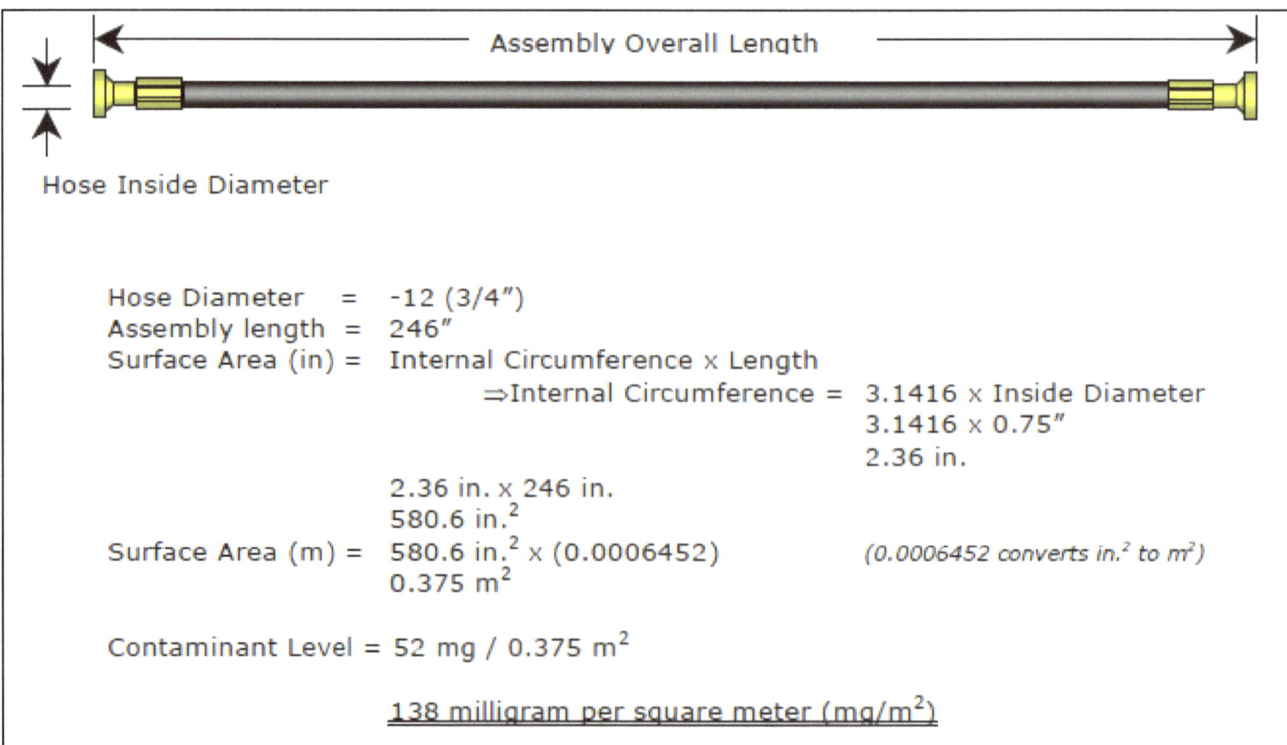

Fig. 3.22- Hose Cleanliness Measurement (Courtesy from Gates)

Reliability Assessment Test (Fig. 3.23):   The SAE and ISO standards for hydraulic hose impulse tests are ran for a finite number of cycles under standardized conditions. If the hose passes this test, it meets the minimum requirements of the applicable standard. Comparable hoses from various manufacturers will usually always pass this test due to the limited number of cycles. Therefore, there is no way to determine from this test which manufacturer will last longer in an application. To do so, a test protocol must be developed using application specific parameters and then test to failure or a specific number of cycles required by the application.

The *Hose Reliability Assessment Test* is an application-based laboratory test program developed by Milwaukee School of Engineering (MSOE). One specific application example was to assess the reliability of hydraulic hoses used on refuse vehicles under various operating conditions. The test program considered investigating the effect of the following operating conditions:
- Four different hose manufacturers and two different hose construction types (2W and 4W spiral) multiple specimens from each manufacturer and hose type.
- Level of continuous pressure: based on vehicle measured data.
- Magnitude and frequency of pulsating pressure: based on vehicle measured data.
- Working temperature: based on vehicle measured data.
- Bend radius: one-half of the minimum specified by the applicable hose construction.
- Twisting angle: 45 Degrees between hose couplings when assembled on the manifold.
- Minimum test length: 2,000,000 cycles.

The test results concluded the following:
- 4W spiral hose lasts up to 5 times longer than the 2W hose in this specific application
- Increased vehicle productivity.
- Reduced environmental costs of cleanups.
- Estimated company annual savings was $10M.

Fig. 3.23- Hose Reliability Assessment Test

3.5-BP-Transmission Lines-05-Transportation and Storage

Referring to Fig. 3.24, the following best practices are guidelines for transportation and storage of transmission lines:

Hose Storage (1): Should be coiled on reels that may rotate manually or using a motor.

Tube/Pipe Storage (2): They should be placed on supports with distances based on the size and length to avoid bending. Pipe manufacturers must be reviewed for such information.

Dust Caps (3): All lines ends should be closed by proper dust caps.

Storage Space: It should be clear of high temperature, corrosive liquids, humidity, ultraviolet light, ozone, oils, fumes, solvents, high humidity, insects, electromagnetic fields or radioactive materials.

Storage Time: Maintain a system of age control that shows a hose must be used before the shelf life is expired and facilitate first-in first-out usage based the date of manufacturing. Follow the manufacturer's shelf life. However, **SAE J517** specifies the shelf life of hydraulic hoses.

Transportation: transmission lines of all types must be sealed with protective caps.

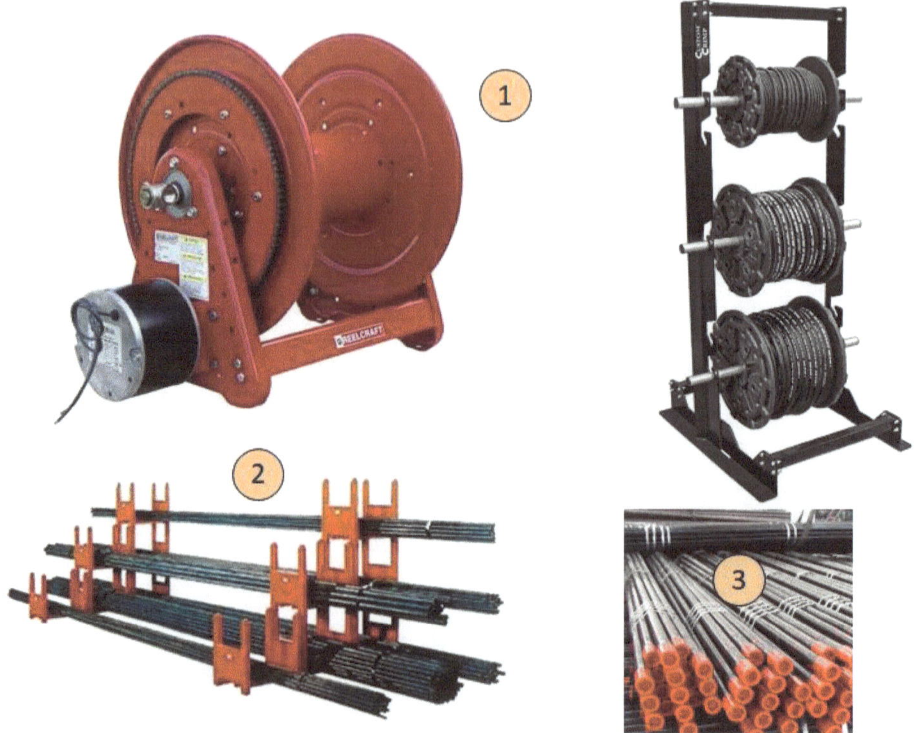

Fig. 3.24- Best Practices for Storage of Transmission Lines

3.6- Oil Injection Avoidance and Treatment

How Oil Injection Happens?
Fine streams of escaping pressurized fluid can penetrate the skin and thus enter the human body causing, as shown in Fig. 3.25, serious injuries, loss of organs, or even death. This is known as Oil Injection.

Challenges of Oil Injections:
- According to the "Occupational Injuries Handbook," oil can penetrate the skin at pressures as low as 100 PSI (7bar).
- The symptoms of oil injection sometimes don't appear until the injected area becomes in critical condition.
- Fluid injected into the skin must be surgically removed within hours or gangrene may occur resulting in amputation of the affected area.
- Some medical care providers are unfamiliar with fluid injection injuries.

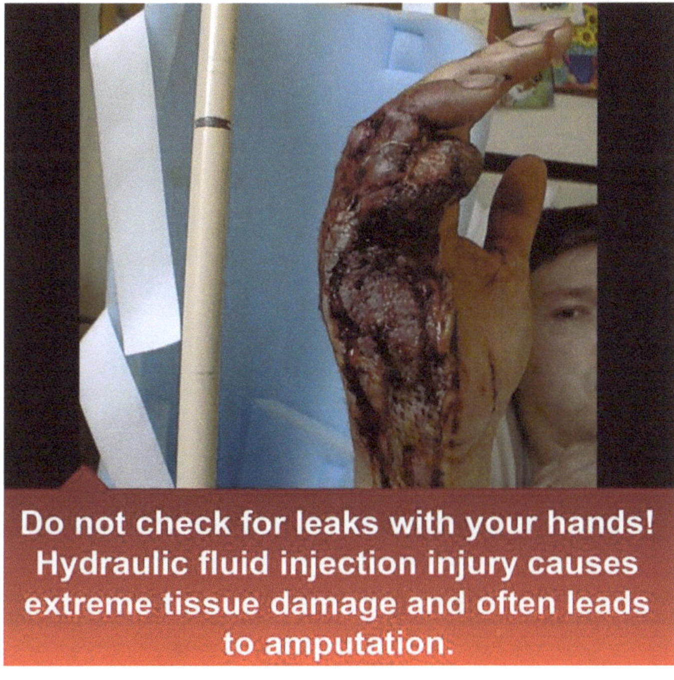

Fig. 3.25 – Risks of Oil Injection
(Courtesy of the International Hydraulic Safety Authority)

Best practices to avoid oil injection:
- Make the people aware of oil injection risks.
- Never use your hand or fingers to search for leaks.
- Wear protective gloves when servicing hydraulic transmission lines.

Best practices if fluid injection occur:
- Report the case immediately to your supervisor.
- DO NOT allow the injured person to drive himself to the medical facility.
- DO NOT give food or drink to the injured person.
- DO NOT treat injections as a simple cut!
- DO NOT delay treatment.
- See a medical specialist doctor immediately.
- Prepare the case information as shown in the safety card provided by the International Fluid Power Society (IFPS), shown in Fig. 3.26

Fig. 3.26 – Safety Focus Card
(Courtesy of the International Fluid Power Society)

Reported Case History (Fig. 3.27):

Action: A technician used his hand to check leakage from a hose spraying oil.

Result: 2 fingers lost.

Reason: High-pressure oil injection not treated immediately

For more information about the subject matter, the Fluid Power Training Institute. released a demonstrative video shown in Fig. 3.27. The video can be ordered from www.fluidpowersafety.com.

Fig. 3.27 – Oil Injection Training Video
(Courtesy of Fluid Power Training Institute)

Chapter 4

Troubleshooting and Failure Analysis of Transmission Lines

Objectives

This chapter discusses hydraulic *transmission lines* inspection, troubleshooting, and failure analysis. In this chapter, a troubleshooting chart for transmission line faults is presented. The chapter also presents examples of defective transmission lines.

Brief Contents

4.1- Hydraulic Transmission Lines Inspection
4.2- Hydraulic Transmission Lines Troubleshooting
4.3- Hydraulic Transmission Lines Failure Analysis

Chapter 4: Troubleshooting and Failure Analysis of Transmission Lines

4.1- Hydraulic Transmission Lines Inspection

Hydraulic *transmission lines* could be pipes, tubes or flexible hoses. They are used to transmit fluid power between hydraulic components. Failure of any transmission line causes consequences that range from minor problem to loss of life.

Table 4.1 shows typical inspection sheet for transmission lines.

Hydraulic Transmission Lines Inspection Sheet	
Manufacturer	
Model #	
Serial #	
Location	
Transmission Line Type	☐ Pipe ☐Tube ☐Hoses
Transmission Line Nominal Size (ID for Hoses or OD for tubes and pipes)	mm:( ) inches:( )
Transmission Line Length	
Transmission Line Ends	
Special features of the line	
Conditions of Transmission line	▪ Hose kinks and twists. ▪ Hose cracks from minimum bend radius exceeded. ▪ Hose brittleness or loss of flexibility. ▪ Hose frayed protective layers. ▪ Hose broken reinforcement layers. ▪ Hose outer cover pulled back from coupling ends. ▪ Hose rusted, broken, or loosen ends joints. ▪ Hose abrasion.
Other Notes	

Table 4.1 – Hydraulic Transmission Lines Inspection Sheet

4.2- Hydraulic Transmission Lines Troubleshooting

Problem of leaking transmission line is beyond just loss of fluid. The real cost of leaking fluid is the cost of:
- Make-up the fluid.
- Cleaning-up the mess.
- Proper disposal of the spilled fluid in accordance with the local state and federal regulations.
- Contamination ingression.
- Possible safety liability.

To get a better understanding of the real cost of leakage, a study has been done to calculate the cost of a fitting that is leaking six drops of oil per minute. The cost was found to be nearly $1000 per year.

Table 4.2 shows troubleshooting guidelines for transmission lines.

T-Transmission Lines-01: Transmission Lines Troubleshooting	
Leaking transmission line? **(See Note 1)**	▪ Line Body (check the following signs): ▪ Line body is damaged. ▪ Line is subjected to mechanical stresses. ▪ Line is over pressurized. ▪ Line wasn't cleaned before assembly **(See Note 2).** ▪ Line minimum bend radius isn't respected. ▪ Hose is twisted, stretched, or fried. ▪ Hose is subjected to abrasion. ▪ Line Fittings (check the following signs): ▪ Line ends are damaged, detached, or corroded. ▪ Line ends are over tightened. ▪ Line ends are nonstandard and of a bad quality. ▪ Tube flare has cracks or embedded dirt. ▪ Tube is not properly aligned with fitting. ▪ Threaded pipe is assembled with no sealant used. ▪ Fitting threads are distorted. ▪ O-Ring leak. ▪ Hydraulic Fluid: ▪ Verify that the fluid is compatible with the line, the outer cover of hoses, line ends, and O-rings.

Table 4.2– Transmission Lines Troubleshooting Chart

Note 1: Transmission Line Leakage Inspection: Figure 4.1 shows an effective method of leakage inspection by blending a small amount of concentrated special dye with the fluid. By applying a light of a specific wavelength using a special lamp, even the smallest leaks become easily visible. Fluid leakage test kit is available.

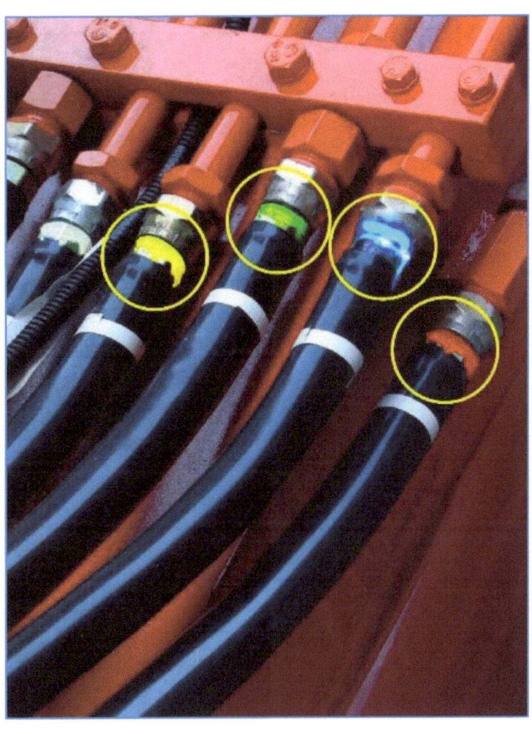

Fig. 4.1– Effective Method for Transmission Line Leakage Inspection (Courtesy of Spectroline)

Note 2: Hose Leakage due to Line Not Cleaned Before Assembly: Contamination can cause several problems for a hydraulic hose assembly. As shown in Fig. 4.2, when cutting a hose, metal particles and debris can settle inside the hose if not properly cleaned. This abrasive debris causes small fractures between fitting and hose assembly, resulting in developing leakage passes. To prevent hose failures due to contamination, the hose must be properly cleaned before inserting the fittings. After the fittings are crimped, make sure to cap the ends in order to keep the hose clean and avoid recontamination during transportation.

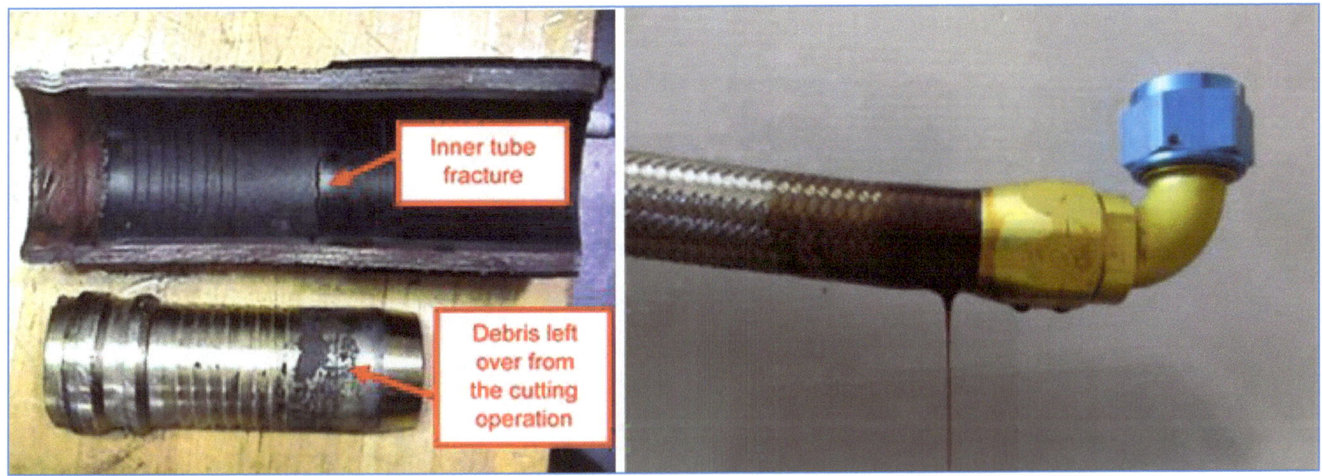

Fig. 4.2– Hose Leakage due to Line Not Cleaned Before Assembly (Courtesy of Parker)

4.3- Hydraulic Transmission Lines Failure Analysis

Hose Cover Abrasion: Figure 4.3 shows hose cover abrasion due to rubbing with a metallic sharp edge. The figure shows best practices of resolving this problem by mounting the hose away from moving elements.

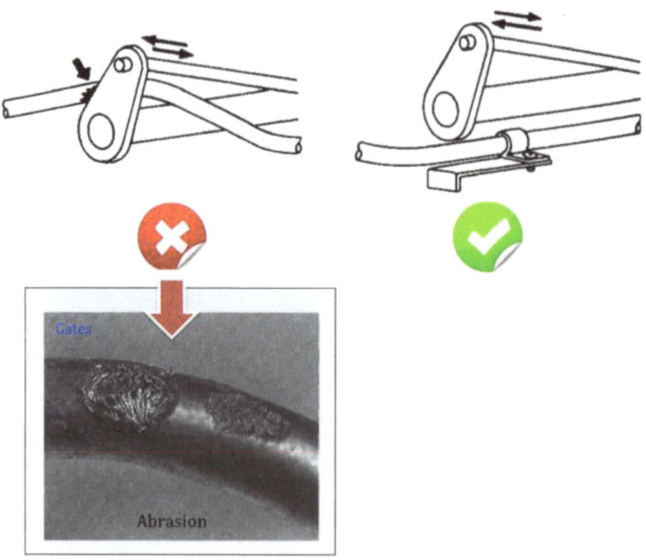

Fig. 4.3– Hose Cover Abrasion (Courtesy of Gates)

Hose Ends Detached due to Short Length: Figure 4.4 shows hose ends detached due to cutting the hose short and not respecting minimum hose length guidelines. When the hose is pressurized, it becomes even shorter causing the hose ends blow off.

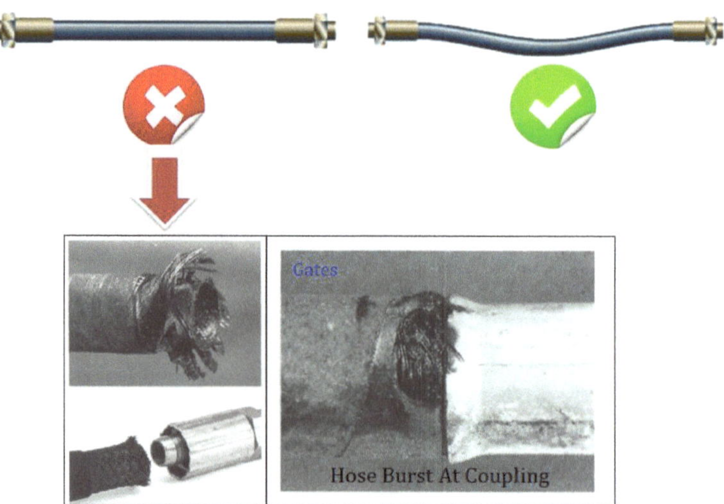

Fig. 4.4– Hose Ends Detached due to Short Length (Courtesy of Gates)

Hose Cover Cracked due to Poor Routing: Figure 4.5 shows hose cover cracked due to improper mounting, too many bends, and not respecting minimum bend radius.

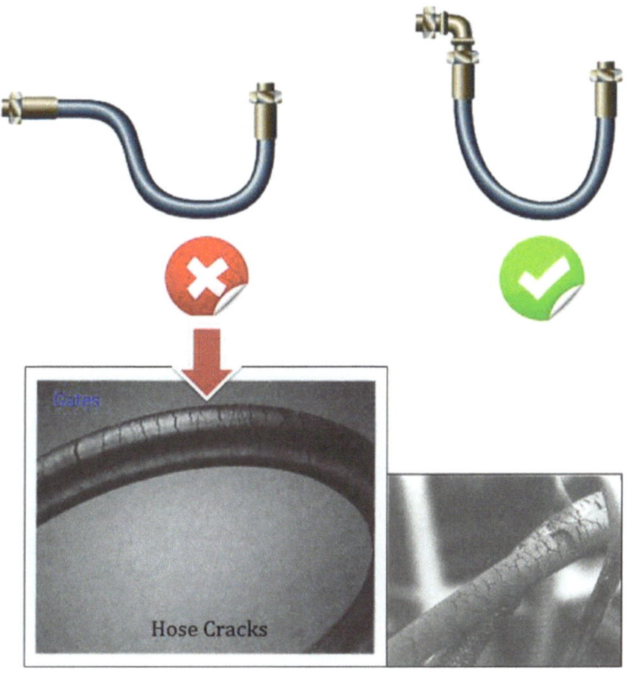

Fig. 4.5– Hose Cover Cracked due to Poor Routing (Courtesy of Gates)

Hose Plastically Deformed due to Improper Assembly: Figure 4.6 shows a hose is twisted due to improper assembly. As a result, over the time, the hose is kinked and plastically deformed causing high pressure losses.

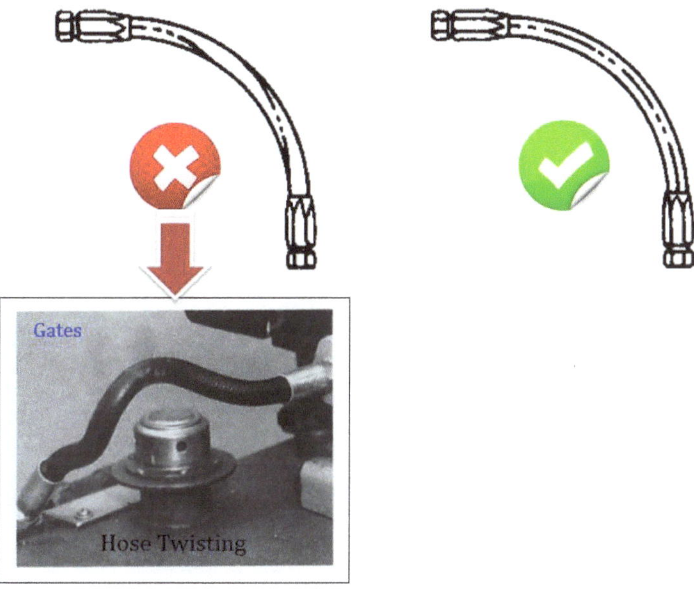

Fig. 4.6– Hose Twisting due to Improper Assembly (Courtesy of Gates)

Hose Burned due to Direct Contact with Heat Sources: Figure 4.7 shows a fried hose due to contact with external heat sources. It is recommended to shield hoses from heat sources.

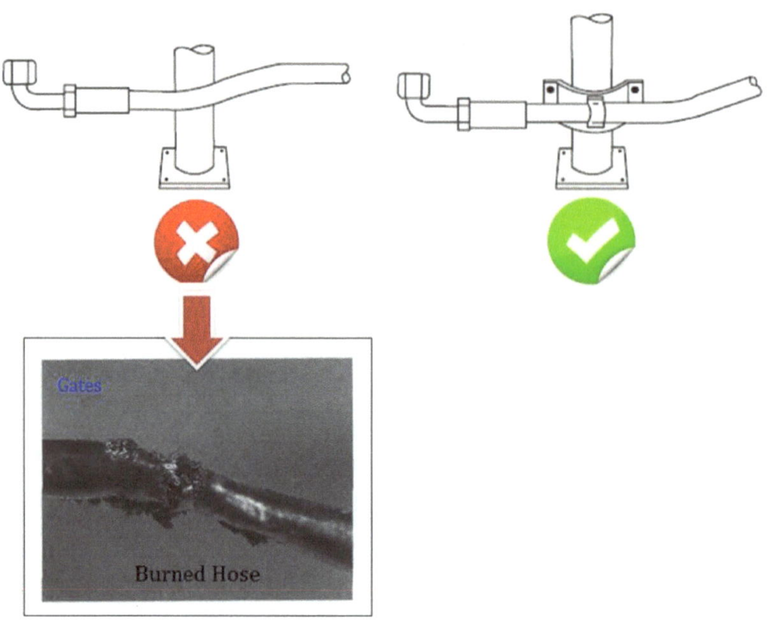

Fig. 4.7– Hose Burned due to Direct Contact with Heat Sources (Courtesy of Gates)

Hose Body Burst due to Exceeding Minimum Bend Radius: Figure 4.8 shows hose burst. The reported cause is due to the kinematical motion of the cylinder, the hose is severely bent exceeding its minimum bend radius. After number of cycles, the hose fatigued and exploded.

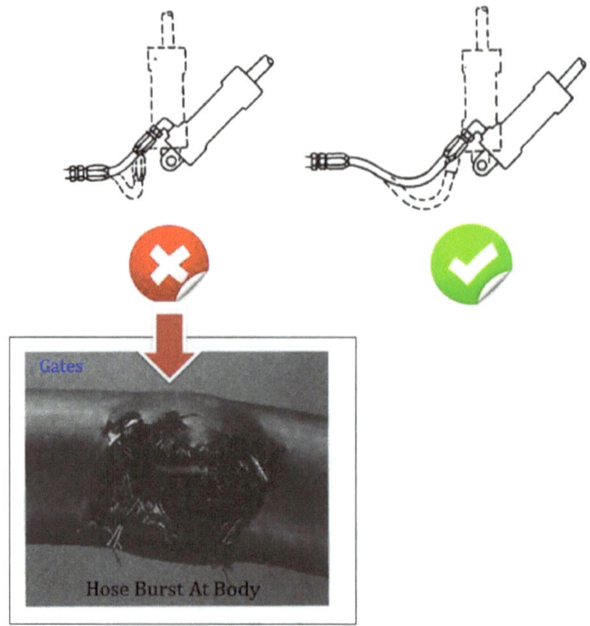

Fig. 4.8– Hose Burst due to Exceeding Minimum Bend Radius (Courtesy of Gates)

Hose Body Burst due to Exceeding lifetime: Figure 4.9 shows hose burst. The reported cause is that the hose was used for a longer time than the recommended lifetime of the hose.

Fig. 4.9– Hose Body Burst due to Exceeding lifetime

Hose Failure due to Fluid Incompatibility: Figure 4.10 shows the result of using an incompatible fluid that causes the inner tube of the hose assembly to deteriorate, swell, and delaminate.

Fig. 4.10– Hose Failure due to Fluid Incompatibility (Courtesy of Parker)

Hose Ends Blow Off due to Short Insertion Depth: Figure 4.11 shows that, when a hose assembly is not properly assembled, it can create very dangerous damages. Fittings need to be pushed on completely to meet the recommended insertion depth. If the hose insertion depth is not met, fittings can blow off, leaving a failed hose assembly. The last grip in the fitting shell is essential to the holding strength.

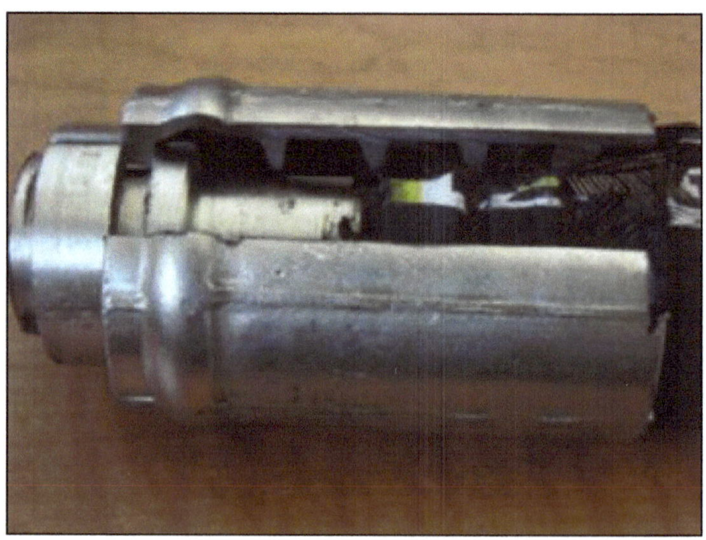

Fig. 4.11– Hose Ends Blow Off due to Short Length (Courtesy of Parker)

Hose Failure due to Overheating: Hose failure can occur from overheating the hose assembly. As shown in Fig. 4.12, overheating will cause the hose to become very stiff. The inner tube is hardened and begin to crack because the plasticizers in the elastomer will break down or harden under high temperatures. In some cases, the cover may show signs of being dried out. The hose assembly may remain in its installed shape after being removed from the application and if flexed, audible cracking can be heard. In order to prevent overheated hydraulic hose assemblies, confirm hoses are rated for the temperatures required by your application. Also, reduce ambient temperatures using good ventilation or use heat guards and shields the hose from nearby high-temperature areas.

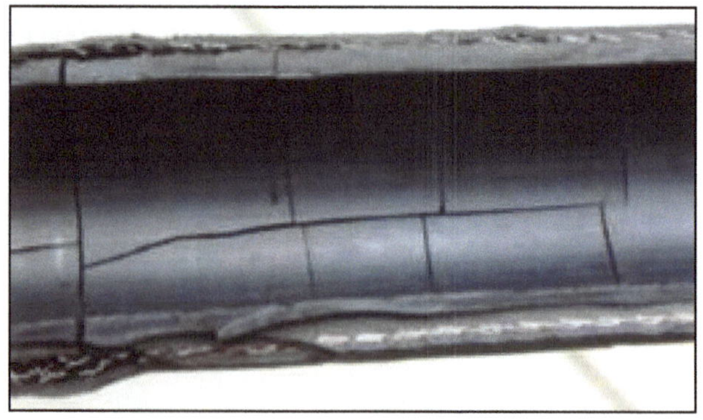

Fig. 4.12– Hose Failure due to Overheating (Courtesy of Parker)

Hose Cracked due to Exposure to Paint Spray: Fig. 4.13 shows a hose is hardened and cracked only on the side which had been exposed to paint overspray during aircraft manufacturing. The paint chemically attacked the hose causing such a damage that reduces the hose lifetime to half.

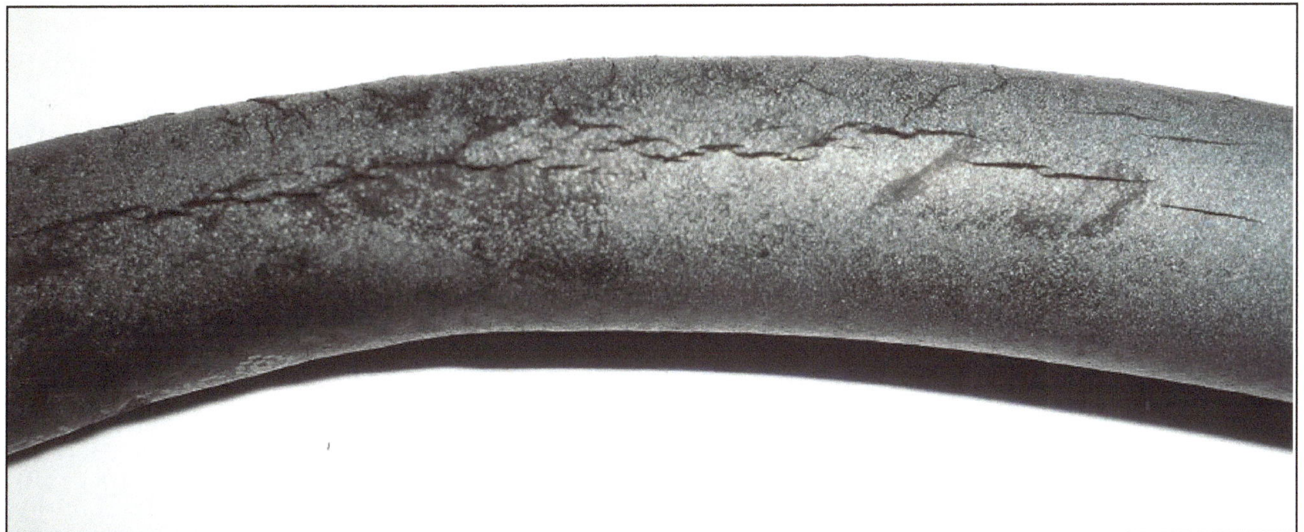

Fig. 4.13– Hose Cracked due to Exposure to Paint Spray

Tube and Pipe Burst due to Overpressure: Exceeding the maximum allowable pressure in transmission lines may result in a line burst. Fig. 4.14, shows tube burst due to overpressure. Fig. 10.14, shows tube burst due to overpressure. Burst pressure supposed to be known through the line manufacturer and the relevant standards.

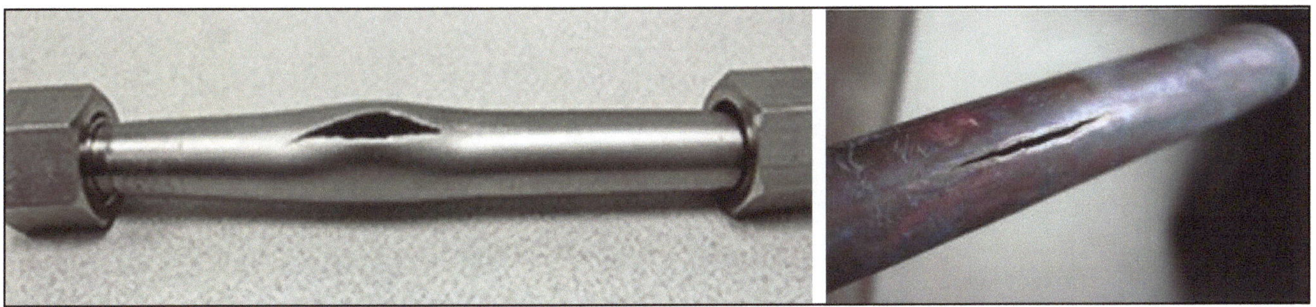

Fig. 4.14– Tube and Pipe Burst due to Overpressure (Courtesy of Parker)

Pipe and Tube Pin Holes due to Poor Material: Fig. 4.15 shows a pin hole in a hydraulic pipe due to poor material. Not any pipe can be used to transmit hydraulic power. Selection of pipe size and material should be in accordance with appropriate industry standards.

Fig. 4.15– Pipe and Tube Pin Holes due to Poor Material

Pipe and Tube are Leaking due to Mechanical Stresses: Fig. 4.16 shows a leaking hydraulic pipe due to mechanical stresses as a result of the pipe weight. Such a pipeline must be hanged to the ceiling to release the mechanical stresses.

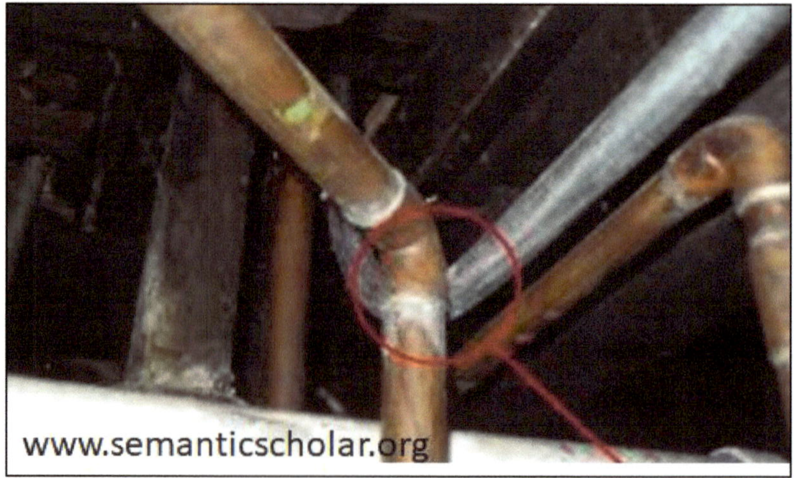

Fig. 4.16– Pipe and Tube are Leaking due to Mechanical Stresses

Chapter 5

Energy Losses in Transmission Lines

Objectives

This chapter discusses the two common flow patterns in hydraulic transmission lines, laminar flow, and turbulent flow. The chapter presents reviews of fundamental concepts conservation of mass and conversation of energy as applied to fluid flow in transmission lines. of power losses in hydraulic transmission line. The chapter presents both mathematical and graphical methods of quantifying the power losses in transmission lines, fittings, and orifices.

Brief Contents

5.1- Flow Patterns in Transmission Lines
5.2- Conservation of Mass in Fluid Flow
5.3- Conservation of Energy in Fluid Flow
5.4- Pressure Drop in Transmission Lines

Chapter 5 - Energy Losses in Transmission Lines

5.1- Flow Patterns in Transmission Lines

In order to analyze the *energy losses* in a hydraulic transmission line, the pattern of the fluid flow inside the line must be identified first. While a fluid is flowing in a transmission line, it behaves based on one of two known flow patterns: *Laminar* or *Turbulent*. In the following section both flow patterns and how to identify them are discussed.

5.1.1- Laminar Flow

As shown in Fig. 5.1, laminar flow is characterized by the following features:
- Streamlines form coaxial layers.
- Fluid layers are steady and do not cross each other.
- Boundary layer, that is adjacent to the inner surface of the line, is standstill.
- Layers following the boundary layer move with speed increasing towards the center.
- The velocity gradient through the line cross section is parabolic.
- Average velocity = 0.5 of the maximum velocity at the center of the line.
- Fluid flow generates fluid-fluid friction due to the relative motion between the fluid layers that are sliding against each other.

5.1.2- Turbulent Flow

As shown in Fig. 5.1, turbulent flow is characterized by the following features:
- Unsteady streamlines.
- Streamlines crossing each other.
- Moving fluid have a good chance to hit the inside walls of the transmission line.
- Layers following the boundary one move with speed increasing towards the center.
- Average velocity = 0.8 of the maximum velocity at the center of the line.
- Fluid flow generates fluid-fluid friction and fluid-wall friction.

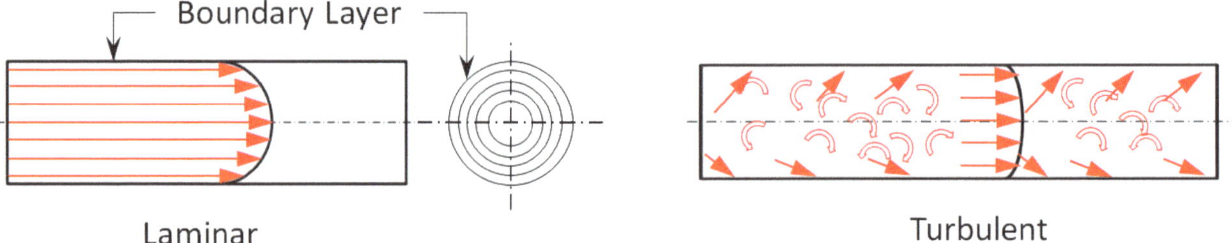

Fig. 5.1- Flow Patterns

5.1.3 - Laminar versus Turbulent Flow

Mistakenly it is believed that turbulent has higher coefficient of friction than laminar flow. The fact is that coefficient of friction in turbulent flow, particularly in smooth transmission lines, is less than in laminar flow. In laminar flow, while there is no fluid-wall friction because the boundary layer is stationary, fluid-fluid friction is high because several layers of fluid slide on each other. In turbulent flow, fluid-fluid friction is small, and fluid-wall friction is high only if a very rough transmission line is used.

The other part of the fact is turbulent flow causes higher overall pressure losses. That is because pressure losses in a transmission line is due to a combined effect of coefficient of friction (*Frictional Losses*) and how fast the fluid is flowing inside the line (*Kinetic Losses*). Obviously, the fluid speed in case of turbulent flow is higher than in case of laminar flow. However, the other reasons why a hydraulic system should be designed to secure laminar flow, as shown in Fig. 5.2, are as follow:

- Better system response due to flow fills 100% of the cross section of the line.
- No separation from the inside walls of the conductor.
- Less formation of eddies, air pockets and air bubbles in the flow.
- One phase flow occurs since no air is mixed with the fluid.
- Better lubrication.
- Higher equivalent bulk modulus.
- Surface roughness of the transmission lines will have no effect on friction coefficient.

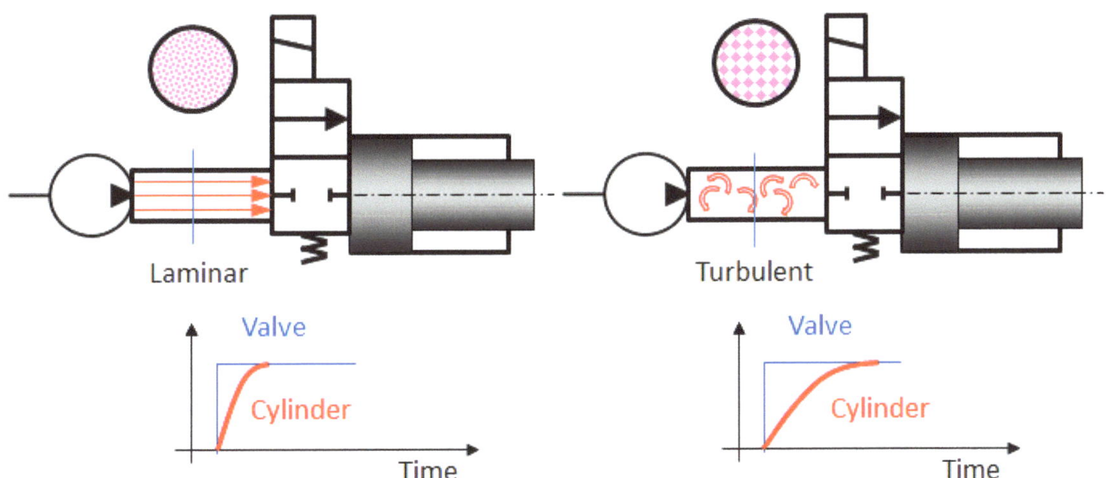

Fig. 5.2 - Laminar versus Turbulent Flow

In some cases, turbulent flow may be needed on purpose. A good example of this case is in heat exchangers. In order maximize the cooling efficiency and heat transfer between hot oil and cooling water, these fluids must have good contact with the wall of the passages through which the fluids are passing. Another example is in flushing systems. To make sure the flushing fluid cleans the transmission lines properly, the flushing fluid must hit the inside walls of the lines.

5.1.4- Reynolds Number

To predict the flow pattern in a transmission line, *Reynolds Number* (**R_e**) must be calculated. As shown in Fig. 5.3, if Reynolds Number is below 2000, then the flow is laminar. If Reynolds Number is above 3500, then the flow is turbulent. In between the flow is in transition status. Reynolds number is a dimensionless number represents the ratio between the accelerating and the decelerating forces that act on a unit fluid element. The accelerating forces are function of how fast the fluid is moving. The decelerating forces are function of the moving mass of the fluid and the molecular bindings that are caused by the fluid viscosity. If the accelerating forces increases while the decelerating forces decreases, Reynolds number increases, and the flow is biased to be turbulent. If the accelerating forces decreases while the decelerating forces increases, then Reynolds number decreases, and the flow is biased to be laminar.

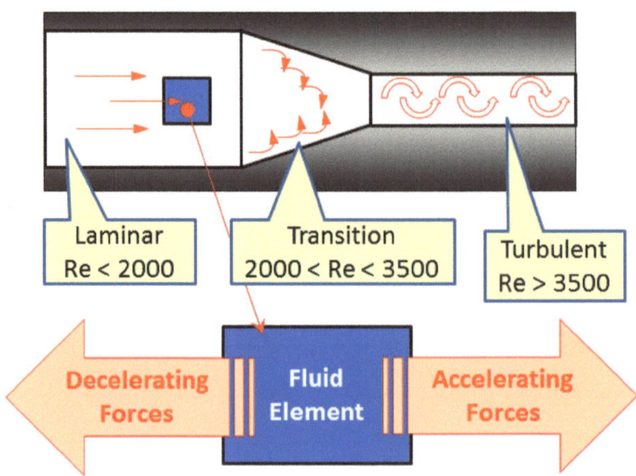

Fig. 5.3- Reynold's Number

Equation 5.1 is used to calculate Reynolds Number. The equation shows that Reynolds Number is function of a combination of design variable **D_h**, operational variable **v** and fluid property **ν**. By a quick look to the equation, first impression reveals that **R_e** is proportional to the line diameter, meaning that it increases with the increase of the line diameter. This is a deceiving impression! It should be remembered that, when the diameter increases, the fluid velocity is decreased simultaneously for the same flow rate passing through the line.

$$\mathbf{R_e} = \frac{\rho v D_h}{\mu} = \frac{v D_h}{\nu} \qquad\qquad 5.1$$

Where:
- ρ = Fluid density.
- v = Fluid velocity.
- D_h = Hydraulic diameter of the transmission lines.
- μ = Dynamic viscosity.
- ν = Kinematic viscosity.

The definition of a *hydraulic diameter* for any conductor is the ratio between four times its cross-section area and its perimeter. For circular cross section conductors, like what are used in hydraulic systems, the hydraulic diameter is same as just the diameter. The reason a term of hydraulic diameter is used in the equation, instead of just a diameter, is because this equation is used for non-circular cross section conductors, such as in air ducts, etc.

For practical use of equation 5.1 in hydraulic system design, Eq. 5.2.A and Eq. 5.2.B have been developed with balanced units in Metric and English system of units, respectively. Now, by looking at these equations, it is clear that R_e is inversely proportional to the diameter.

$$R_e = \frac{21231\ Q[l/min]}{v\ [cSt] \times D[mm]} \qquad\qquad 5.2.A$$

$$R_e = \frac{3164\ Q[gpm]}{v\ [cSt] \times D[in]} \qquad\qquad 5.2.B$$

It is to be noted that transmission lines (tubes, pipes, and hoses) are not selected first then checked for flow pattern. During system design, conductors are sized primarily to secure laminar flow from the very beginning. With the knowledge of the flow rate and the fluid viscosity, Eq. 5.2.A and 5.2.B can be rewritten to solve for the line diameter. So, Eq. 5.3.A and Eq. 5.3.B are used to size transmission lines to secure laminar flow (Reynolds Number < 2000).

$$D[mm] = \frac{21231\ Q\left[\frac{l}{min}\right]}{v\ [Cst] \times R_e} \qquad\qquad 5.3.A$$

$$D[mm] = \frac{3164\ Q[gpm]}{v\ [Cst] \times R_e} \qquad\qquad 5.3.B$$

In the example below, Fig. 5.4 shows a fluid line with a turbulent flow. System designer has one of three following choices to secure laminar flow as follows:

- Increase the line diameter.
- Increase the fluid viscosity.
- Reduce the flow rate.

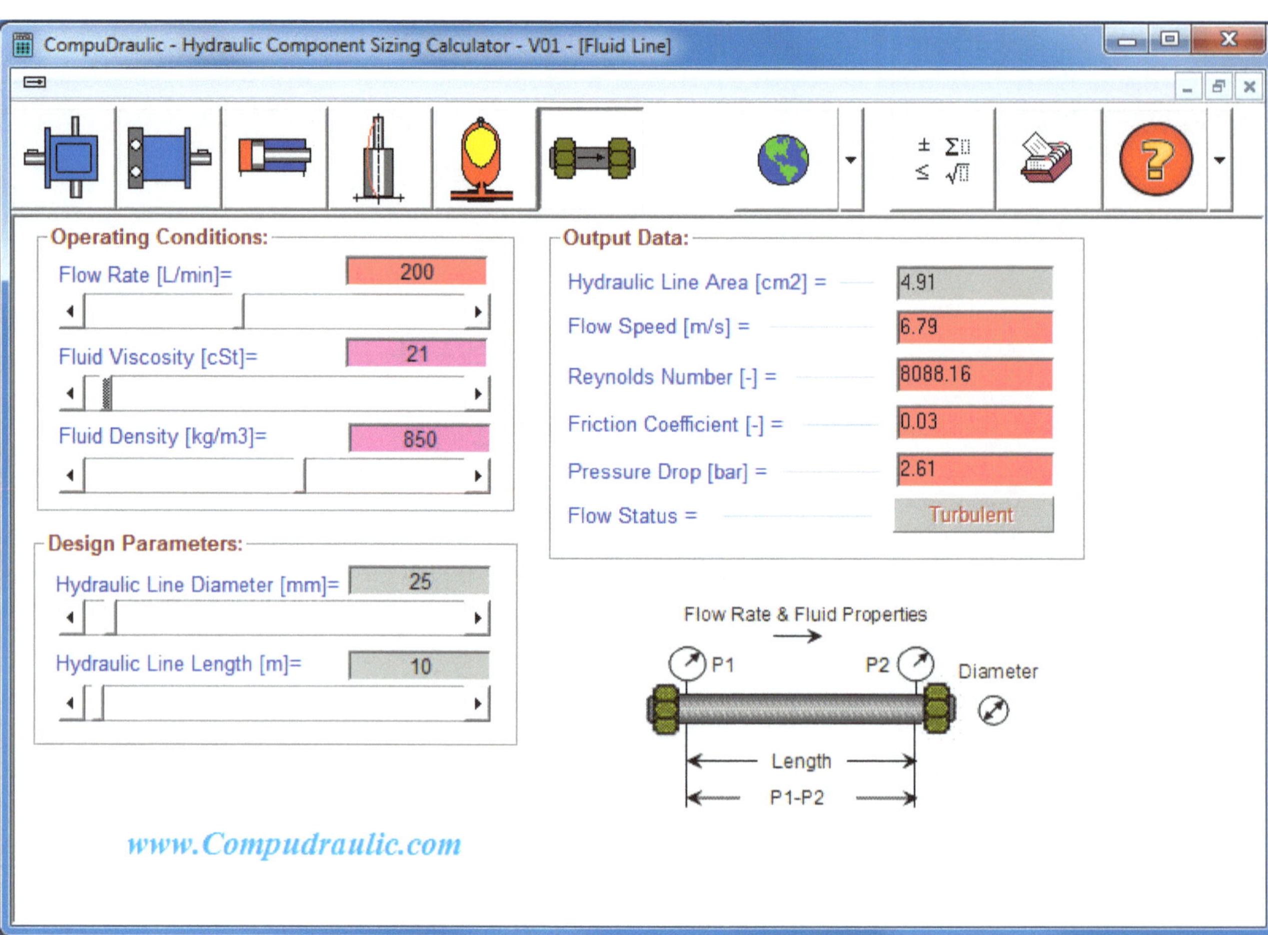

Fig. 5.4 - Fluid Line Design

5.1.5- Typical Fluid Speed in a Hydraulic Conductor

For a transmission line of a certain diameter, the fluid velocity is proportional to flow rate. Basically, with the common range of viscosity of hydraulic fluids, fluid velocity should be kept below 5 m/s (16.4 ft/s) to avoid generating turbulent flow in the line.

However, in addition to the issue of the flow pattern, there are some considerations in sizing hydraulic transmission lines. As shown in Fig. 5.5, a hydraulic system has the following basic transmission lines: pressure (delivery) line, suction (intake) line and return (drain) line. The following rules of thumb are applicable for sizing these transmission lines as follows:

Suction Line: recommended speed = (0.6-1) m/s = (2-4) ft/s. This is to minimize the losses and avoid pump cavitation.

Return Line: recommended speed = (1.2-2.1) m/s = (4-7) ft/s. This is to minimize the turbulence in the tank. If the return oil gets back to the tank with high momentum, that generates turbulence in the tank. The oil must settle down in the tank before it is sucked back by the pump. During this settling period, the fluid dissipates the heat through the reservoir's side wall, settles the contaminants at the bottom, and get rid of the air bubbles at the surface.

Delivery Line: recommended speed = (2.1-4.7) m/s = (7-15) ft/s. There are no limitations in a pressure line but making sure the flow is laminar.

It is to be noted that, in mobile applications, such rules of thumb for return and pressure lines may be voided because they may result in large lines to be assembled in limited space. So, the speed of fluid in a delivery line can reach 7.6 m/s (25 ft/s).

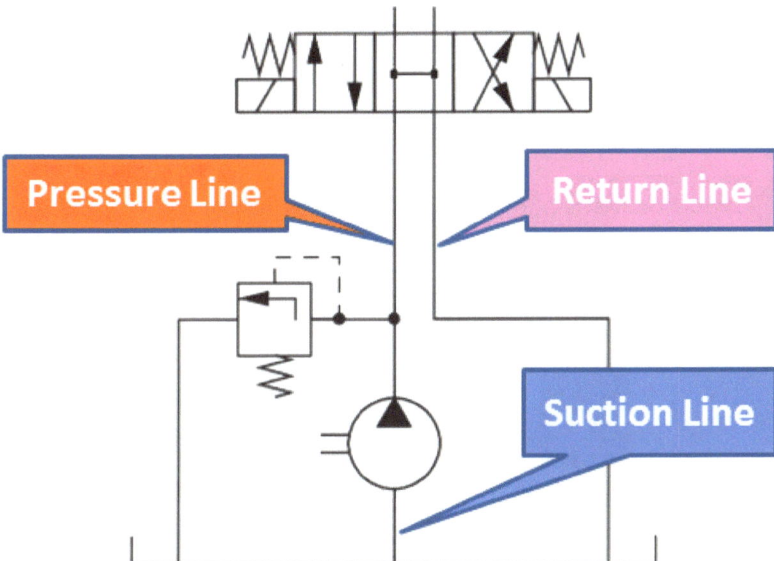

Fig. 5.5- Recommended Fluid Speed inside a Conductor

5.2- Conservation of Mass in Fluid Flow

5.2.1- Compressible Flow

A fluid, to some extent, is a compressible substance that acts like a spring. The fluid property that represents the fluid compressibility is named *Bulk Modulus.* This physical property for a fluid is equivalent to spring *Stiffness* and *Modulus of Elasticity* of a piece of metal. They are all represent stress/strain ratio. Eq. 5.4 defines the bulk modulus **B** as the ratio between the incremental increase in the pressure **Δp** that acts on a confined volume of a fluid and the resulted volumetric strain ΔV/V of the fluid. A fluid with a high bulk modulus is less compressible.

$$\mathbf{B} = -\frac{\Delta \mathbf{p}}{\frac{\Delta \mathbf{V}}{\mathbf{V}}} \qquad\qquad 5.4$$

The practical reading of Eq. 5.4 is how much incremental increase of the pressure **Δp** is required to reduce an original volume of oil **V** by a defined incremental volume **ΔV**. Obviously bulk modulus has units of pressure. The negative sign in the equation indicates that an increase in the pressure will cause a decrease in the volume of oil and vice versa. For a typical petroleum-based hydraulic fluid, under the recommended working conditions, an increase of 70 bar (1000 psi) causes 0.5% decrease in oil volume. Typical values of bulk modulus at ambient conditions are 220 000 psi (≈ 15 000 bars) for petroleum fluid and 300 000 psi (≈ 20 000 bars) for water.

The bulk modulus is affected highly by the working conditions, namely, temperature and pressure. Therefore, fluid compressibility effect on system performance will be more noticeable in some applications or under some conditions such as those shown in Fig. 5.6:
- If a large oil volume is exposed to high pressure, e.g., in hydraulic presses.
- In applications of precise motion control, e.g., servo-controlled hydraulic actuators.
- If a volume of oil will be subjected to pressure spikes, e.g., landing gears.
- If the working temperature exceeds the advisable higher limit, e.g., steel mills.

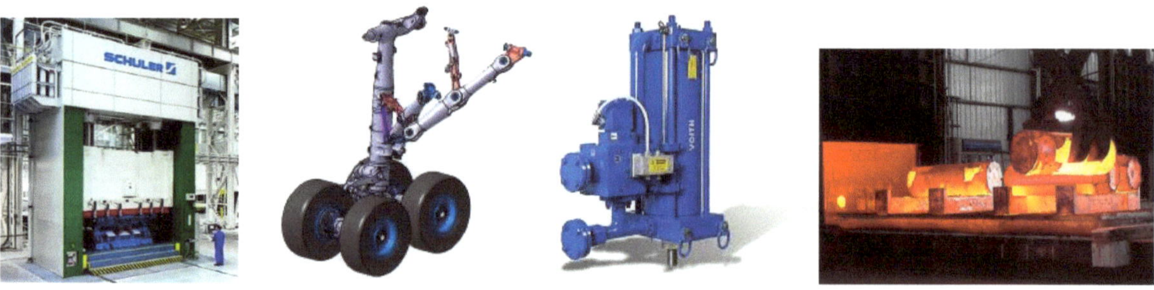

Fig. 5.6- Examples where Fluid Compressibility are Considered in System Design

The question now is how does oil compressibility affects system performance and how this effect can be mathematically quantified?

To answer this question, and to ease the understanding of the fluid compressibility, consider the mass-spring system shown in Fig. 5.7. If a mass m_1 is being pushed by a step force F, will the mass m_2 move at the same moment with the same speed of m_1? Obviously, the answer is "No" because of the presence of the spring between the two masses. The second mass will be delayed a bit until the spring is compressed enough to transfer the force to mass m_2. The response of the mass m_2 is proportional to the spring constant k. So, the spring stiffness affects the transient response of m_2 before it moves in steady state condition with m_1. If the spring is very tough, an assumption can be made that the two masses are attached by a shaft.

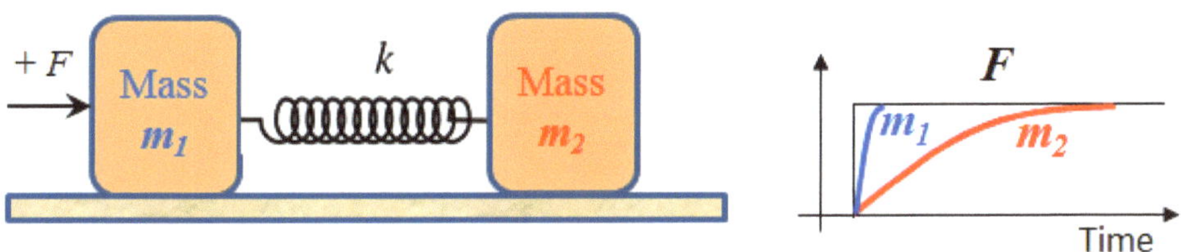

Fig. 5.7- Mass-Spring System

In an example similar to the mass-spring system, Fig. 5.8 shows a hydraulic cylinder is used to push a mass over a frictional surface. If a compressible fluid is supplied to the piston chamber, will the mass m move at the same moment at which the valve is opened with a speed proportional to the supplied flow? Obviously, the answer is "NO" because of the fluid compressibility. The mass will be delayed a bit until the fluid pressure builds up in the piston chamber. A fluid with large bulk modulus results in fast response of the hydraulic actuator. So, the fluid bulk modulus affects the transient response of the load before it follows the supplied flow in steady state motion. In another way, it will affect the pressure change in a closed volume based on the flow in and out of that closed volume. If the bulk modulus is very large, an assumption of incompressible flow can be made.

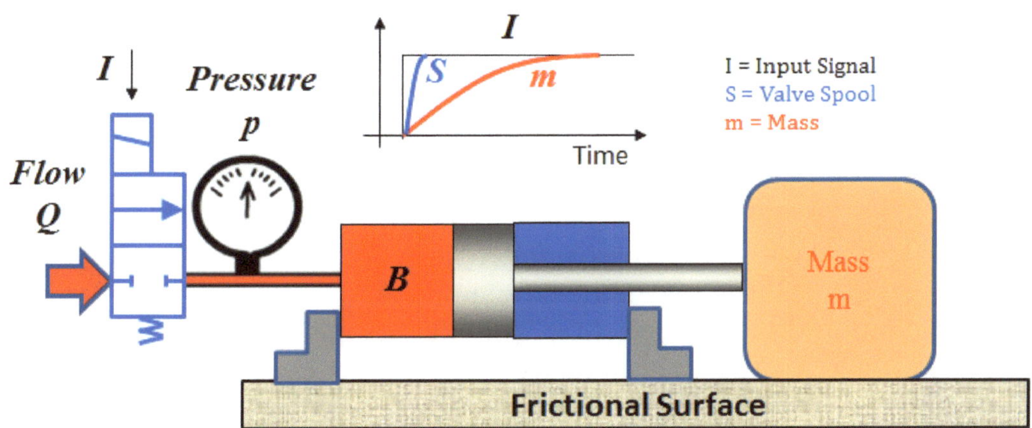

Fig. 5.8- Effect of Oil Compressibility

Now, how the effect of fluid compressibility can be mathematically quantified? As shown in Fig. 5.9, a control volume of a piece of a hydraulic conductor receives flow from multiple sources and supplies flow to multiple sources. Based on the sum of the flow in, the sum of the flow out, and the fluid compressibility, the transient pressure change inside the control volume can be calculated.

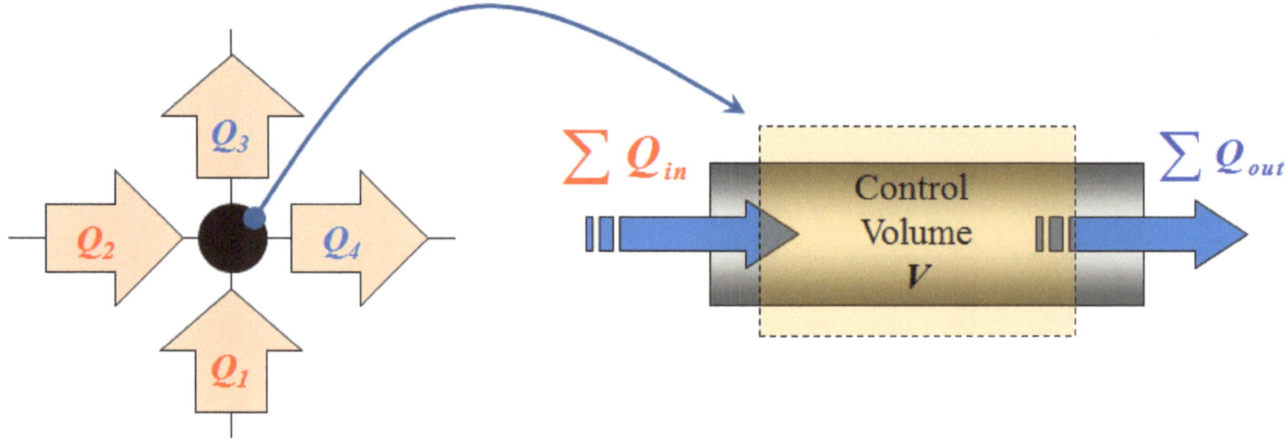

Fig. 5.9- Effect of Oil Compressibility on Pressure Change

In Eq. 5.4, the incremental change in the oil volume, ΔV, has been defined as the reduction in the oil volume due to the pressure increase Δp. It can be also defined as the amount of the accumulated oil in a closed volume that resulted in an incremental increase in the pressure. Then, starting back by Eq. 5.4, Eq. 5.5 describes the transient change in the pressure of a confined volume of oil as follow:

$$\text{Bulk Modulus } B = -\frac{\Delta p}{\frac{\Delta V}{V}} \rightarrow \Delta p = -B\frac{\Delta V}{V}$$

Then

$$\Delta p = \int \frac{B}{V}\left[\sum Q_{in} - \sum Q_{out}\right] dt \;\; OR \;\; p = p_{initial} + \int \frac{B}{V}\left[\sum Q_{in} - \sum Q_{out}\right] dt \qquad 5.5$$

Equation 5.5 shows that, for a compressible flow, the pressure change in a closed volume is a time-variant variable. If the sum of the flow supplied to inside the control volume equals the sum of the flow withdrawn out of the control volume, then the pressure will not change and stays at a steady state value. So, Eq. 5.5 can be used to simulate the dynamics of the pressure including both the transient and the steady state, for example, when a valve is closed or opened. With the knowledge of the pressure change in any closed volume, dynamics of the rest of the system can be analyzed.

5.2.2- Incompressible Flow

If the spring in Fig. 5.7 is replaced by a solid shaft, the two masses will move almost simultaneously assuming the shaft has infinite modulus of elasticity. Similarly, if the fluid used in Fig. 5.8 is assumed incompressible fluid, with infinite bulk modulus, then the actuator will respond instantaneously when the valve opened.

The assumption of an incompressible flow can be made under conditions of a small volume of fluid is used, under moderate pressure, and relatively lower working temperature.

By differentiating the two sides of Eq. 5.5, the flow rate versus the rate of change of pressure is shown in Eq. 5.18.

$$\sum Q_{in} - \sum Q_{out} = \frac{V}{B} \times \frac{dp}{dt} \qquad\qquad 5.6$$

An incompressible fluid has an infinite bulk modulus. Then the right-hand side of Eq. 5.6 equals zero. This practically means that, in cases of incompressible flow, the sum of the flow supplied to a control volume must equal the sum of the flow out. This is known as the *Continuity Equation* and can be mathematically expressed as shown in Eq. 5.7.

$$\sum Q_{in} - \sum Q_{out} = 0 \qquad OR \qquad \sum Q_{in} = \sum Q_{out} \qquad\qquad 5.7$$

There are two practical observations of the Continuity Equation. First, at any branching point, flow distribution can be determined. For example, at a branching point, there are inflow sources of 10 [lit/min] total, one outflow has 7 [lit/min], and a second outflow is unknown. Then, per the Continuity Equation, the unknown outflow is 3 [lit/min].

In order to explain the second practical observation of the Continuity Equation, the difference between *flow rate* and *flow speed* should be clarified. Flow rate is the volumetric flow rate. Flow speed means how fast the fluid is passing through that same cross section. They are related to each other as presented in Eq. 5.8.A and Eq. 5.8.B in Metric and English systems of units, respectively.

$$v \left[\frac{cm}{s} \right] = \frac{1000\ Q[l/min]}{60 \times A[cm^2]} \qquad\qquad 5.8.A$$

$$v\ [fps] = \frac{0.321\ Q[gpm]}{A[in^2]} \qquad\qquad 5.8.b$$

Based on Eq. 5.8 and in order for the Continuity Equation to be valid, Fig. 5.10 shows inverse proportionality between the flow speed and the conductor's cross-sectional area.

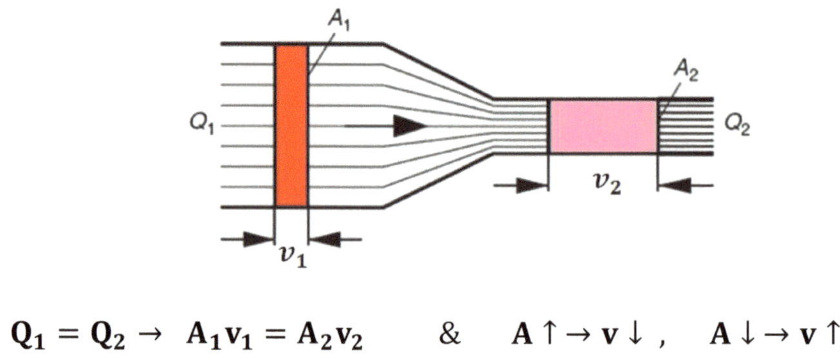

$$\mathbf{Q_1 = Q_2 \rightarrow A_1 v_1 = A_2 v_2} \quad \& \quad \mathbf{A \uparrow \rightarrow v \downarrow}, \quad \mathbf{A \downarrow \rightarrow v \uparrow}$$

Fig. 5.10- Continuity Equation

So far, it can be concluded that a change in a conductor area will result in a change in the flow speed. This conclusion will be discussed also when discussing the conservation of energy in fluid flow.

This relationship between the flow rate and the flow speed can be observed in the following example. Figure 5.11 shows two identical pumps driving two identical motors. The result is that the two motors rotate at the same speed because they receive the same flow rate. The outer circuit has a conductor with an inner diameter double the inner diameter of the conductor in the inner circuit. Therefore, the flow speed in the outer circuit is almost one fourth of the flow speed in the inner circuit because the area is function of the square of the diameter. Despite that the two motors rotate at the same speed; inner circuit may suffer from turbulent flow in the transmission lines.

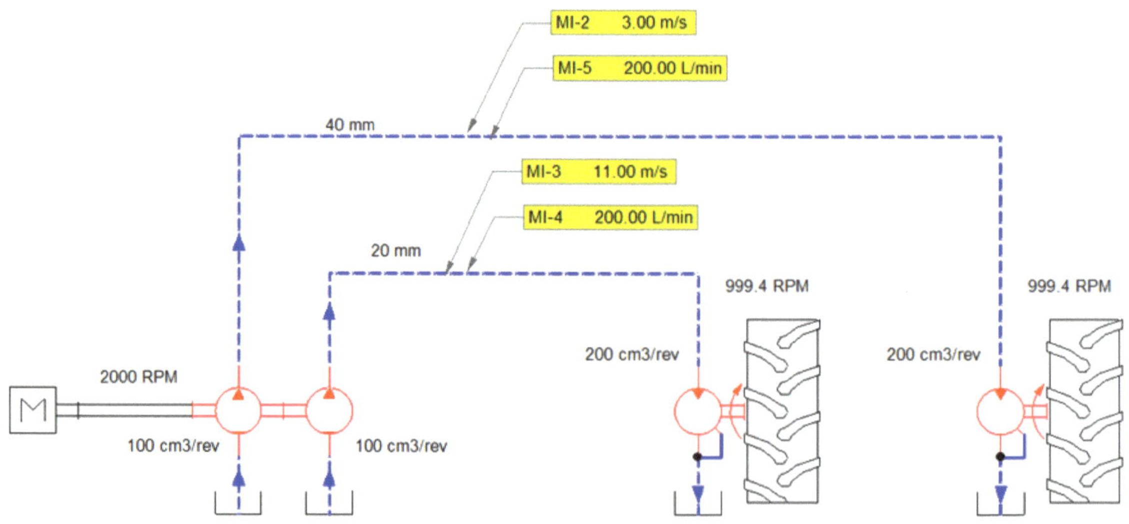

Fig. 5.11- Flow Rate and Flow Speed

5.3- Conservation of Energy in Fluid Flow

5.3.1- Bernoulli Equation for Ideal Fluid Flow

Bernoulli's analysis for ideal fluid flow has been built based on two considerations as follows:

❖ **First** is the concept that a unit weight of fluid flowing in a transmission line gains a total energy that consists of three portions as follows:

Pressure Energy = pV
    That is because the unit weight of fluid flows within a pressure field.

Kinetic Energy = mv²/2
    That is because the unit weight of fluid is a moving mass.

Potential Energy = mgz
    That is because the elevation of the unit weight of fluid varies along the stream of flow with respect to a certain line of reference.

If each portion of these energies is divided by the unit weight of the fluid element, then it can be expressed by a term called *"Head"* as follows:

- Pressure Head = **p / ρg**
- Kinetic Head = **v² / 2g**
- Potential Head = **z**

Where:
- **p** = pressure surrounding the fluid unit weight.
- **V** = volume of the fluid unit weight.
- **m** = mass of the fluid unit weight.
- **v =** speed of the fluid element
- **g** = gravitational acceleration.
- **z** = the elevation of the fluid unit weight with respect to a line of reference.
- **ρ** = fluid density.

❖ **Second:** The flow is ideal so that the overall energy gained by a unit weigh of fluid is constant along the stream of fluid flow. As shown in Fig. 5.12, fluid flows in a conductor that has the upstream portion is of a higher elevation and a larger cross section area than the downstream portion of it. Equation 5.9 formulates mathematically the aforementioned two considerations of Bernoulli's analysis for ideal fluid flow.

$$(\frac{p}{\rho g} + \frac{v^2}{2g} + z)_1 = (\frac{p}{\rho g} + \frac{v^2}{2g} + z)_2 = \text{Constant} \qquad\qquad 5.9$$

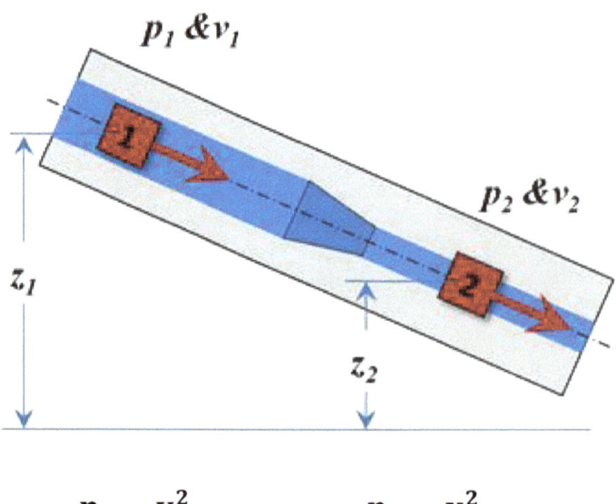

$$(\frac{p}{\rho g} + \frac{v^2}{2g} + z)_1 = (\frac{p}{\rho g} + \frac{v^2}{2g} + z)_2$$

Fig. 5.12- Ideal Fluid Flow

Out of the general conditions expressed by equation 5.9, three special cases can be discussed.

First special case: As shown in Fig. 5.13, applying equation 5.9 on a vertical conductor between points 1 and 2 will result in Eq. 5.10. The equation shows that, in such a situation, the static pressure at point 1 is proportional to the potential head H.

$$\Delta p = p_1 - p_2 = \rho g\,(z_2 - z_1) = \rho g H \qquad\qquad 5.10$$

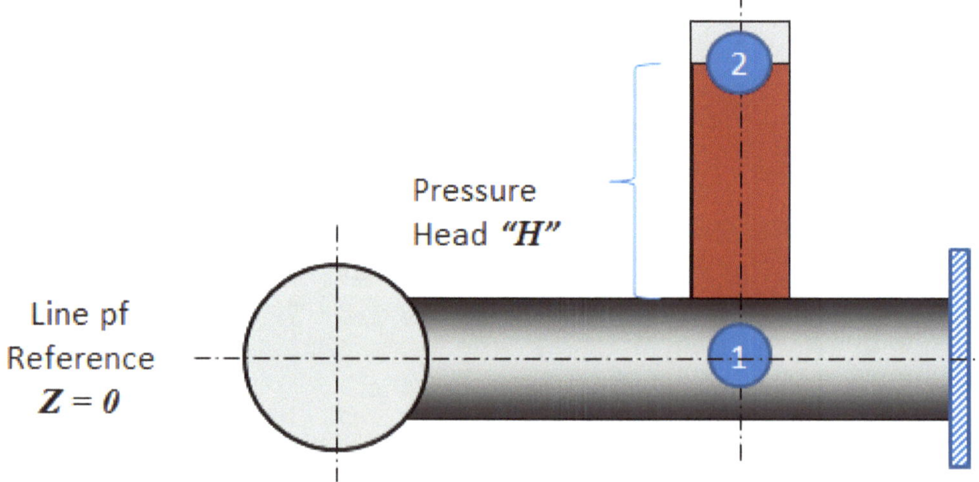

Fig. 5.13- Pressure Head

Second special case: As shown in Fig. 5.14, a conductor is being placed horizontally and has different cross section areas at locations 1 and 2. Elevation will be the same and consequently the potential head will be same everywhere in the conductor. Therefore, any change in the cross-section area will cause a change in the flow speed according to the conservation of mass; and consequently, a change in the pressure according to conservation of energy. Applying equation 5.9 between points 1 and 2 will result in Eq. 5.11. Since the potential head is the same at point 1 and 2, a reduction in the pressure head is substituted by an increase in the kinetic head.

$$\Delta p = p_1 - p_2 = \frac{\rho}{2}\left(v_2^2 - v_1^2\right) \qquad\qquad 5.11$$

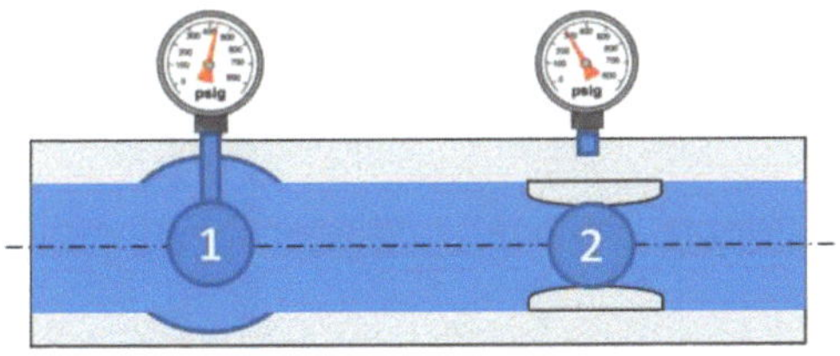

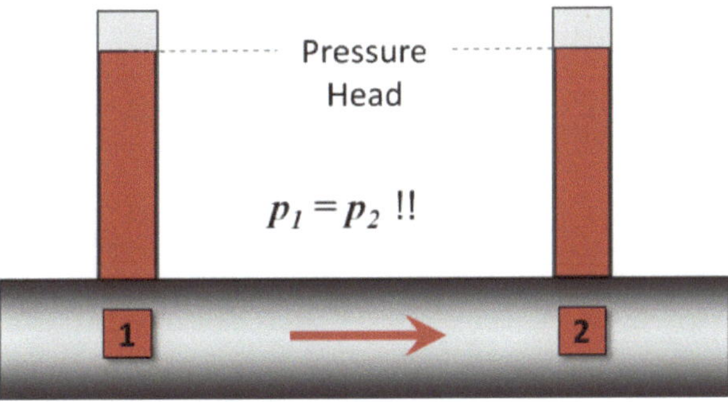

Fig. 5.14- Kinetic Head

One point to make here, it has been made clear now that reducing the area of a conductor will result in an increased flow speed and consequently reduced pressure. Therefore, a suction line mustn't be sized less than the size of the intake port of the pump, which is sized primarily to avoid severe reduction in suction pressure and consequently cavitation.

Third special case: As shown in Fig. 5.15, a conductor of a constant cross section area is positioned horizontally. As per Bernoulli's analysis so far, since the area is unchanged, the flow speed and the pressure are unchanged. Applying equation 5.9 between points 1 and 2 will result in equal pressure everywhere in the conductor. This is an impractical observation that contradicts what has been mentioned in the first section of this chapter. Practically, to generate flow in a pipeline, there must a differential pressure. What is missing? see the following section.

Fig. 5.15- Practical Observation of Continuity Equation and Bernoulli's Equation

5.3.2- Bernoulli Equation for Real Fluid Flow

The previous analysis was based on an ideal fluid that does not experience any energy losses while flowing in a conductor. So, to include the line losses for a real fluid, Eq. 5.9 must be rewritten in the form of Eq. 5.12. Fig. 5.16 shows a special case in which a conductor of a seamless cross-section area is placed horizontally. While both potential head and kinetic head are constants, line losses from point 1 to point 2 are seen as a reduction in the pressure head. Applying Eq. 5.12 for such a special case will result in Eq. 5.13.

$$(\frac{p}{\rho g} + \frac{v^2}{2g} + z)_1 - (\frac{p}{\rho g} + \frac{v^2}{2g} + z)_2 = H_{1 \to 2} \qquad\qquad 5.12$$

$$\Delta p = p_1 - p_2 = \rho g H_{1 \to 2} \qquad\qquad 5.13$$

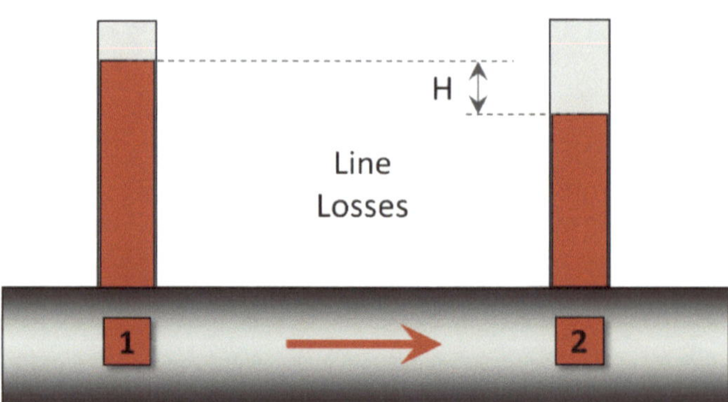

Fig. 5.16- Real Fluid Flow

5.4- Pressure Drop in Transmission Lines

The term line losses refer to the total pressure drop between the inlet and outlet of a transmission line. As show in Fig. 5.17, total pressure drop in the hydraulic fluid are due to frictional losses and local losses.

Frictional losses Δp_F it is called *Frictional Losses* despite that it refers to the differential pressure needed to overcome the fluid-fluid friction, fluid-wall friction, and kinetic losses.

Local losses Δp_L it is called *Local Losses* because it refers to the differential pressure needed to overcome the line resistivity due to the change of the magnitude and/or the direction of the fluid speed in a specific local point in the line.

Quantifying the line losses is an important step in system design for two reasons. First, power losses in transmission lines generates heat. So, quantifying the generated heat is required for proper sizing of a heat exchanger. Second, power losses in transmission lines is required for proper sizing of the prime mover.

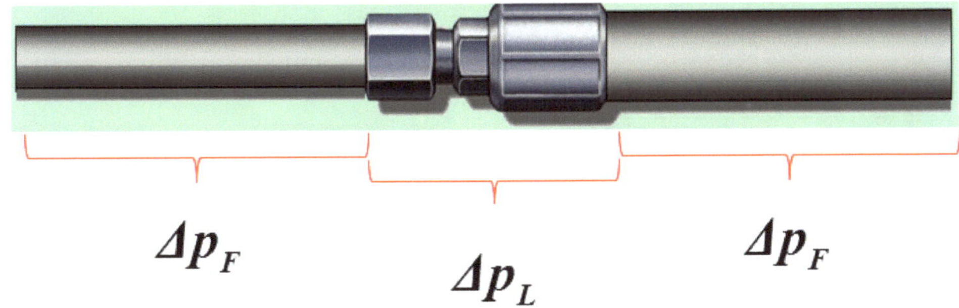

Fig. 5.17- Line Losses

5.4.1 – Frictional and Kinetic Pressure Losses in Hydraulic Lines

Equation 5.14 is used to calculate the frictional pressure losses in a hydraulic line.

$$\Delta p_F = p_1 - p_2 = \rho g h_{L1 \to 2} = \rho g \lambda \frac{L}{D} \frac{v^2}{2g} = \lambda \frac{L}{D} \frac{\rho v^2}{2} \qquad\qquad 5.14$$

Where:
- ρ = Fluid density.
- v = Fluid velocity.
- D = Line diameter.
- L = Length of the line.
- λ = Coefficient of friction.
- μ = Dynamic viscosity.

The equation indicates that the pressure drop in a transmission line is function of:
- Dimensional parameters as line diameter D and line length L.
- Fluid property as fluid density ρ.
- Operational conditions as frictional coefficient λ and speed of fluid in the line v.

So, in addition to the line dimensions and fluid density, pressure drop in a transmission line depends on the friction (fluid-to-fluid and fluid-to-wall) and how fast the fluid is moving in the line. It is to be reminded that turbulent flow has lower friction coefficient and much higher kinetic losses because the speed of the fluid in the equation is squared.

For a practical use of equation 5.14 in hydraulic system design, Eq. 5.15.A and Eq. 5.15.B have been developed in the Metric and English systems of units, respectively.

$$\Delta p_F(bar) = \frac{2254 \times \lambda \times L(m) \times SG \times [Q(\frac{lit}{min})]^2}{[D(mm)]^5}$$

5.15.A

$$\Delta p_F(psi) = \frac{\lambda \times L(ft) \times SG \times [Q(gpm)]^2}{74.3 \times [D(in)]^5}$$

5.15.b

Pretty much every parameter is assumed to be known to the system designer except the coefficient of friction λ, which is based on the pattern of flow as follows.

In case of laminar flow, Eq. 5.16 is used to calculate the coefficient of friction.

$$\lambda = \frac{64}{R_e}$$

5.16

Substituting Eq. 5.16 in Eq. 5.14 will result in Eq. 5.17 as follows:

$$\Delta p_F = \frac{128\,\mu\,L}{\pi\,D^4}\,Q = RQ$$

5.17

It is to be noted that Eq. 5.17:
- Is valid only for the conditions of laminar flow.
- Shows linear relationship between the frictional pressure losses and the volume flow rate in a hydraulic line. The proportionality factor is **R**.

In case of turbulent flow, Eq. 5.18 is used to calculate the coefficient of friction based on an assumption that the inside surface of the hydraulic line is smooth. This assumption is feasible unless otherwise been stated.

$$\lambda = \frac{0.3164}{R_e^{0.25}}$$

5.18

The coefficient of friction found out of Eq. 5.18 is to be substituted in the general equation 5.14 in order to calculate the corresponding frictional losses in case of turbulent flow in a smooth line.

Moody Diagram, shown in Fig. 5.18, is an alternative method to find the coefficient of friction regardless the flow pattern inside the transmission line. The first step to use the diagram is to locate the value of the Reynolds's number on the logarithmic x-axis then draw a vertical line from there. If the flow is laminar, the vertical line intersects with the solid line. Then, from the intersection point go left to find the coefficient of friction. For example, if R_e = 1,000 then coefficient of friction = 0.065. If the flow is in transition state, the vertical line intersects with the dashed line, and then go left to find the coefficient of friction. For example, if R_e = 3,000 then coefficient of friction = 0.0215. If the flow is turbulent, the vertical line will intersect with one of the *Relative Roughness* curves. Relative roughness is defined as the ratio between the inside surface roughness ε of the fluid conductor and its nominal diameter **D**. The lowest relative roughness curve represents smoothest fluid conductor. For example, R_e = 100,000 then coefficient of friction for a smooth pipe = 0.0185 approximately.

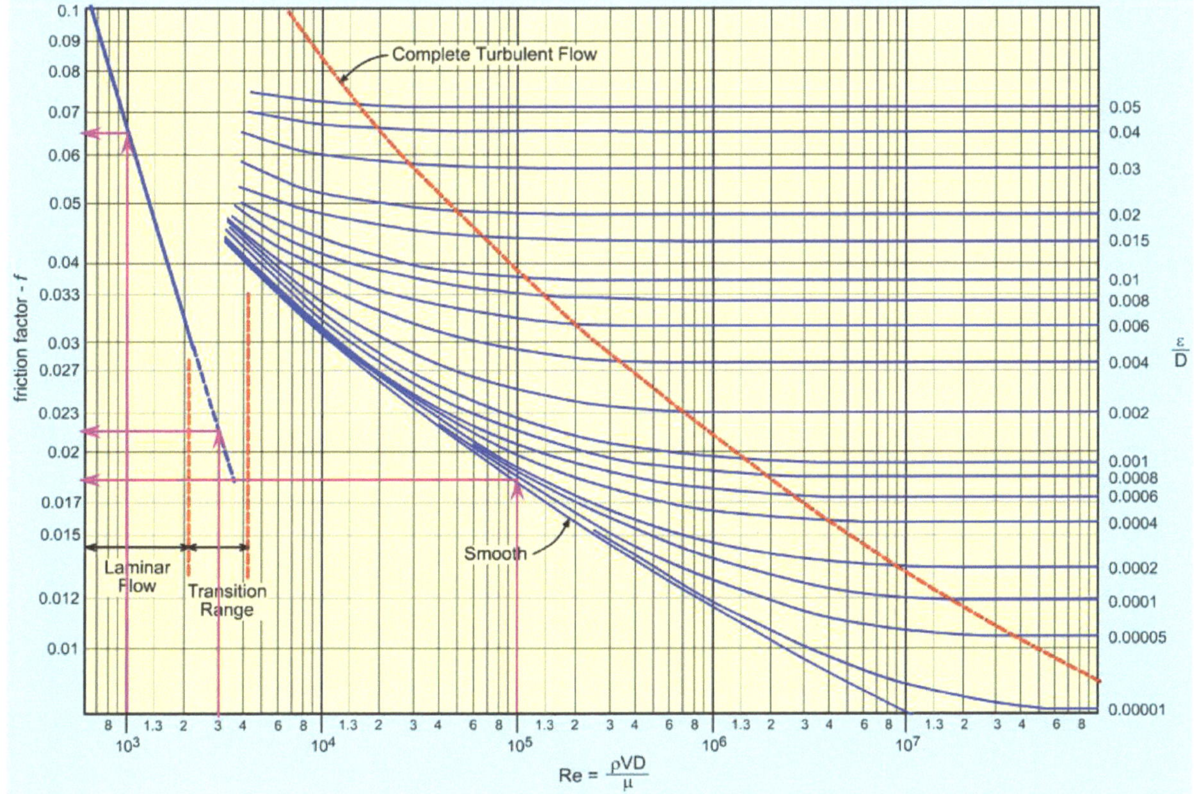

Fig. 5.18- Moody Diagram

Two observations can be made from the use of the Moody diagram. First is to confirm that laminar flow isn't necessarily generate low friction as is usually wrongly understood. IT depends on the transmission line internal surface roughness. A turbulent flow may generate less friction if the fluid flows in a smooth pipe. Additionally, the fluid-fluid friction in laminar flow is much more than in turbulent flow because the coaxial layers in laminar flow move relative to each other.

Second is that laminar flow neutralizes the effect of line inside surface conditions. That is because the boundary layer in laminar flow is at standstill and is not in friction with the inside wall. That is why the effect of inside surface roughness is considered with turbulent flow only.

Example 1 (Frictional Pressure Losses Calculation in Metric System of Units):

Given: A flow in a smooth pipe has the following variables.

$$Q\left(\frac{\text{lit}}{\text{min}}\right) = 76, \quad v\,(\text{cSt}) = 64, \quad D(\text{mm}) = 25, \quad SG = 1.055 \quad \text{and } L\,(\text{m}) = 3.$$

Solution: $\qquad \text{Eq. 2.15.A} \rightarrow R_e = \dfrac{21231 \times 76}{64 \times 25} = 1009$

$$\text{Laminar Flow} \rightarrow \text{Moody Diagram} \rightarrow \lambda \approx 0.06$$

$$\text{Eq. 1.15.A} \rightarrow \Delta p_F(\text{bar}) = \frac{2254 \times 0.06 \times 3 \times 1.055 \times 76^2}{25^5} = 0.27$$

Figure 5.19 shows the validation of the solution shown above using the "Hydraulic Component Sizing Calculator".

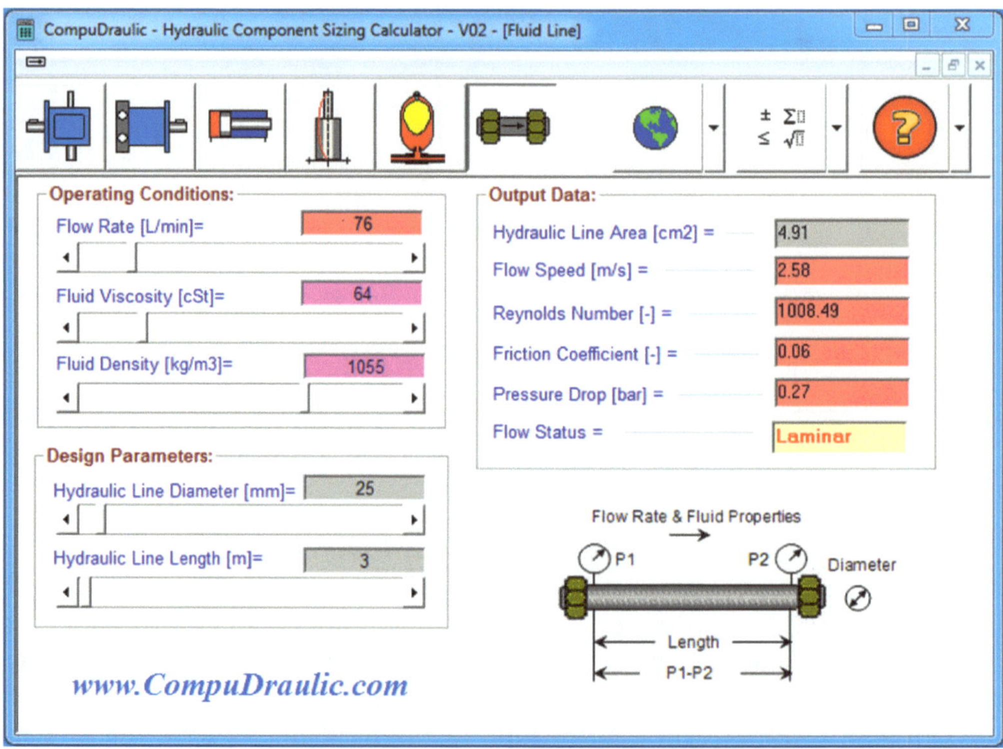

Fig. 5.19- Frictional Pressure Losses Calculation in Metric System of Units

Example 2 (Frictional Pressure Losses Calculation in English System of Units):

Given: A flow in a smooth pipe has the following variables.

$$Q\,(\text{gpm}) = 20.08, \quad \nu\,(\text{cSt}) = 64, \quad D(\text{in}) = 0.98, \quad SG = 1.055 \quad \text{and } L(\text{ft}) = 9.84.$$

Solution:

$$\text{Eq. 2.15.A} \rightarrow R_e = \frac{3164 \times 20.08}{64 \times 0.98} = 1013$$

$$\textbf{Laminar Flow} \;\rightarrow\; \textbf{Moody Diagram} \;\rightarrow\; \lambda \approx 0.06$$

$$\text{Eq. 1.15.B} \rightarrow \Delta p_F(\text{psi}) = \frac{0.06 \times 9.84 \times 1.055 \times 20.08^2}{74.3 \times 0.98^5} = 3.82$$

Figure 5.20 shows the validation of the solution shown above using the "Hydraulic Component Sizing Calculator".

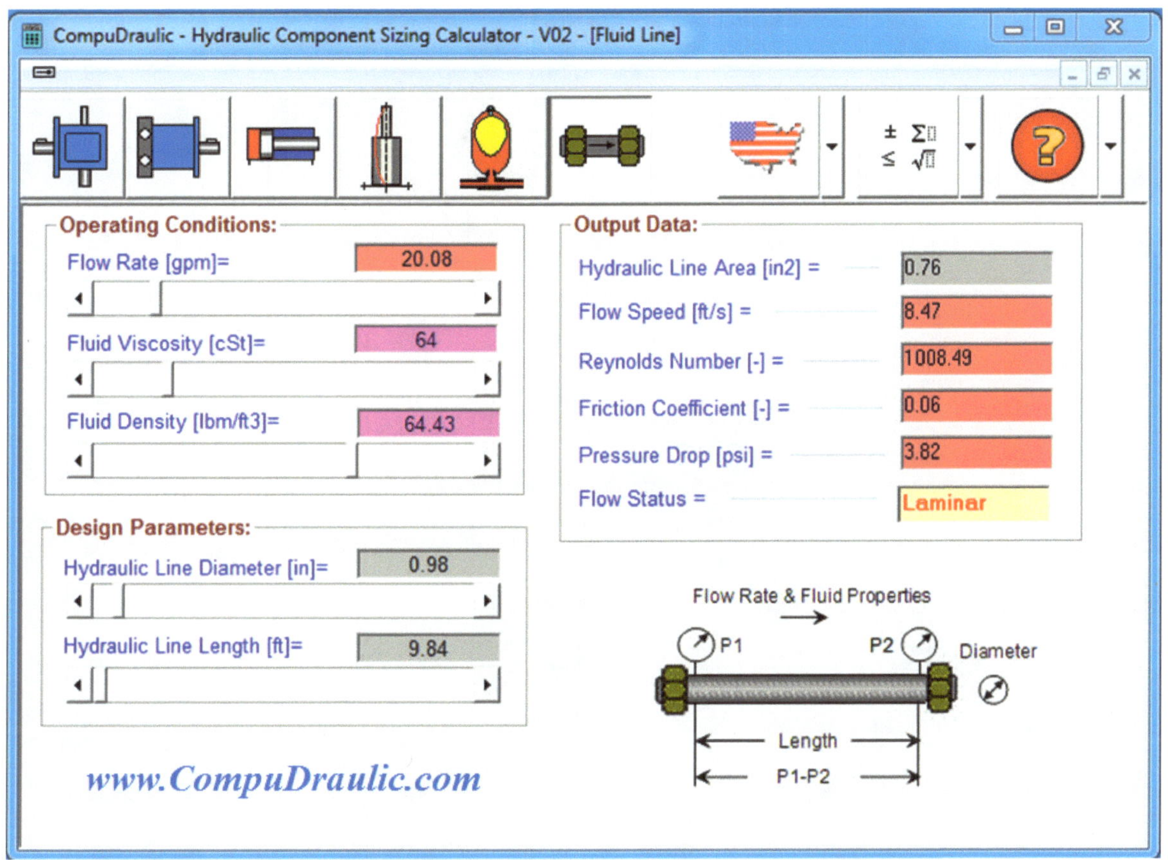

Fig. 5.20- Frictional Pressure Losses Calculation in English System of Units

5.4.2 – Local Losses in Fittings

As it has been mentioned previously, local losses $\Delta \mathbf{p_L}$ refers to the differential pressure needed to overcome the line resistivity due to the change of the magnitude and/or the direction of the fluid speed in a specific local point in the line. This happens most likely at the place of adapting two conductors with a fitting. The first two examples shown in Fig. 5.21 shows local losses due to change in the magnitude of the fluid speed resulted from sudden contraction or expansion of the fluid conductor's diameter. In case of fluid conductor sudden contraction or expansion, fluid should not be assumed to flow adjacent to the inside wall of the fluid conductor. It will rather separate from the inside walls causing turbulences and pressure drop in the spot of separation. The third example shows a seamless elbow in which the area is constant, consequently the magnitude of the fluid speed is also constant. Because the changing in the fluid direction, there will be a place of separation because of the centrifugal force and a place of more friction because of the fluid inertia.

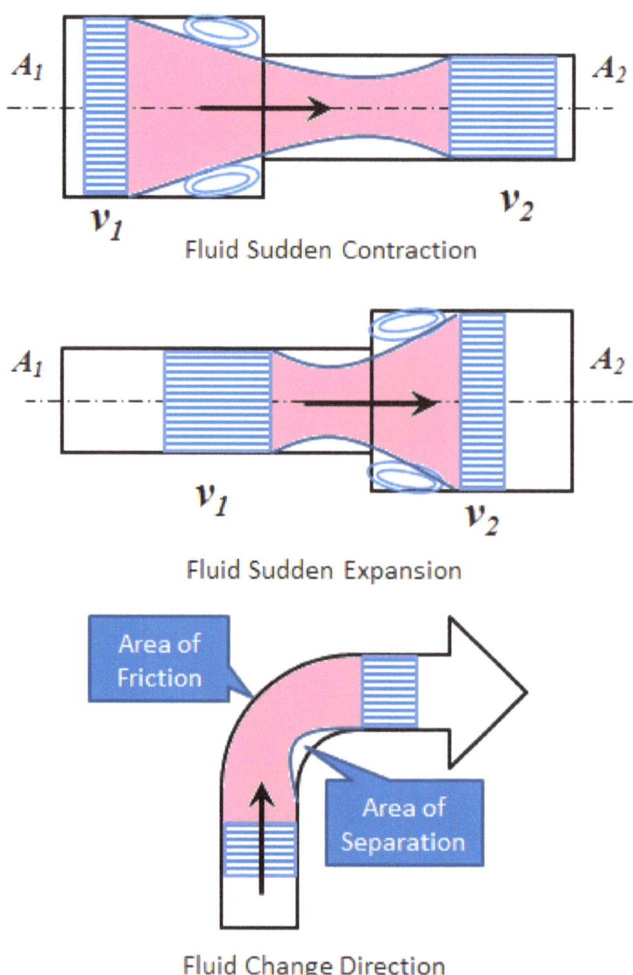

Fig. 5.21- Local Losses in Fittings

The easiest and most practical way of quantifying the local losses is to find the manufacturer test results that records pressure drop across the fitting versus the flow rate. If test results are not found, it can be performed at an appropriate test lab. Mathematically Eq. 5.19 is used to quantify local losses. The factor **K** is the Losses Factor that is equivalent to the value **(λL/D)** in Eq. 5.14. For practical use of Eq. 5.19 in hydraulic system design, Eq. 5.20.A and Eq. 5.20.B have been developed in Metric and English system of units, respectively.

$$\Delta p_L = p_{in} - p_{out} = K\,\frac{\rho v^2}{2} \qquad\qquad 5.19$$

$$\Delta p_L(bar) = \frac{2.254 \times K \times SG \times [Q(\frac{Lit}{min})]^2}{[D(mm)]^4} \qquad\qquad 5.20.A$$

$$\Delta p_L(psi) = \frac{K \times SG \times [Q(gpm)]^2}{891.3 \times [D(in)]^4} \qquad\qquad 5.20.B$$

The question now is how to get the value of the losses factor. Fluid Mechanics references are good resource for that. Table 5.1 shows losses factor for some fittings.

Fitting	90° bend	90° angle	T-Piece	Double angle	Valve
K	0.5 - 1	1.2	1.3	2	5 - 15
Sudden Contraction	Well-Rounded	Slightly Rounded	Sharp Edged	Shoulder	
K	0.04	0.2	0.5	0.8	
Sudden Expansion	Well-Rounded	Slightly Rounded	Sharp Edged	Shoulder	
K	1	1	1	1	

Table 5.1 – Losses factor in Fittings

Figure 5.22 shows empirical data based on a test performed to measure the losses factor in an elbow. A family of curves has been generated based on various values of relative roughness.

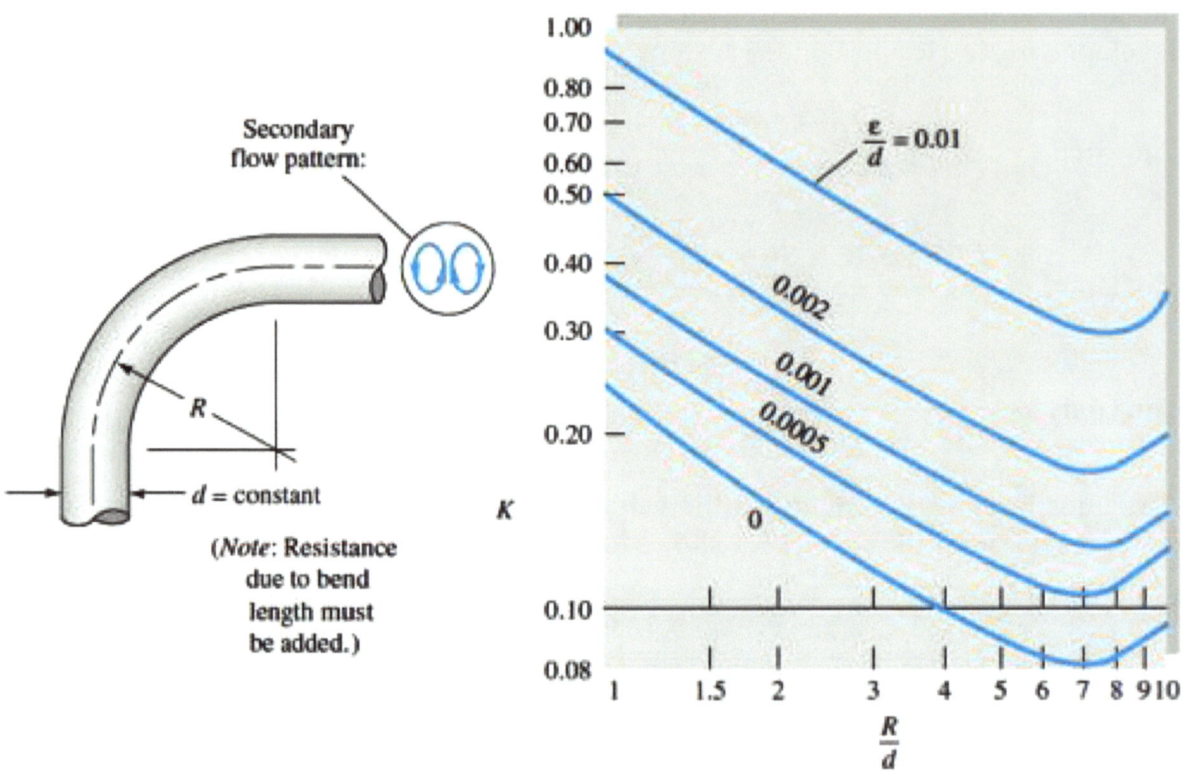

Fig. 5.22- Local Losses in Elbow, Empirical Data

5.4.3 – Local Pressure Losses in Orifices

From the theory of fluid mechanics, the very well-known Eq. 5.21 represents the pressure drop across orifices. Design parameters are defined on Fig, 5.23.

$$Q = C_D A \sqrt{\frac{2(p_1 - p_2)}{\rho}} \qquad\qquad 5.21$$

C_D is called *Discharge Factor*. It depends on the shape of the orifice. For a sharp-edged short orifice, shown in Fig. 5.40, the discharge factor equals 0.611.

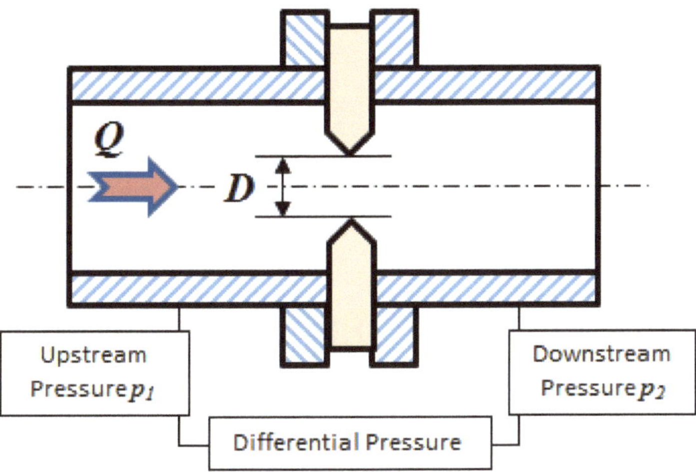

Fig. 5.23- Local Losses in Orifices

For a practical use of Eq. 5.20, it has been rewritten in several forms. One of these forms is to combine orifice design parameters and fluid properties in one factor called *Throttling Coefficient* k_{th} as shown in Eq. 5.22.

$$k_{th} = \frac{\Delta p}{Q^2} = \frac{\rho}{2C_D^2 A^2} \qquad\qquad 5.22$$

The value of the throttling coefficient varies from zero to infinity. Zero value represents a fully opened throttle that experiences no differential pressure and maximum flow. Infinity value represents fully closed throttle that experience maximum differential pressure and no flow. In between the orifice is partially closed.

For practical use of Eq. 5.22 in hydraulic system design, Eq. 5.23.A and Eq. 5.23.B have been developed with balance of units in Metric and English system of units, respectively.

$$k_{th}\left[\frac{bar}{(Lit/min)^2}\right] = \frac{2.254 \times SG}{C_D^2 [D(mm)]^4} \qquad\qquad 5.23.A$$

$$k_{th}\left[\frac{psi}{(gpm)^2}\right] = \frac{5.34 \times 10^3 \times SG}{C_D^2 [D(in)]^4} \qquad\qquad 5.23.b$$

Chapter 6
Modeling and Simulation of Hydraulic Transmission Lines

Objectives

This chapter presents concepts of modeling transmission lines, fittings and orifices. Model for a transmission line considers compressible fluid so that effect of line capacitance can be investigated. Developed models were validated based on other software.

Brief Contents

7.1- Modeling of Seamless Hydraulic Transmission Lines

7.2- Modeling of Hydraulic Fittings

7.3- Modeling of Hydraulic Orifices

7.4- Modeling Hydraulic Transmission Line Assembly

Chapter 6: Modeling and Simulation of Hydraulic Transmission Lines

6.1- Modeling of Seamless Hydraulic Transmission Lines

A seamless hydraulic transmission line shown in Fig. 6.1 has resistive effect, inductive effect and capacitive effect. Therefore, modeling of transmission line can be built on concept of finite elements as follows.

1. **Lumped Resistance**: In this portion of the line, pressure drops due to <u>fluid-fluid and fluid-wall friction</u>, but flow will be the same as Qin.

2. **Lumped Inductance**: In this portion of the line, pressure drops due to <u>fluid inertia</u>, but flow will be the same as Qin.

3. **Lumped Capacitance:** In this portion of the line, outlet flow changes based on the fluid compressibility and the rate of change of the differential pressure across the line.

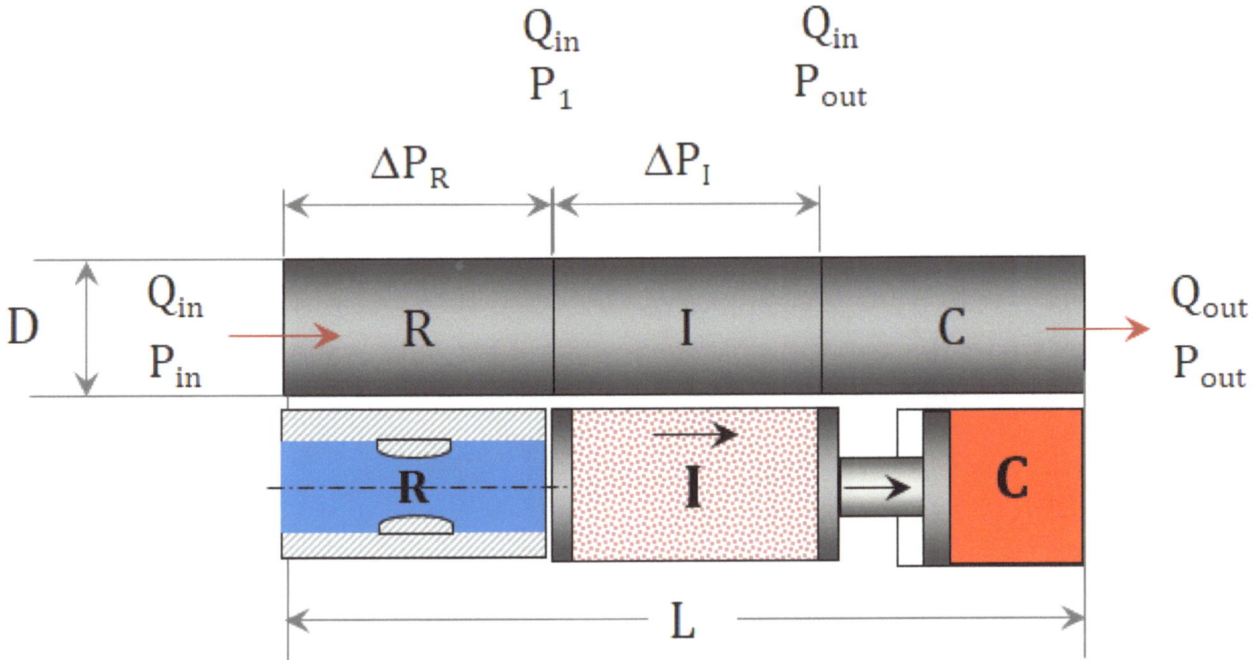

Fig. 6.1 – Modeling Pressure Drop in Hydraulic Transmission Lines

Figure 6.2 shows the stream of calculations to model a transmission line as follows:

Step 1: With the knowledge of inlet flow, fluid properties, and the line dimensional parameters, the model calculates the line pressure drop due to line *resistance.*
Step 2: With the knowledge of inlet flow, fluid properties, and the line dimensional parameters, the model calculates the line pressure drop due to line *inductance.*
Step 3: With the knowledge of outlet pressure and the overall pressure drop across the line, the model calculates the inlet pressure.
Step 4: With the knowledge of the inlet flow and the rate of change in the outlet pressure, the model calculates the change in the outlet flow due to the line *capacitance.*

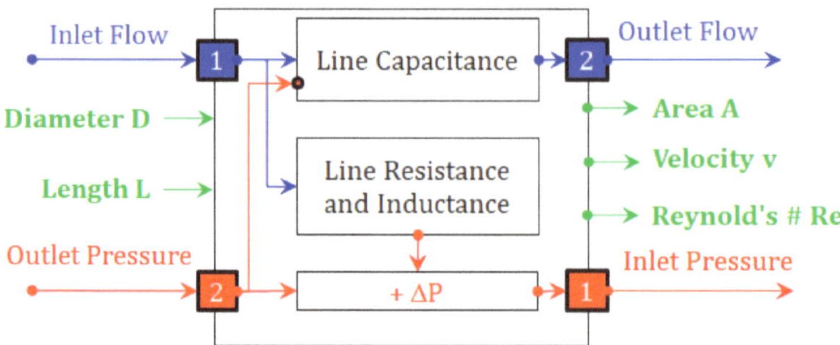

Fig. 6.2- Structure of a Generic Model for a Hydraulic Transmission Line

The challenge to build a Simulink model is to make sure units are balanced. Therefore, the following set of variables with the assigned units are used to build the model.

- **ρ (kg/m³)** = Fluid Density. Fluid properties must be interactively corrected based on working temperature and pressure. Therefore, it will be delivered to this model through *Global Variables* from hydraulic fluid model.
- **v (m/s)** = Fluid Velocity.
- **D (mm)** = Line Diameter.
- **A [cm²)** = Line Area.
- **L (m)** = Line Length.
- **R_e (-)** = Reynold's Number
- **λ (-) =** Coefficient of Fluid Friction.
- **ν (cSt = mm²/s)** = Kinematic Viscosity.
- **Q_{in} (lit/min)** = Inlet Flow.
- **Q_{out} (lit/min)** = Outlet Flow.
- **P_{in} (bar) =** Inlet Pressure.
- **P_{out}(bar) =** Outlet Pressure.
- **β_E =** Equivalent Bulk Modulus of the hydraulic fluid.

It is to be noted that some equations in this chapter are excerpted from Chapter 5 and used with different equation number just to follow the stream of calculation and the relevant Simulink model.

6.1.1- Modeling Pressure Losses due to Line Resistivity

The following set of equations shows the sequence based on which the model is structured.

$$A\,[\text{cm}^2] = [\pi \times D^2(\text{mm})]\,/(4 \times 100) \tag{6.1}$$

$$v\,[\text{m/s}] = \frac{Q_{\text{in}}(\text{lit/min}) \times \dfrac{1}{60 \times 1000}}{A\,(\text{cm}^2) \times 10^{-4}} = \frac{Q_{\text{in}}(\text{lit/min})}{A\,(\text{cm}^2) \times 6} \tag{6.2}$$

$$R_e\,[-] = \frac{v\,[\text{m/s}] \times 1000 \times D\,[\text{mm}]}{v\,(\text{Cst} = \text{mm}^2/\text{s})} \tag{6.3}$$

The *Coefficient of Fluid Friction* λ depends on the flow pattern (Laminar or Turbulent) inside the transmission line that is determined based on the *Reynolds's Number* as follows:

For Laminar and Transitional Flow (Reynolds's Number <= 3500):

$$\lambda = \frac{64}{R_e} \tag{6.4}$$

For Turbulent Flow (Reynolds's Number > 3500):
Equation 6.5 calculates the coefficient of friction based on an assumption that the inside surface of the hydraulic line is smooth. This assumption is feasible unless otherwise stated.

$$\lambda = \frac{0.3164}{R_e^{0.25}} \tag{6.5}$$

An alternative way to determine the coefficient of fluid friction λ is to use the *Moody Diagram* shown in Fig. 6.3. The first step is to locate the Reynolds's Number on the logarithmic x-axis then draw a vertical line.

If the flow is laminar, the vertical line intersects with the solid line, and then go left to find the coefficient of friction. For example, if $\mathbf{R_e}$ = 1,000 then coefficient of friction = 0.065.

If the flow is in transition, the vertical line intersects with the dashed line, and then go left to find the coefficient of friction. For example, if $\mathbf{R_e}$ = 3,000 then coefficient of friction = 0.0215.

If the flow is turbulent, the vertical line will intersect with one of the *Relative Roughness* curves. Relative roughness is defined as the ratio between the inside surface roughness ε of the transmission line and its nominal diameter **D**. The lowest relative roughness curve represents the smoothest line. For example, $\mathbf{R_e}$ = 100,000 then coefficient of friction for a smooth line = 0.0185 approximately. This curve of lowest relative roughness is used to develop the previous equation, Eq. 6.5.

Moody diagram can be converted into a lookup table to avoid using Eqs. 6.4 and 6.5 in the model.

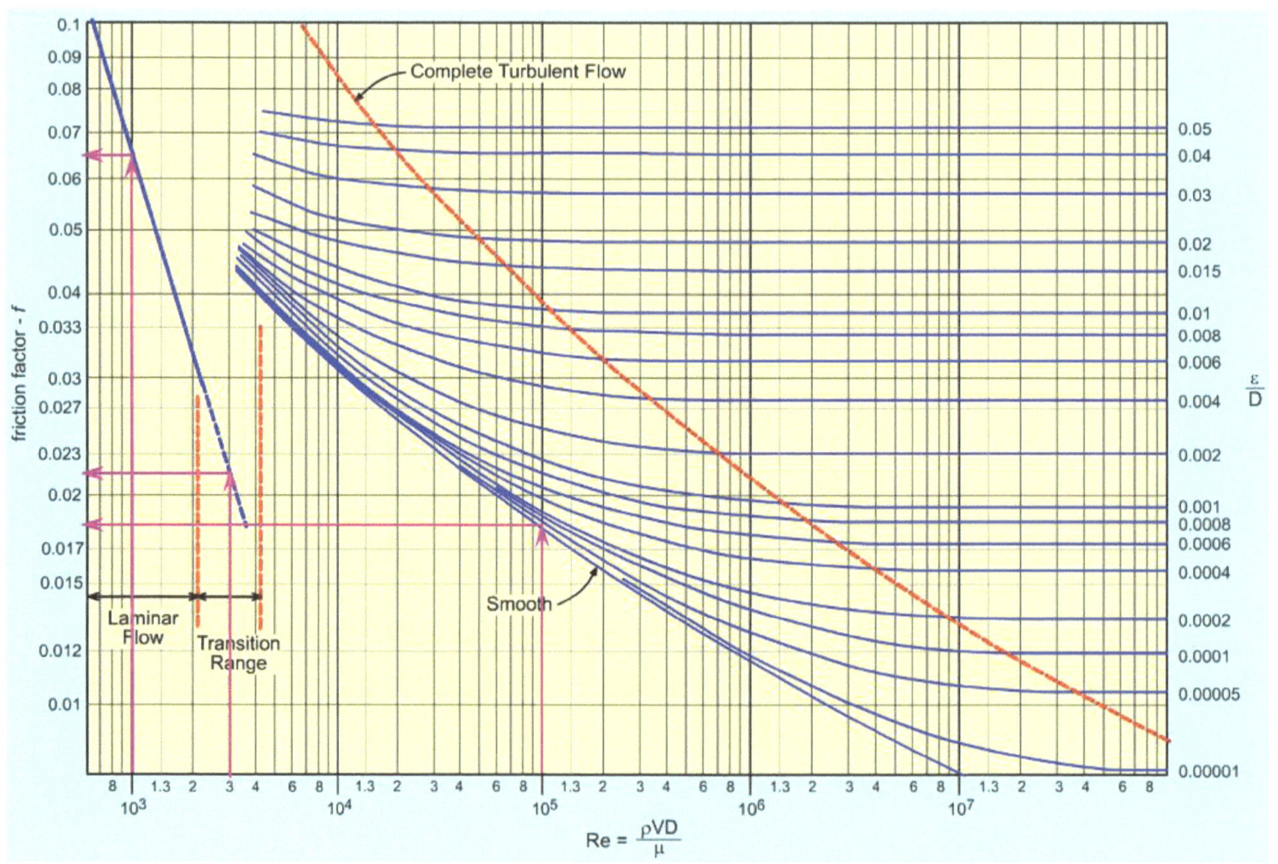

Fig. 6.3- Moody Diagram

Finally, Eq. 6.6 calculates the pressure drop across a hydraulic transmission line due to its resistivity. This pressure drop is conventionally known as *"Frictional Losses"*.

$$\Delta P_R = \lambda \frac{L}{D} \frac{\rho v^2}{2} = \lambda \frac{L}{D} \frac{SG \times \rho_W \times v^2}{2} \qquad 6.6$$

Equation 6.7 presents Eq. 6.6 with the units being balanced as follows:

$$\Delta P_R(\text{Pascal}) = \frac{\lambda \times L(m)}{D(mm) \times 10^{-3}} \frac{SG \times 1000\ (kg/m^3) \times v^2(m/s)^2}{2}$$

$$\Rightarrow \Delta P_R(\text{Pascal}) = \frac{10^6 \times \lambda \times SG \times L \times v^2}{2 \times D} \left[\frac{kg.\,m}{s^2} \frac{1}{m^2} = \frac{N}{m^2}\right]$$

$$\Rightarrow \Delta P_R(\text{bar}) = \frac{\Delta p_R(\text{Pascal})}{10^5} = \frac{5 \times \lambda \times SG \times L(m) \times [v(m/s)]^2}{D(mm)} \qquad 6.7$$

Figure 6.4 shows the structure of a Simulink model that solves Eq. 6.1 through 6.6. For better understanding of the structure, colors on the figure are made to match the colors on the text.

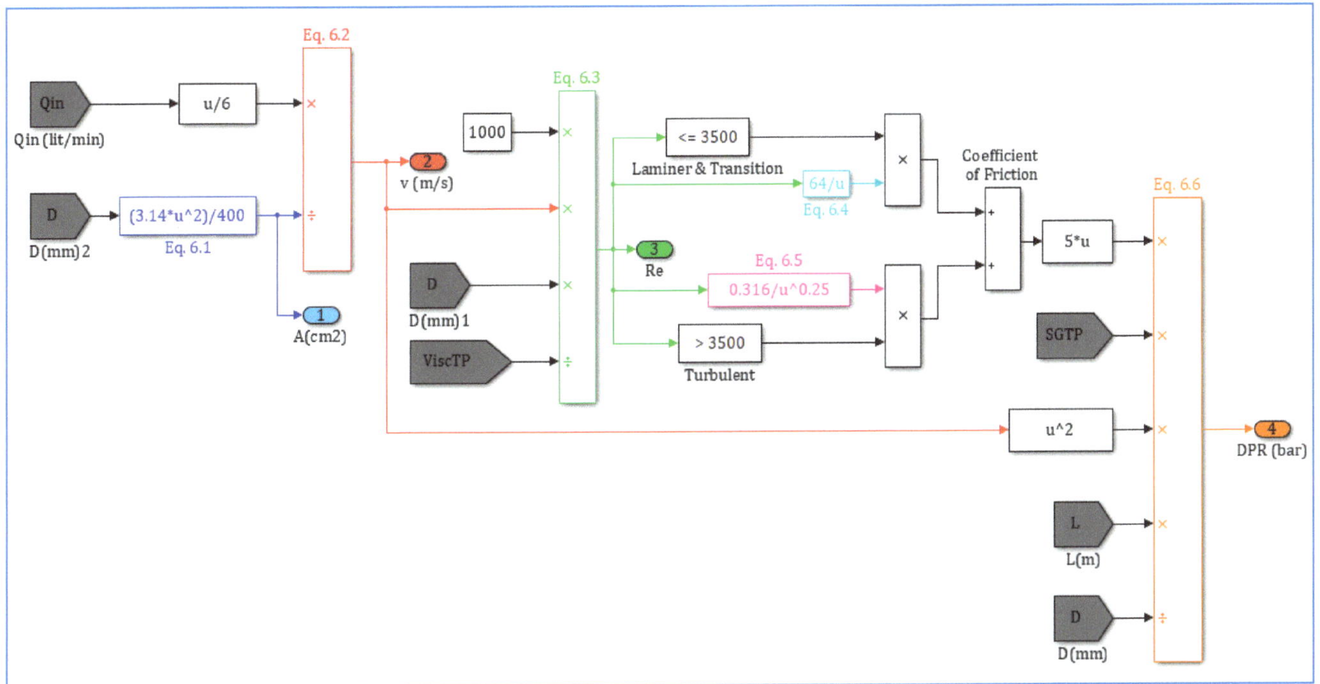

Fig. 6.4- Modeling Pressure Losses due to Line Resistivity

6.1.2- Modeling Pressure Losses due to Line Inductance

Equation 6.8 is used to calculate the pressure drop across a transmission line due to its inductance. The equation is developed based on the concept that a pressure drop is required to accelerate a mass of fluid in motion inside a transmission line of known length and diameter.

$$\Delta P_I = \frac{\rho L}{A} \times \frac{dQ_{in}}{dt} \qquad \qquad 6.8A$$

Equation 6.8B presents Eq. 6.8A with the units being balanced as follows:

$$\Delta P_I \text{ (Pascal)} = \frac{\rho \left[\frac{kg}{m^3}\right] L[m]}{A[cm^2] \times 10^{-4}} \times \frac{dQ_{in}\left[\frac{lit}{min}\right] \times \frac{1}{60 \times 1000}}{dt[s]}$$

$$\Rightarrow \Delta P_I \text{ (Pascal)} = \frac{\rho L}{A \times 10^{-4}} \times \frac{dQ_{in} \times \frac{1}{60 \times 1000}}{dt} \left[\frac{kg \times m \times m^3}{m^3 \times m^2 \times s^2}\right]$$

$$\Rightarrow \Delta P_I \text{ (Pascal)} = \frac{\rho L}{A \times 6} \times \frac{dQ_{in}}{dt} \left[\frac{kg.\,m}{s^2}\frac{1}{m^2} = \frac{N}{m^2}\right]$$

$$\Rightarrow \Delta P_I \text{ (bar)} = \frac{\rho \left[\frac{kg}{m^3}\right] L[m]}{A[cm^2] \times 6 \times 10^5} \times \frac{dQ_{in}(lit/min)}{dt} \qquad \qquad 6.8B$$

Figure 6.5 shows the structure of a Simulink model that solves Eq. 6.8.

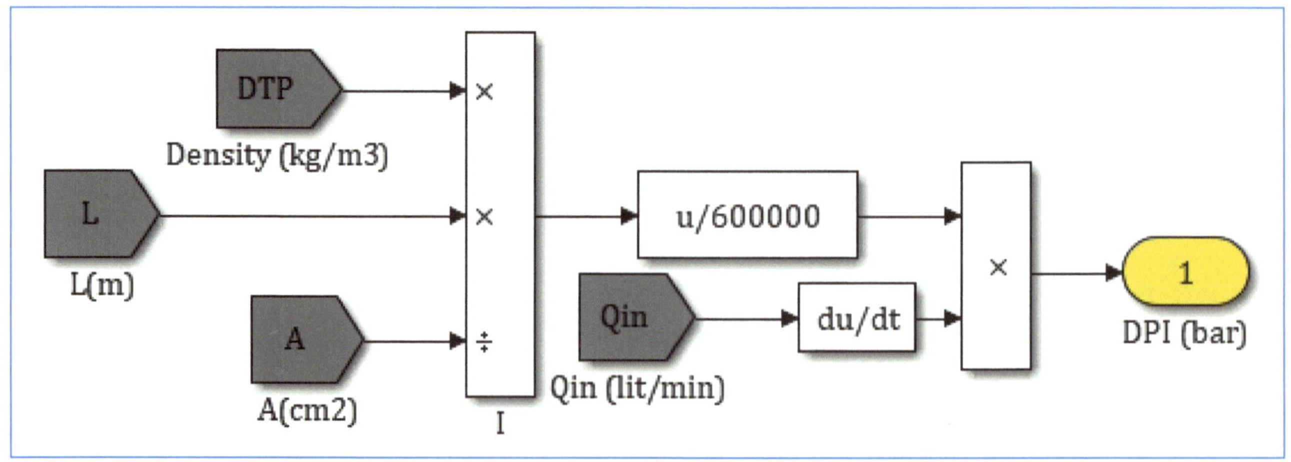

Fig. 6.5- Modeling Pressure Losses due to Line Inductance

6.1.3- Modeling Inlet Pressure of Transmission Line

$$\mathbf{P_{in}\ (bar)\ =\ P_{out}\ (bar)\ +\ \Delta P_R\ (bar) + \Delta P_I\ (bar)} \qquad\qquad 6.9$$

6.1.4- Modeling Outlet Flow due to Line Capacitance

Referring to Eq. 1.5D in Chapter 1,

$$\Delta \mathbf{P_{out}}\ =\ \frac{\textbf{Equivalent Bulk Modulus }(\beta_E)}{\textbf{Container Volume }(V)} \int \mathbf{Q\,d} \implies \frac{\mathbf{dP_{out}}}{\mathbf{dt}} = \frac{\beta_E}{V}[\mathbf{Q_{in} - Q_{out}}]$$

$$\implies \mathbf{Q_{out} = Q_{in} - \frac{V}{\beta_E}\frac{dP_{out}}{dt} = Q_{in} - \frac{A \times L}{\beta_E}\frac{dP_{out}}{dt}} \qquad\qquad 6.10A$$

Equation 6.10B presents Eq. 6.10A with the units being balanced as follows:

$$\mathbf{Q_{out}\left[\frac{lit}{min}\right] = Q_{in}\left[\frac{lit}{min}\right] - \frac{A\,[cm^2]\times\dfrac{1}{10000}\times L[m]}{\beta_E\,[bar]}\frac{dP_{out}[bar]}{dt\,[s]\times\dfrac{1}{60}}\times 1000}$$

$$\mathbf{Q_{out}\left[\frac{lit}{min}\right] = Q_{in}\left[\frac{lit}{min}\right] - \frac{6\times A\,[cm^2]\times L[m]}{\beta_E\,[bar]}\frac{dP_{out}[bar]}{dt\,[s]}} \qquad\qquad 6.10B$$

Figure 6.6 shows the structure of a Simulink model that solves Eq. 6.10.

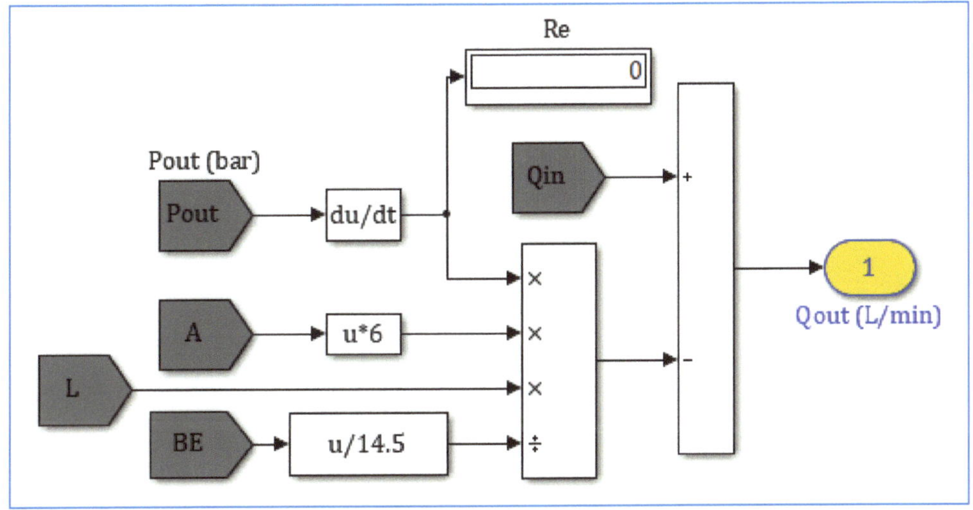

Fig. 6.6- Modeling Outlet Flow due to Line Capacitance

6.1.4- Case Studies for Modeling a Hydraulic Transmission Line

Figure 6.7 shows simulation model (HSV7CH06Model01) for a hydraulic transmission line. The following features were considered in the model to improve model capabilities:

- Model of hydraulic fluid is added to get an instantaneous update on hydraulic fluid properties used in the transmission line model.
- Most of the variables are assigned to global variables so that it can be accessed from anywhere in the model.
- Two Simulink blocks are extensively used ("GOTO" and "From") to reduce the crossing linking lines.
- In order to help a modeler to use the model to investigate various scenarios, inlet flow and outlet pressure can be selected as a constant or ramping value.

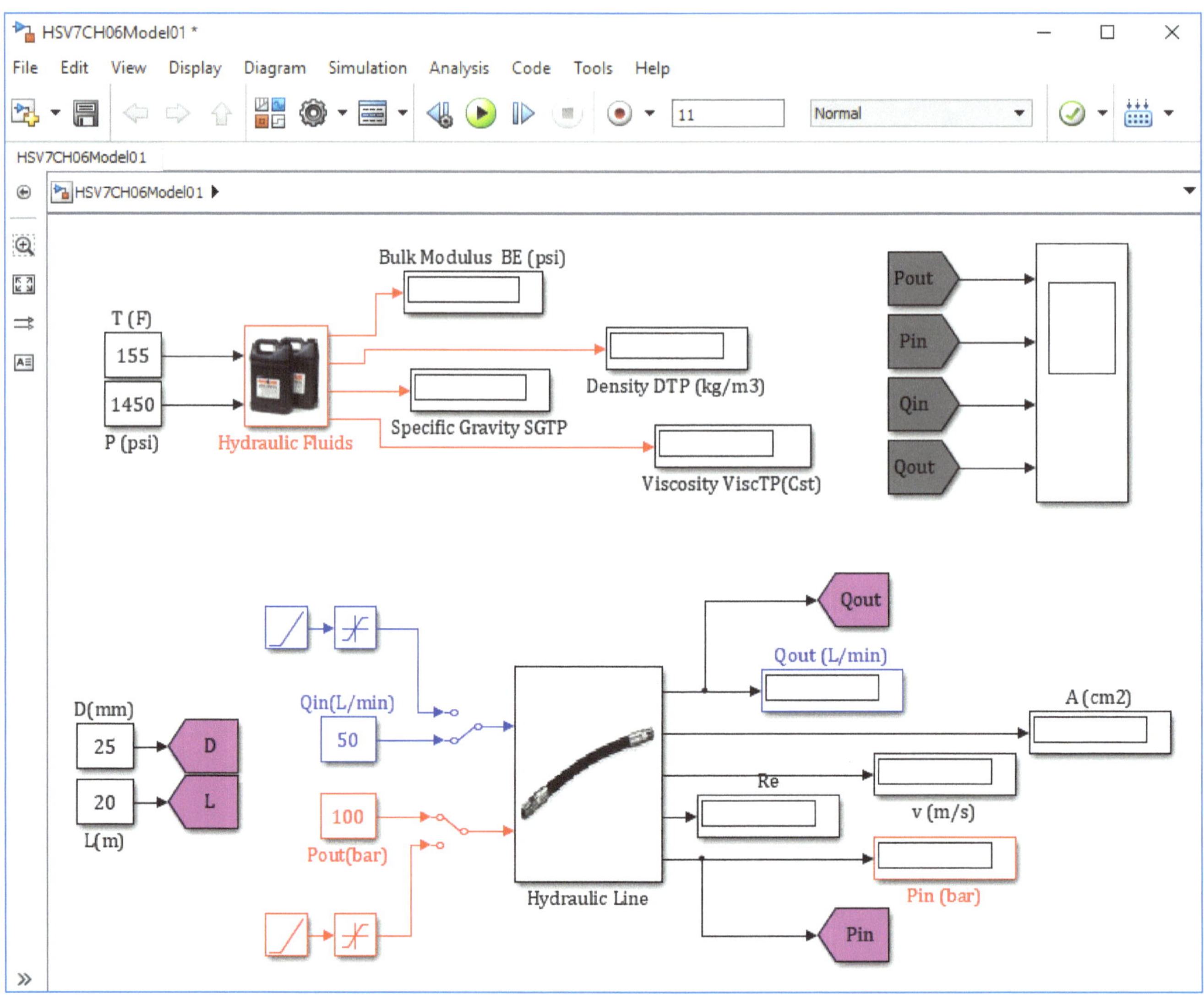

Fig. 6.7- Simulation Model for Hydraulic Transmission Line

Figure 6.8 shows the details of the model under the mask. The figure shows combining the three previous sub-models.

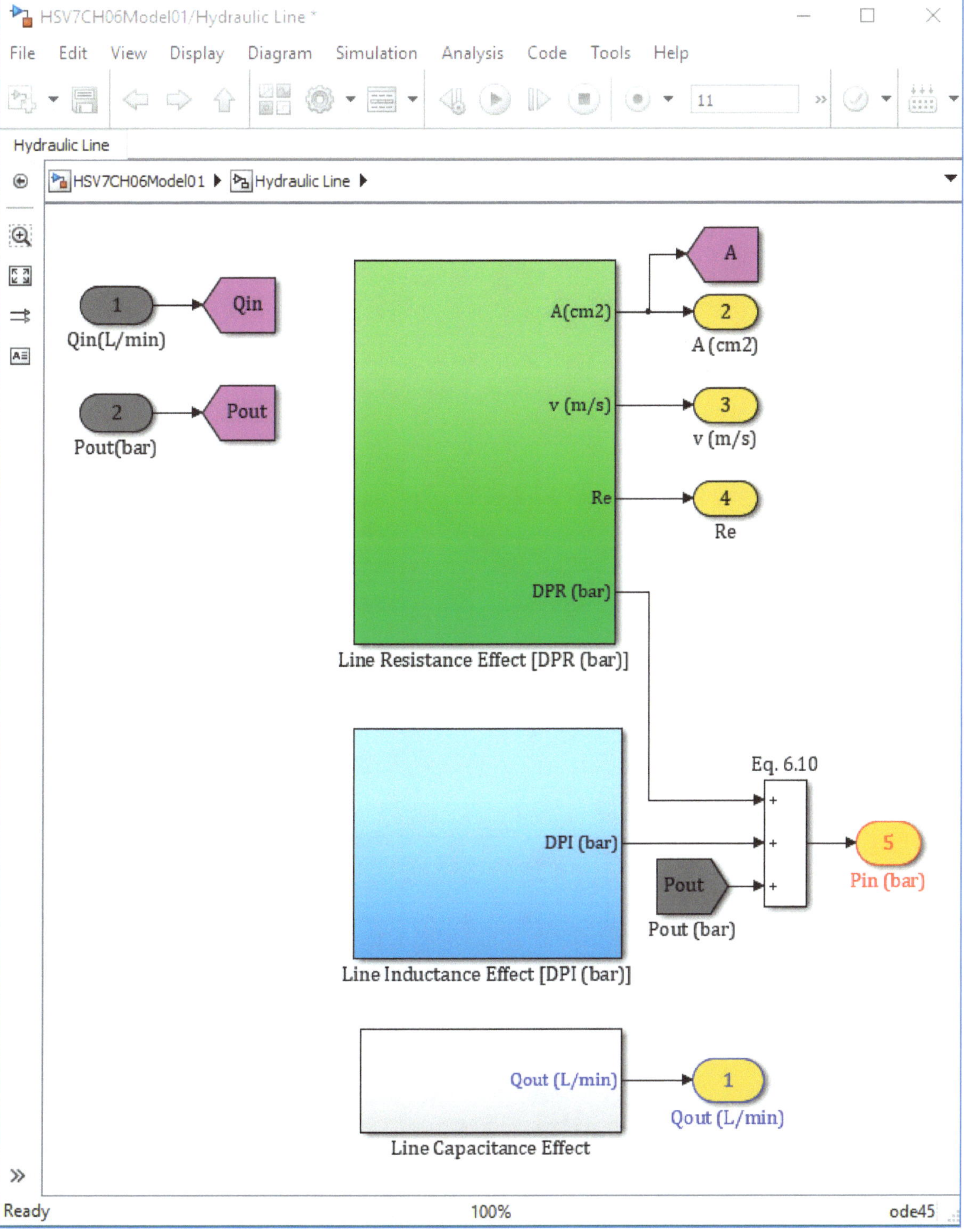

Fig. 6.8- Model Structure under the Mask of Transmission Line Model

6.1.4.1- Case Study 1: Steady State Condition

The upper part of Fig. 6.9 shows the results of simulating a hydraulic transmission line based on constant values of inlet flow and outlet pressure. The lower part is developed by running other software "Hydraulic Component Sizing Calculator - HCSC". The results from HCSC validates the results from the simulation model.

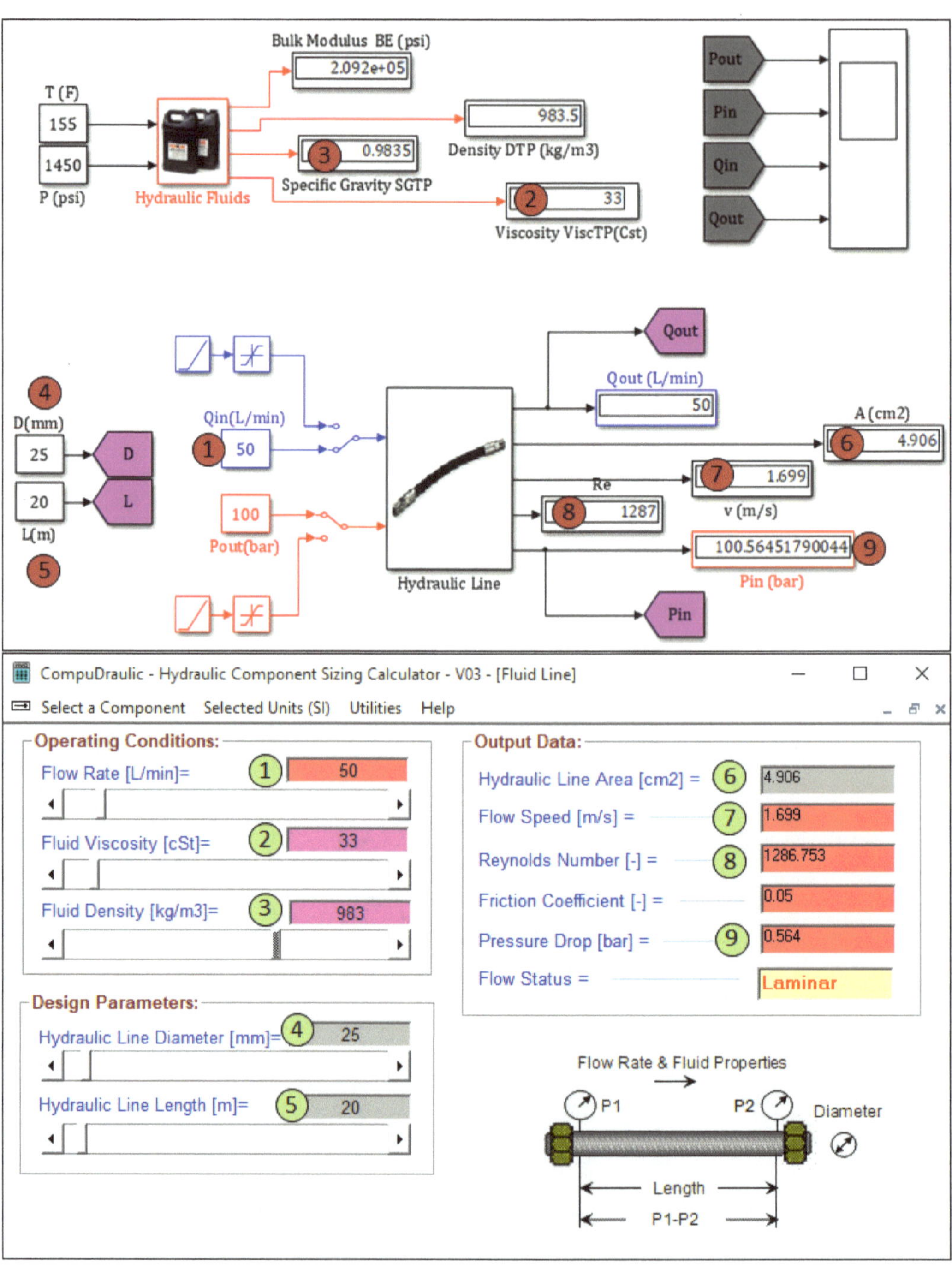

Fig. 6.9- Results of Case Study 1

6.1.4.2- Case Study 2: Surge Inlet Flow

During a machine operation, a hydraulic line may subject to a surge inlet flow. Figure 6.10 shows simulating this case by sudden increase of the inlet flow. The figure shows that, due to surge inlet flow, inlet pressure spikes up before returning to steady state condition.

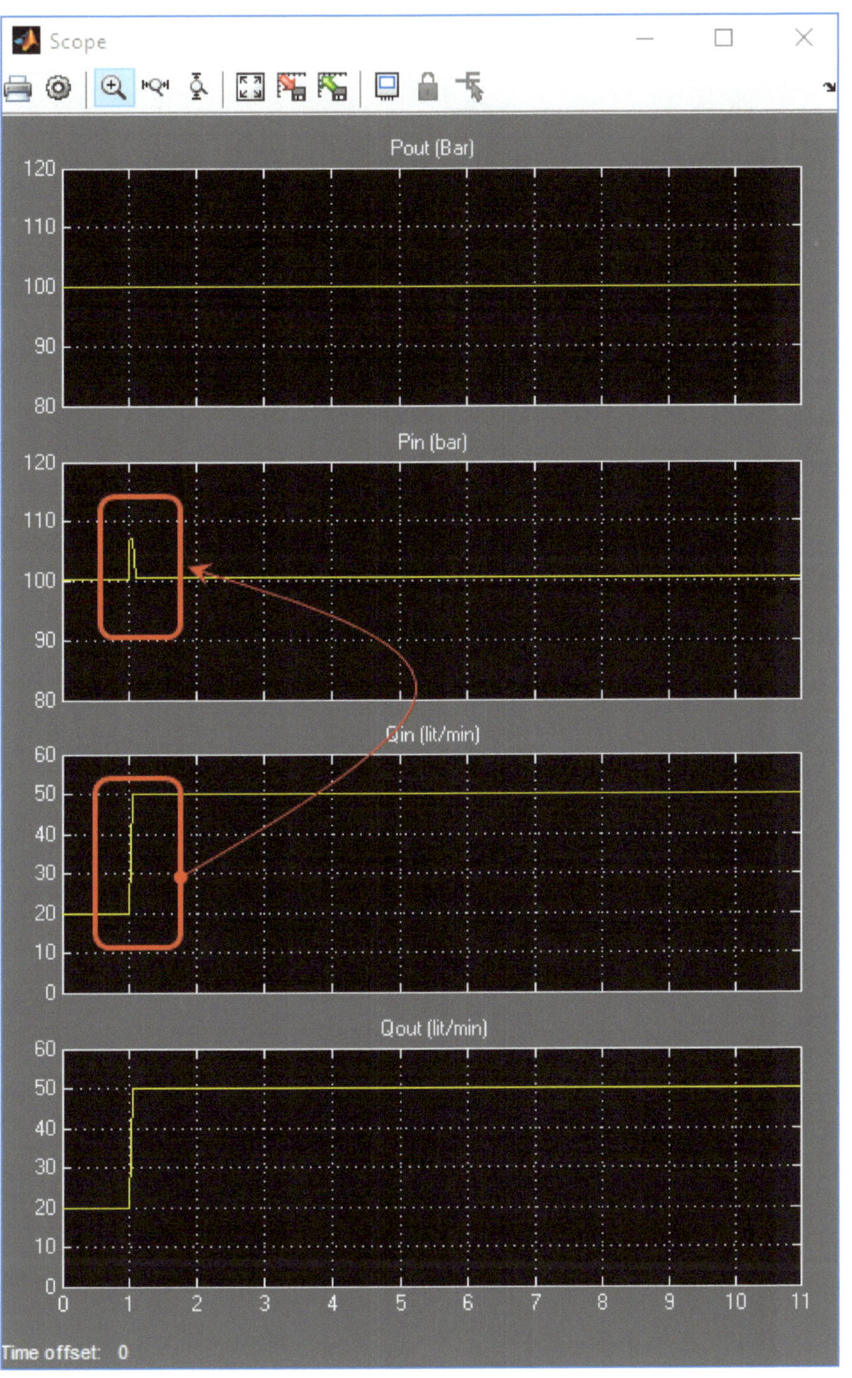

Fig. 6.10- Results of Case Study 2

6.1.4.3- Case Study 3: Surge Outlet Pressure

During a machine operation, a hydraulic line may subject to a surge outlet pressure, let's say during overrunning load or sudden closure of a valve. Figure 6.11 shows simulating this case by sudden increase in the outlet pressure. The figure shows that, due to surge outlet pressure, outlet flow surges backward before returning to steady state condition.

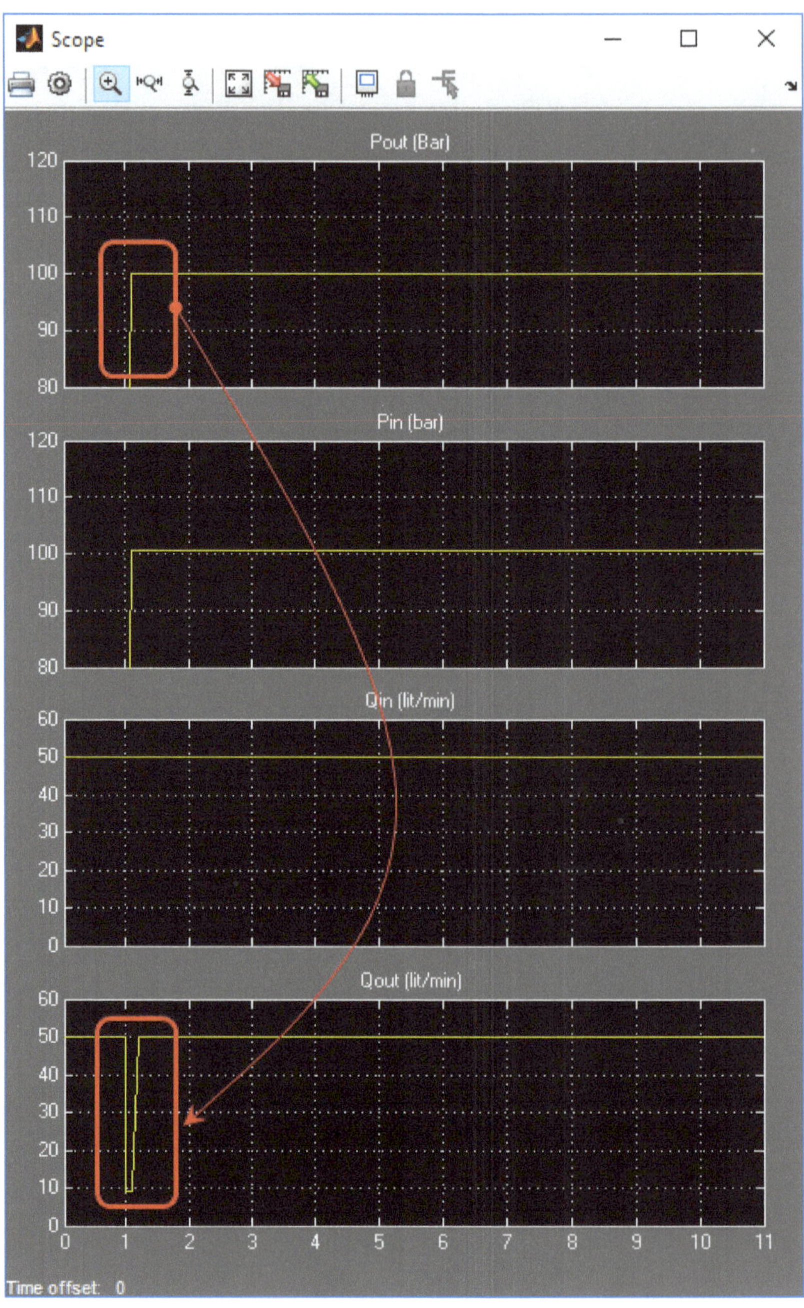

Fig. 6.11- Results of Case Study 3

6.2- Modeling of Hydraulic Fittings

Figure 6.12 shows the stream of calculations in a generic fitting model structure. The model uses the inlet flow to calculate the pressure drop across the fitting. This pressure drop is added to the outlet pressure to get the inlet pressure. Outlet flow will be considered equal to inlet flow because the piece of fitting has small dead volume, so effect of fluid capacitance can feasibly be ignored.

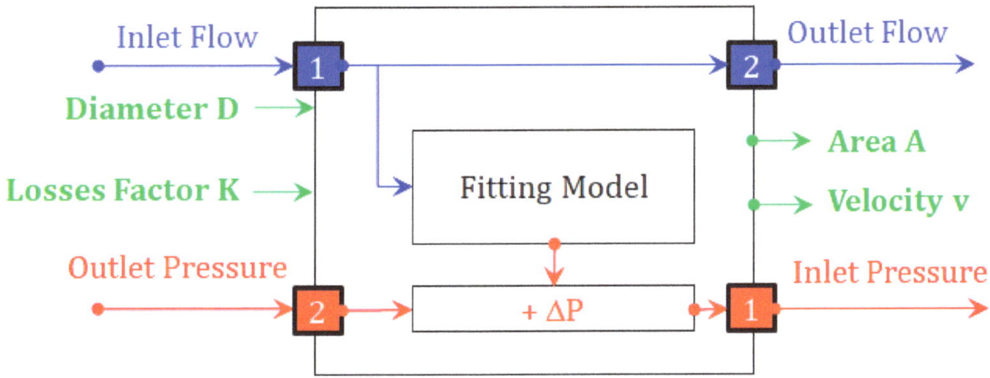

Fig. 6.12- Structure of a Generic Model for a Hydraulic Fitting

Modeling the Pressure Drop: Pressure drop in fittings is referred to as *Local Losses*. This term is used whenever the fluid velocity changes its magnitude (e.g. when a line diameter is changed as in fittings) or its direction (e.g. as in elbows). This pressure drop is mathematically expressed as follows:

$$\Delta p_L = K \frac{\rho v^2}{2} = K \frac{SG \times \rho_W \times v^2}{2} \qquad 6.11A$$

Equation 6.11B presents Eq. 6.11A with the units being balanced as follows:

$$\Delta p_L (\text{bar}) = \frac{K \times SG \times 1000 \, (\text{kg/m}^3) \times v^2 (\text{m/s})^2}{2 \times 10^5} = \frac{K \times SG \times v^2 (\text{m/s})^2}{200} \qquad 6.11B$$

Where **K** is a lump sum resistivity factor and reported numerically for every specific shape of fitting as shown in Fig. 6.13. As shown in Fig. 6.14, sometimes the **K** factor is reported graphically. In such a case, a lookup table should be built to generate the proper **K** factor depending on the given data.

Modeling the Inlet Pressure: Inlet pressure of a fitting is calculated as follows:

$$p_{in} (\text{bar}) = p_{out} (\text{bar}) + \Delta p_L (\text{bar}) \qquad 6.12$$

Other Modeling Method: An alternative easy method to model a hydraulic fitting is to convert the local losses in the fitting into a lookup table as function of the inlet flow.

Fitting	90° bend	90° angle	T-Piece	Double angle	Valve
K	0.5 - 1	1.2	1.3	2	5 - 15
Sudden Contraction	Well-Rounded	Slightly Rounded	Sharp Edged	Shoulder	
K	0.04	0.2	0.5	0.8	
Sudden Expansion	Well-Rounded	Slightly Rounded	Sharp Edged	Shoulder	
K	1	1	1	1	

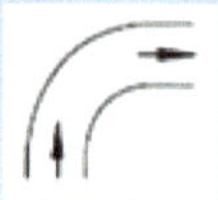

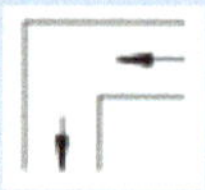

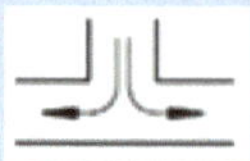

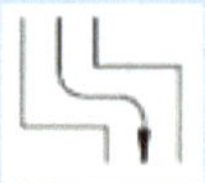

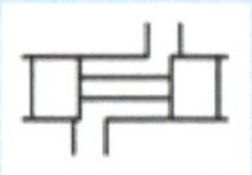

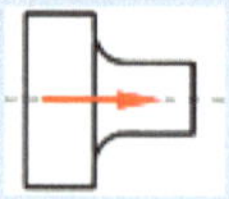

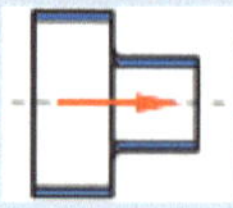

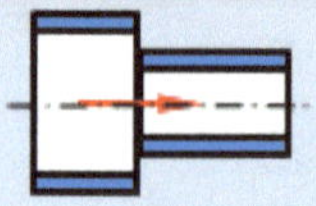

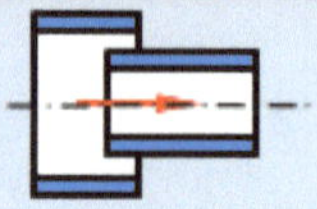

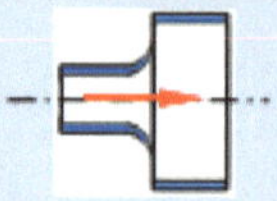

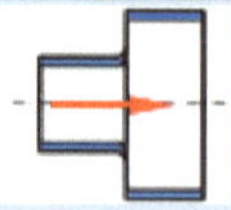

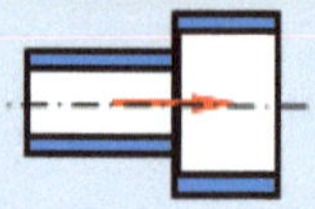

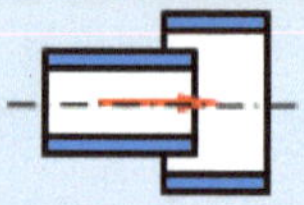

Fig. 6.13- Losses Factor in Fittings

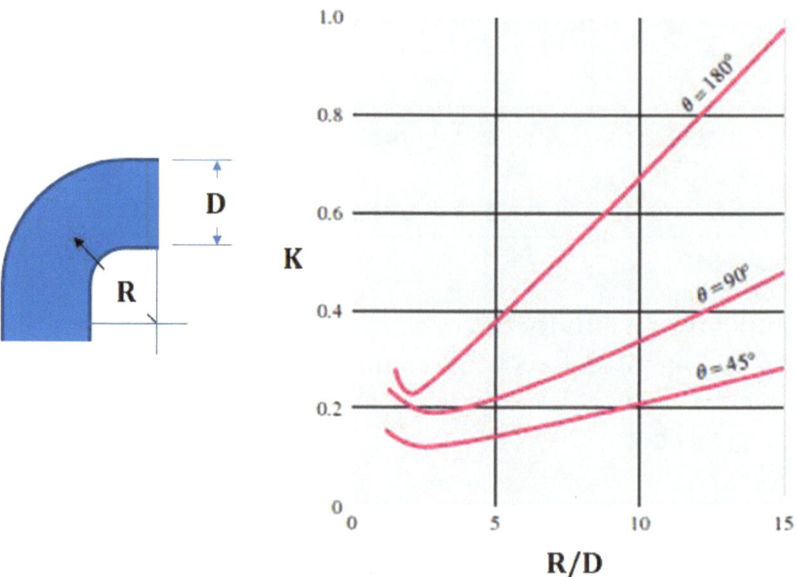

Fig. 6.14- Losses Factor in Elbows

Simulation Model: Figure 6.15 shows the structure of a simulation model (HSV7CH06Model02). The model is built to simulate the local losses with the choice of using a lookup table or the previously described set of equations.

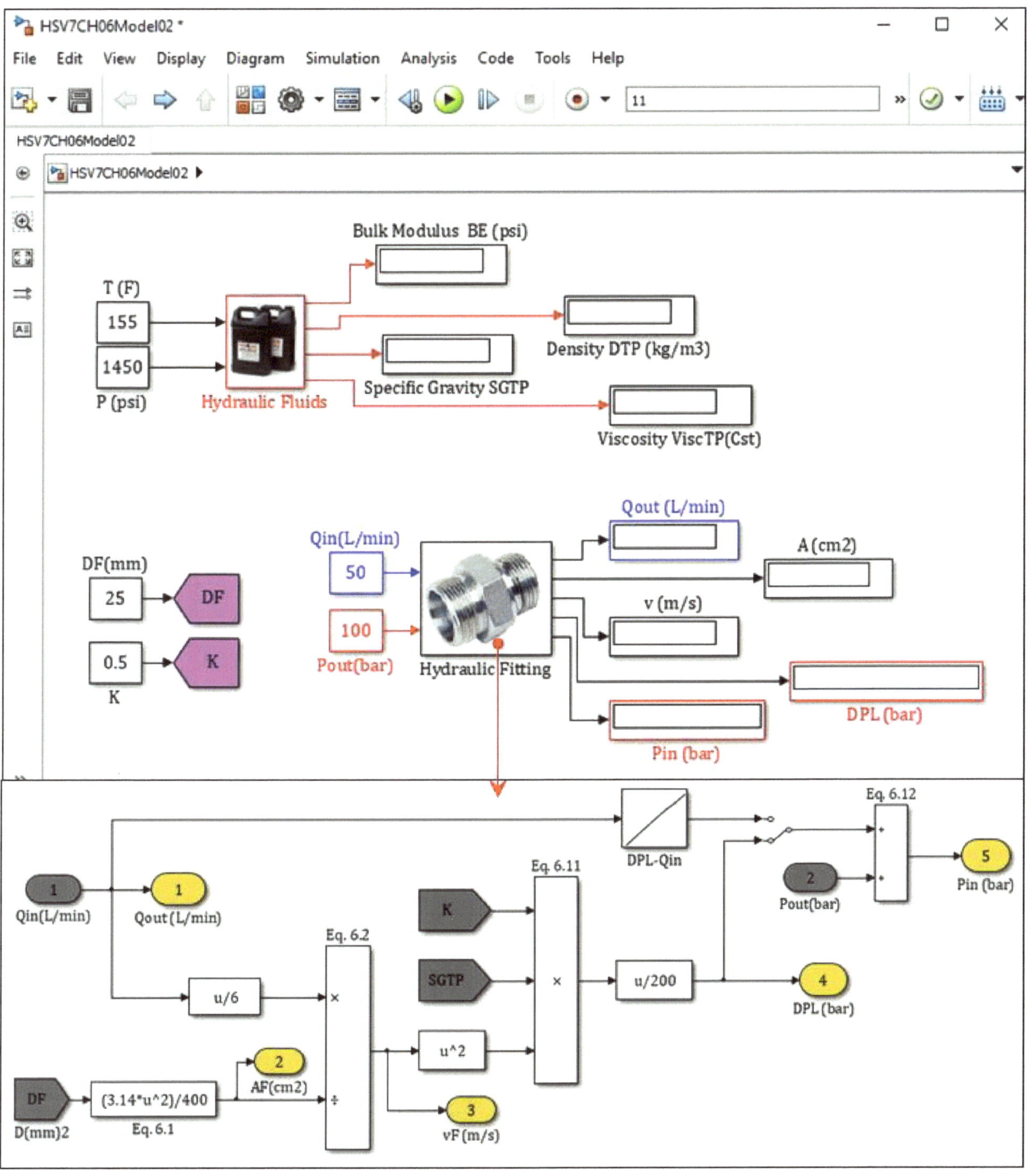

Fig. 6.15- Simulation Model for Hydraulic Fitting

Model Validation: As shown in Fig. 6.16, a sudden contraction is assumed in the hydraulic line shown in Fig. 6.9 from 25 mm diameter to 10 mm. Simulation results shows 0.38 bar for a losses factor of 0.7. Simulation results were validated based on other software "Automation Studio".

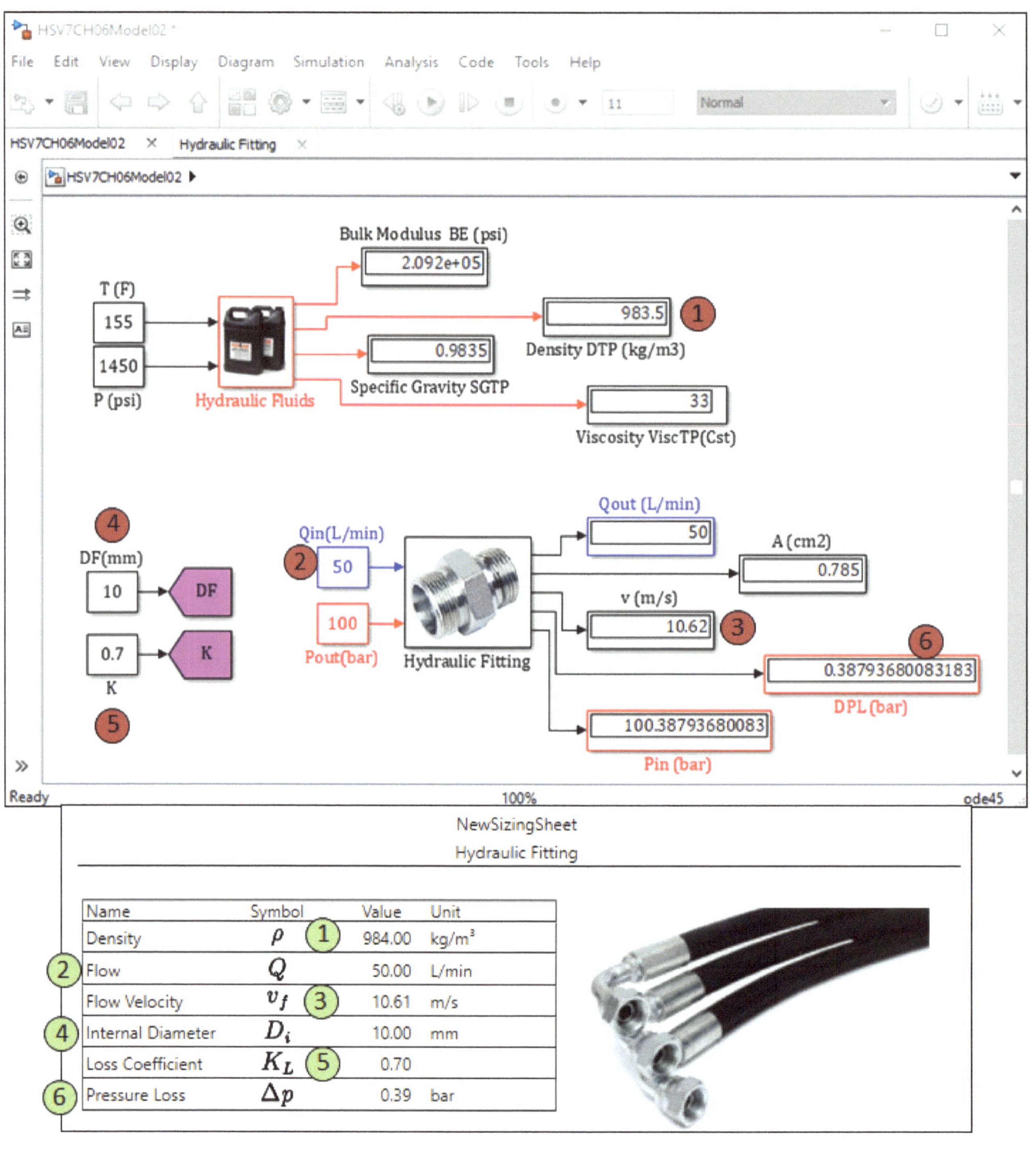

NewSizingSheet

Hydraulic Fitting

Name	Symbol		Value	Unit
Density	ρ	(1)	984.00	kg/m³
Flow	Q		50.00	L/min
Flow Velocity	v_f	(3)	10.61	m/s
Internal Diameter	D_i		10.00	mm
Loss Coefficient	K_L	(5)	0.70	
Pressure Loss	Δp		0.39	bar

Fig. 6.16- Validation of Hydraulic Fitting Model

6.3- Modeling of Hydraulic Orifices

Figure 6.17 shows the stream of calculations in an orifice generic model structure. The model uses the inlet flow to calculate the pressure drop across the orifice. This pressure drop is added to the outlet pressure to get the inlet pressure. Outlet flow will be considered equal inlet flow because the orifice has small dead volume, so effect of fluid capacitance can feasibly be ignored.

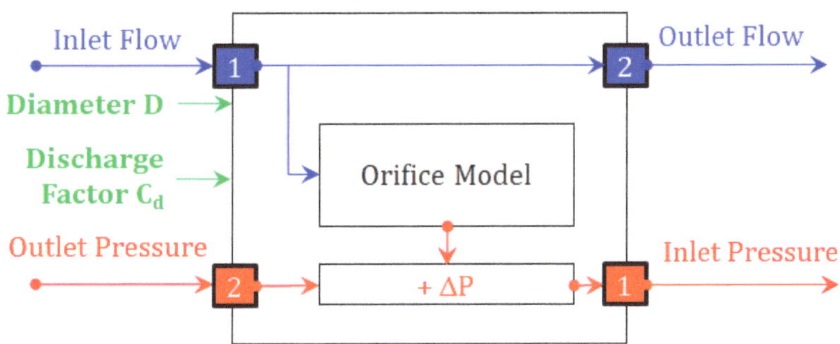

Fig. 6.17- Structure of a Generic Model for a Hydraulic Orifice

Modeling the Pressure Drop: Equation 6.13 shows that the flow through an orifice is a nonlinearly function of the pressure drop across the orifice as follows.

$$Q_{in} = C_d A \sqrt{\frac{2 \times \Delta P_O}{\rho}} \qquad\qquad 6.13$$

Where C_d is the *Discharge Coefficient* that depends on the orifice shape and equals 0.611 for sharp edged orifices. Equation 6.13 is rewritten in a reverse form to solve for the pressure as follows:

$$\Delta P_O = \frac{\rho}{2}\left[\frac{Q_{in}}{C_d A}\right]^2 = \frac{SG \times \rho_w}{2}\left[\frac{Q_{in}}{C_d A}\right]^2 \qquad\qquad 6.14A$$

Equation 6.14B presents Eq. 6.14A with the units being balanced as follows:

$$\Delta P_O (\text{Pascal}) = \frac{SG \times 1000 \,(\text{kg/m}^3)}{2}\left[\frac{Q_{in}[\text{lit/min}] \times \frac{1}{60 \times 1000}}{C_d \times A\,[\text{mm}^2] \times \frac{1}{10^6}}\right]^2$$

$$\Delta P_O (\text{Pascal}) = \frac{SG \times 1000 \times [Q_{in}(\text{lit/min})]^2}{2 \times C_d^2 \times A^2\,[\text{mm}^2]}\left[\frac{100}{6}\right]^2 \left[\frac{\text{kg.m}^6}{\text{m}^3} \frac{1}{\text{s}^2} \frac{1}{\text{m}^4}\right]$$

$$\Delta P_O (\text{Pascal}) = \frac{10^7 \times SG \times [Q_{in}(\text{lit/min})]^2}{72 \times C_d^2 \times A^2\,[\text{mm}^2]}\left[\frac{\text{kg.m}}{\text{s}^2} \frac{1}{\text{m}^2} = \frac{N}{\text{m}^2}\right]$$

$$\Delta P_O(\text{bar}) = \frac{\Delta P_O(\text{Pascal})}{10^5} = \frac{1000 \times SG \times [Q_{in}(\text{lit/min})]^2}{720 \times C_d^2 \times A^2 \, [\text{mm}^2]} \qquad\qquad 6.14B$$

Modeling the Inlet Pressure: Inlet pressure of the orifice is calculated as follows:

$$P_{in}\,(\text{bar}) \;=\; P_{out}\,(\text{bar}) \,+\, \Delta P_O\,(\text{bar}) \qquad\qquad 6.15$$

Other Modeling Method: An alternative easy method to model a hydraulic orifice is to convert the local losses in the orifice into a lookup table as function of the inlet flow.

Simulation Model: Figure 6.18 shows the structure of a simulation model (HSV7CH06Model03). The model is built to simulate the local losses with the choice of using a lookup table or the previously descried set of equations.

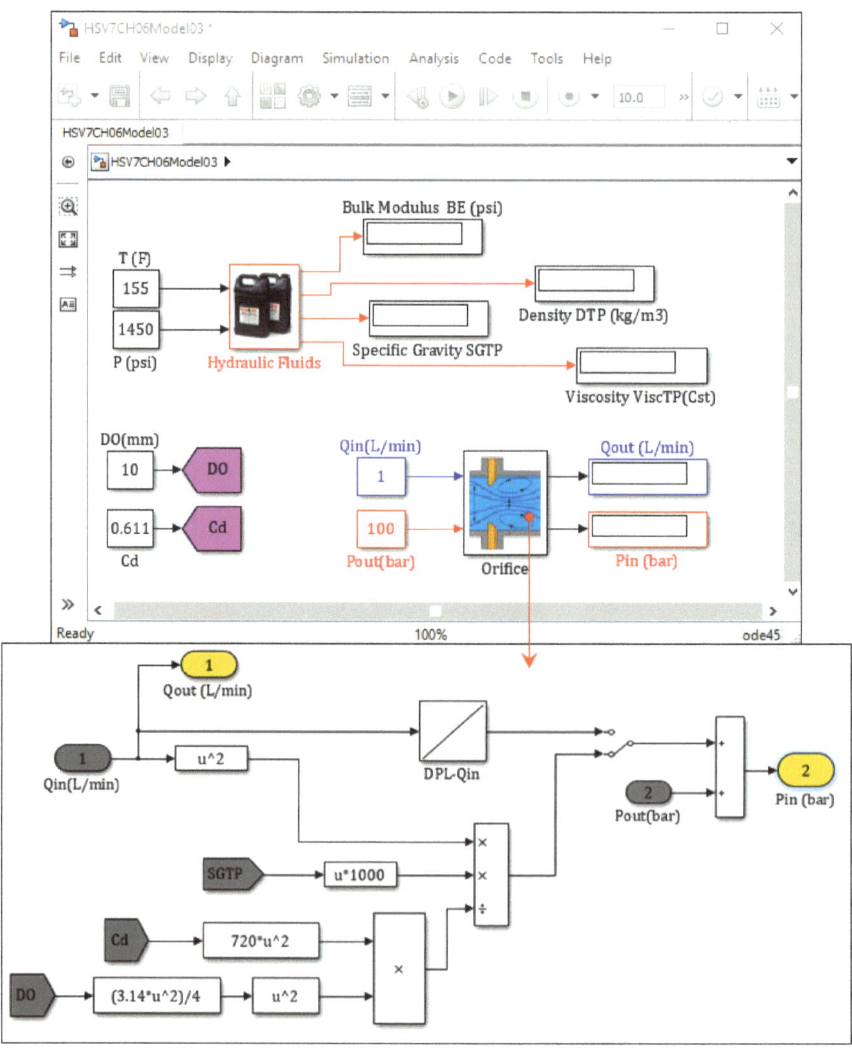

Fig. 6.18- Simulation Model for Hydraulic Orifice

Model Validation: As shown in Fig. 6.19, an orifice of 5 mm diameter is used. Simulation results shows 23.8 bar pressure drop across the orifice. Simulation results were validated based on other software "Automation Studio".

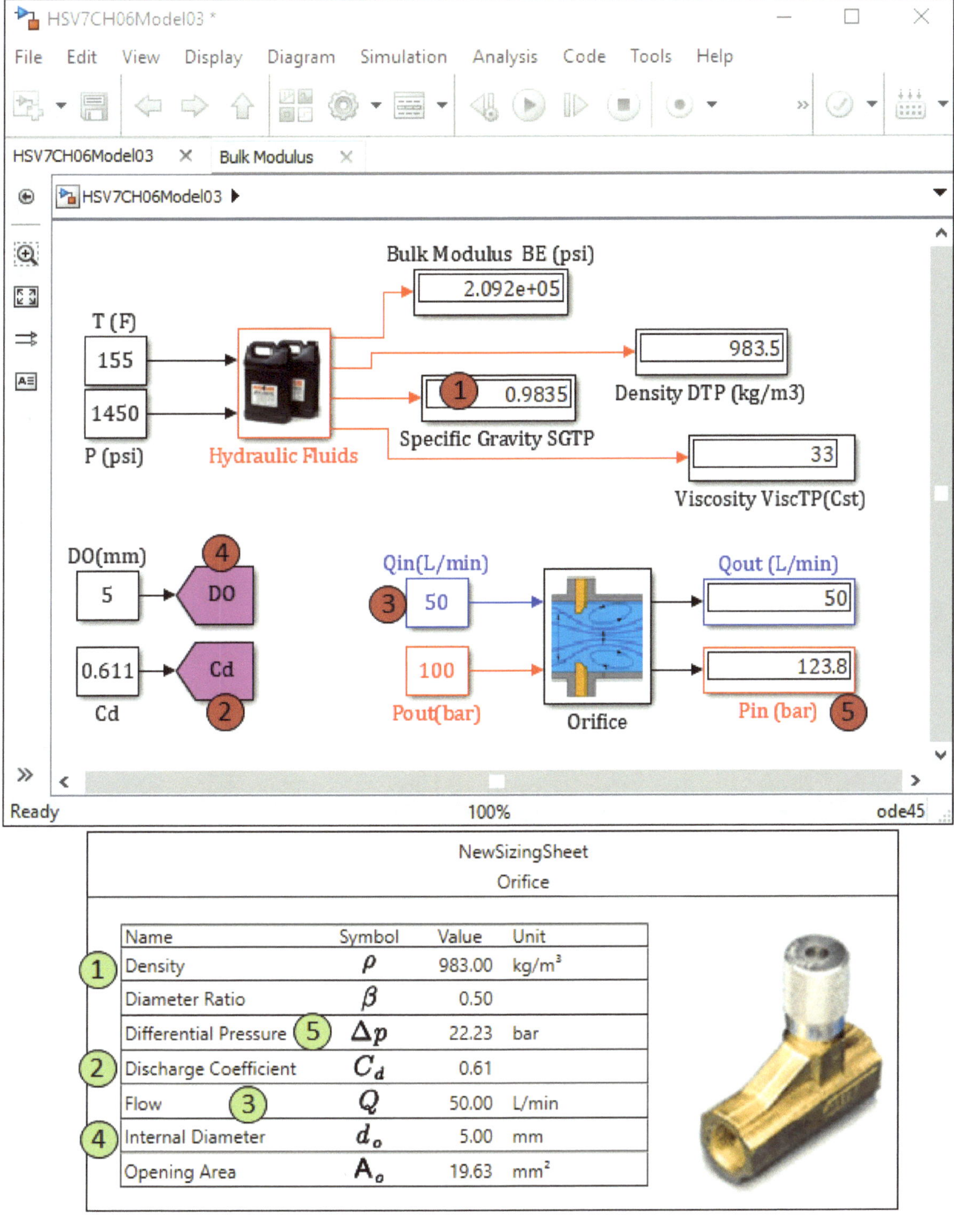

NewSizingSheet
Orifice

	Name	Symbol	Value	Unit
1	Density	ρ	983.00	kg/m³
	Diameter Ratio	β	0.50	
5	Differential Pressure	Δp	22.23	bar
2	Discharge Coefficient	C_d	0.61	
3	Flow	Q	50.00	L/min
4	Internal Diameter	d_o	5.00	mm
	Opening Area	A_o	19.63	mm²

Fig. 6.19- Validation of Hydraulic Fitting Model

6.4- Modeling Hydraulic Transmission Line Assembly

As shown in Fig. 6.20, in model (HSV7CH06Model04), a transmission line is assembled with a piece of fitting followed by an orifice. As shown in figure, for a 100-bar outlet pressure at the orifice, inlet pressure at the transmission line is 124.7 bar. An overall pressure drop is 24.7 bar including pressure drop in the three elements.

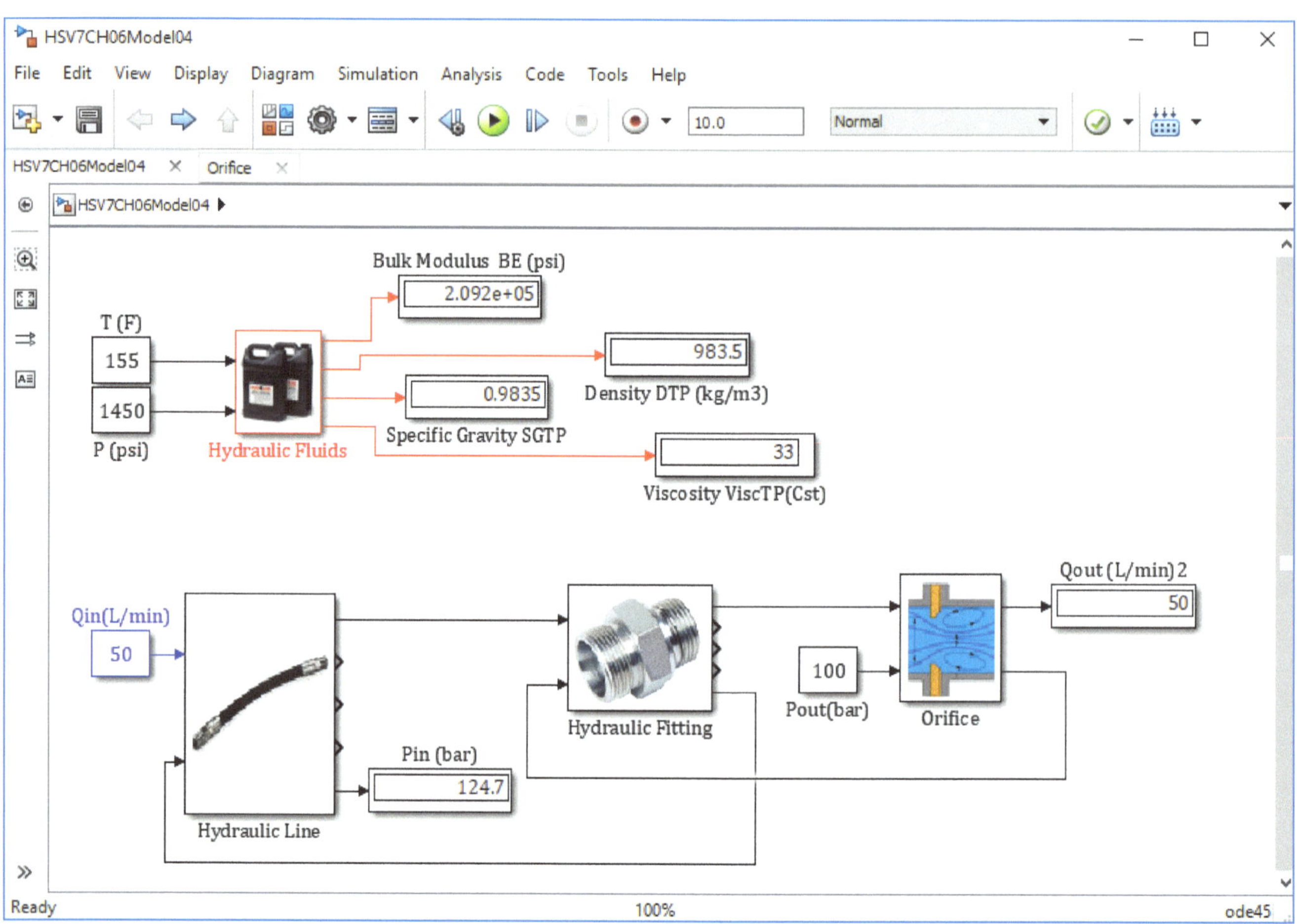

Fig. 6.20- Example of Modeling a Transmission Line Assembly

Chapter 7

Water Hammer and Noise Control

Objectives

This chapter presents a summarized idea about water hammer, its effect, and some techniques to minimize its damaging effects as applicable to hydraulic control systems. This chapter also presents the construction and operating principle of using shock suppressors to reduce the noise that are transmitted by hydraulic transmission lines.

Brief Contents

7.1- Water Hammer
7.2- Noise Control in Hydraulic Transmission Lines

Chapter 7 – Water Hammer and Noise Control

7.1- Water Hammer

What is Water Hammer?

Water Hammer is a shock wave transmitted through fluid contained in a transmission line. As shown in Fig. 7.1, it occurs when fluid in motion is suddenly forced to stop moving, such as when suddenly closing a valve or load is blocked downstream a pump.

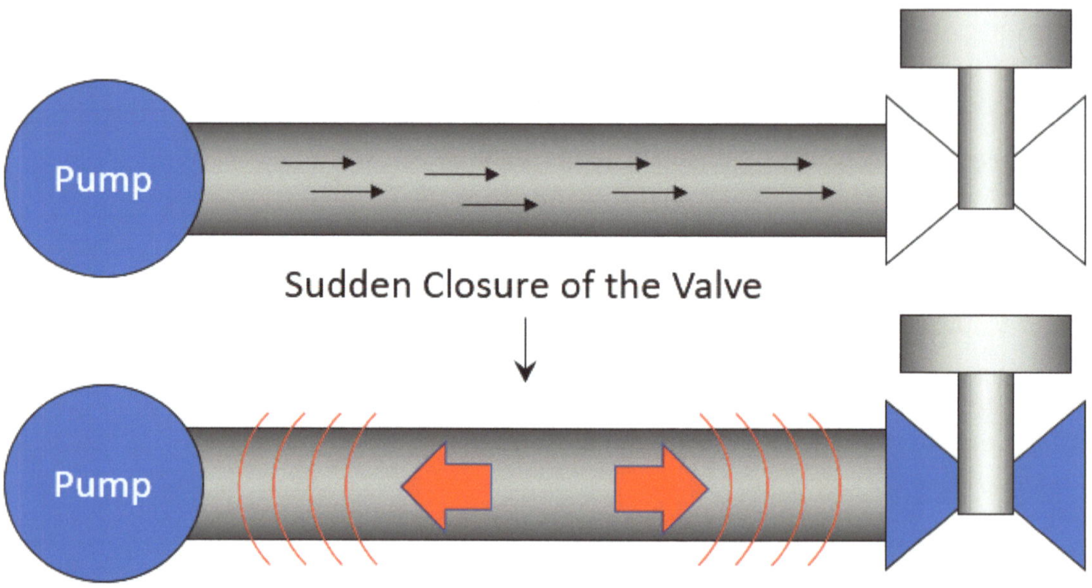

Fig. 7.1 – How Water Hammer is Created in Transmission Lines

What is the Effect of Water Hammer on a Hydraulic System?

Effect of water hammer is more noticeable in long and large water pipelines where the mass of the fluid in motion is huge. However, it still has some damaging effects on hydrostatic control systems. If it occurs very frequently, it may result is several problems such as:

- Line vibration and noise.
- Fatigue on pipelines and components due to pressure spikes act repeatedly.
- Leakage due loosening fittings under vibration.
- Instrumentations damage such as bourdon-tube pressure gauges.

How to Reduce the Effect of Water Hammer?
Effect of water hammer can be reduced by one or combination of the following:

- **Low Flow Speed:** Reduce the flow speed either by reducing the flow rate and/or increasing the pipeline diameter, whichever possible. As a results, the momentum in the fluid in motion will be reduced and the pressure spikes will be reduced.

- **Reduce the Overall System Stiffness:** When a system is built to be very rigid, the pressure waves move back and forth very fast when water hammer starts. So, one way to reduce the effect of water hammer is to increase the system elasticity by one or combination of the following:
  - **Low Bulk Modulus:** When the fluid has lower *Bulk Modulus*, the shock wave created by water hammer is suppressed faster because the fluid is spongey.
  - **Hose instead of Tubes or Pipes:** Wall elasticity of hoses is larger than hard transmission lines. As a result, the equivalent bulk modulus will be less.
  - **Accumulator:** Add a small diaphragm accumulator will help shaving the pressure spikes resulted from the water hammer.

- **Reduce Valve Closure Time:** Sudden closure of a valve or sudden load stopping downstream a pump cause water hammer. So, a system designer should consider cushioning of cylinders, gradual closure of valves downstream the pump, and using responsive pressure relief valves to limit maximum pressure.

How to Calculate Minimum Valve Closure Time to Minimize Effects of Water Hammer?
Equations 7.1 and 7.2 shows the two steps of calculating the minimum valve closure. Calculations made under the assumption that the pipe is perfectly rigid. Hydraulic pipes and tubes have some wall elasticity. If hoses are used, wall elasticity is even larger than pipes and tubes. Since elasticity reduces the effect of water hammer, then this calculation will be satisfactory no matter what type of hydraulic transmission lines are used.

$$\text{Celerity (m/s)} = \sqrt{\frac{10^5 \times \text{B (bar)}}{\rho\ (\text{kg/m}^3)}} \qquad\qquad 7.1$$

$$\text{T}_{\text{VCMIN}}\ (\text{s}) = \frac{2 \times \text{L}}{\text{Celerity}} \qquad\qquad 7.2$$

Where (referring to Fig. 7.2):

- **Celerity**: *Celerity* is the speed by which the pressure wave is transmitted through the transmission line.
- **B** = Hydraulic Fluid *Bulk Modulus.*
- ρ = Hydraulic Fluid *Density.*
- **L** = Length of the transmission line.
- T_{VCMIN} = Minimum Valve Closure Time.

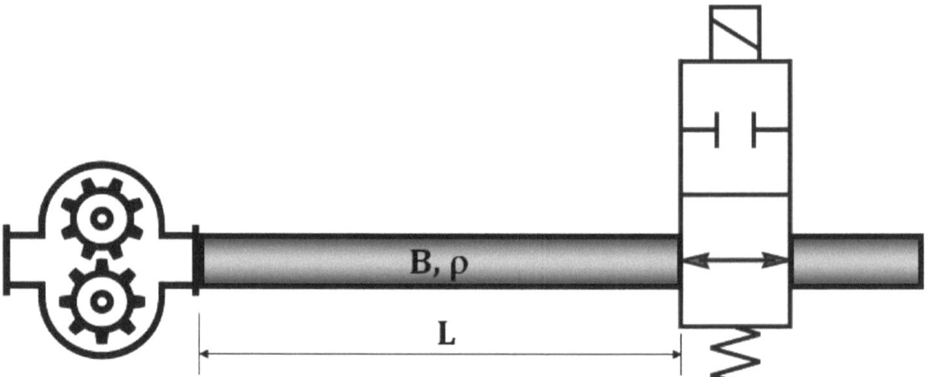

Fig. 7.2 – How Water Hammer is Created in a Transmission Lines

Case Study:

Given Data: L = 2 m, B = 15000 bar, and ρ = 850 kg/m³.

Eq. 7.1 →

$$\text{Celerity} \left(\frac{\mathbf{m}}{\mathbf{s}}\right) = \sqrt{\frac{10^5 \times \mathbf{B}\ (\mathbf{bar})}{\rho \left(\frac{\mathbf{kg}}{\mathbf{m^3}}\right)}} = \sqrt{\frac{10^5 \times 15000}{850}} = 1328$$

Eq. 7.2 →

$$T_{VCMIN} = \frac{2 \times \mathbf{L}}{\mathbf{Celerity}} \quad \frac{2 \times 2}{1328} = 0.003\ \mathbf{s} = 3\ \mathbf{ms}$$

Note: Solenoid operated valves usually requires much more than 3 milliseconds to close. So, for this case study, maximum length of the pipe can be used before thinking of controlling the valve closure time is 12 meter if a valve is closing let say in 18 ms.

7.2- Noise Control in Hydraulic Transmission Lines

7.2.1- Introduction

As known, noise in a hydraulic control system is generated from various sources in the system such as the pump, the prime mover, etc. *Noise Control* in a hydraulic system is a broad subject beyond the scope of this chapter. In this chapter the focus is on the noise control techniques that are applicable on transmission lines.

Transmission lines generate or radiate noise as generated by other sources as follows:
- **Structural-Born Noise:** The is the noise generated by the body of the transmission line, especially pipes, due to the material of the line and the way it is clamped. Hoses are the least transmission lines in generating noise because of the elasticity of the walls. However, all lines must be clamped on distances to suppress the noise.
- **Operational Conditions:** Sounds generated due to turbulent flow or cavitation. Transmission lines must be initially designed to secure laminar flow and not to contribute to developing pump cavitation.
- **Pressure Pulsation from Positive Displacement Pumps:** Transmission lines vibrates and radiate noise as they subject to positive displacement pumps pressure ripples.

In any case, as transmission lines transmit the fluid power, they also transmit the noise to the rest of the system's components. That explains why the reported pump noise is less than the actual noise of a power unit. One technique to minimize the noise of hydraulic power units, in either industrial or mobile applications, is to use and online device called *Shock Suppressor*. Figure 7.3 shows that shock suppressors are available in different sizes. A shock suppressor act like a muffler in automotive engineering to absorb sounds and vibration. It is no longer necessary to build sound enclosures around power units to isolate noise and meet customer noise specifications. Since the noise out of pumps is proportional to pump speed, using shock suppressor help running pumps a higher speed getting more power with the noise specifications are met. Additionally, using shock suppressors reduces pump wear and possibility of line leakage.

Fig. 7.3 – Shock Suppressors (Courtesy of Parker)

7.2.2- Construction of Shock Suppressors

Figure 7.4 shows a longitudinal sectional view in Parker's Pulse-Tone shock suppressor. It consists of the following three different noise baffles or diffusers separated by a 0.25-in. gap:

- Inner radial chamber with a series of 0.5-in. diameter holes.
- A compressed coil spring surrounds the inner chamber.
- An outer radial chamber dotted with 0.03in. diameter holes. This 0.03-in. diameter holes maximize flow but keep the bladder from extruding through the outer chamber.

An elastomeric bladder surrounds the outer chamber. A compressed Nitrogen is charged around the bladder with 50% to 60% of the hydraulic operating pressure.

Shock suppressors are available in sizes 3/8" – 2" for pipe connections and in sizes ¾" – 3" for split flanges connections.

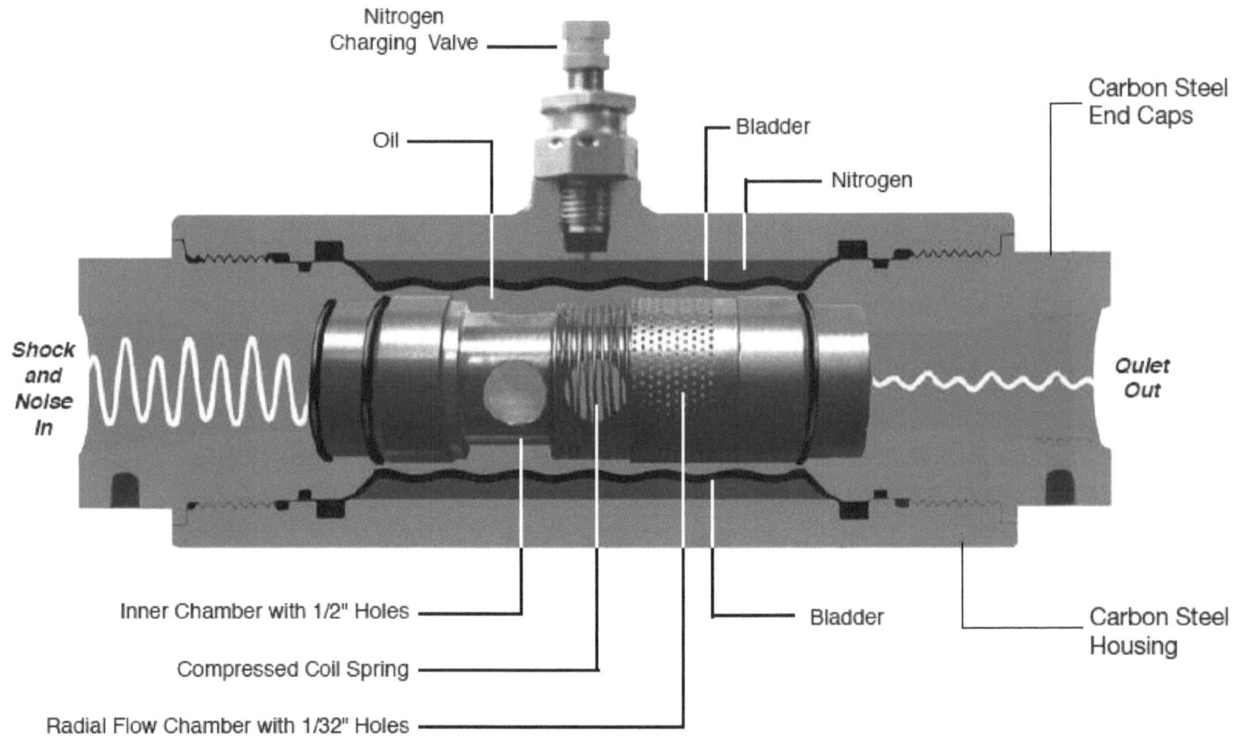

Fig. 7.4 –Suppressor Longitudinal Sectional View (Courtesy of Parker)

7.2.3- Operation of Shock Suppressors

As shown in Fig. 7.5, in operation, oil flows subsequently through the three different noise baffles. Pulsations pass through the holes, strike and deflect the nitrogen-charged bladder. This deflection of the bladder reduces shock and noise. Field tests have shown that the Suppressor will effectively reduce noise by up to 60% and high frequency pulsations over 600 Hertz.

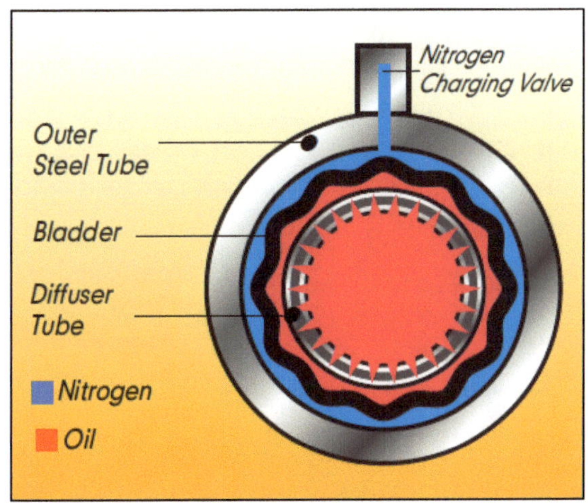

Fig. 7.5 –Suppressor Cross Sectional View (Courtesy of Parker)

Figure 7.6 shows that using stainless steel make the shock suppressor is usable in water transmission lines.

Fig. 7.6 – Stainless Shock suppressor Cross Sectional View (Courtesy of Parker)

Figure 7.7 shows how pressure ripples out of positive displacement pumps are suppressed by using shock suppressors. Results show that shock suppression is more effective with pumps that run at higher speed.

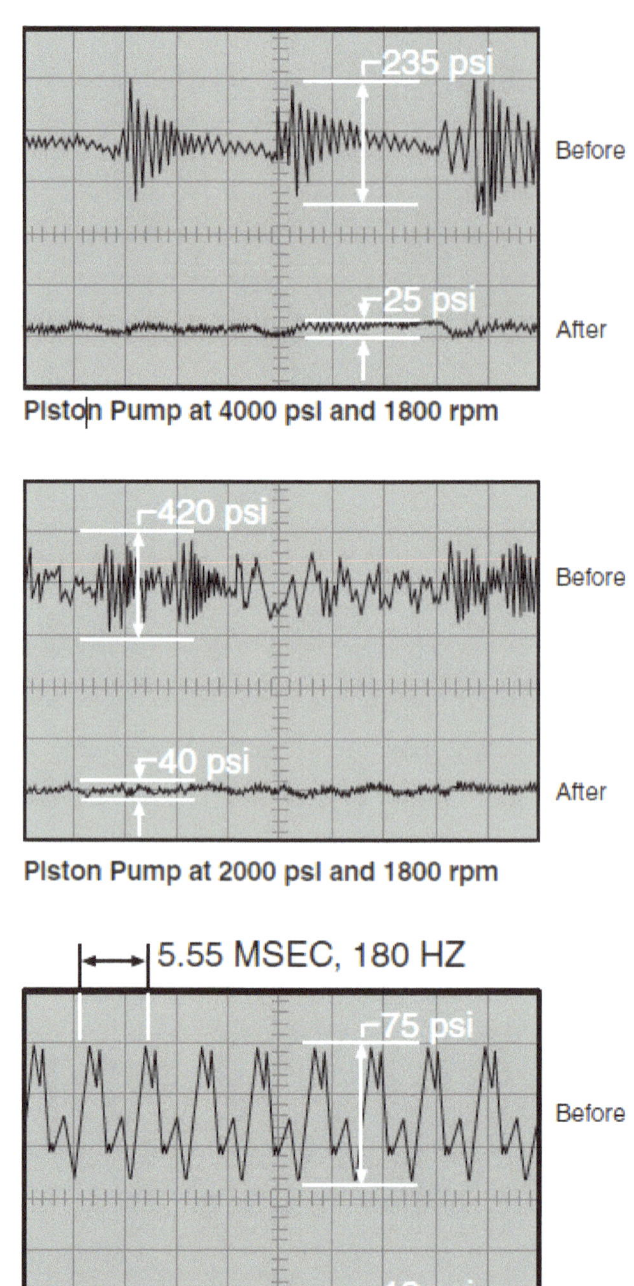

**Fig. 7.7 – Examples of Minimizing Piston Pump Noise by using a Shock Suppressors
(Courtesy of Parker)**

APPENDIXES

APPENDIX A: LIST OF FIGURES

Chapter 1- Basic Types of Hydraulic Transmission Lines

Fig. 1.1 - Types of Hydraulic Transmission Lines
Fig. 1.2 - Example 1 of using Charts for Sizing a Transmission Lines
Fig. 1.3 - Example 2 of using Charts for Sizing a Transmission Lines (Courtesy of Gates)
Fig. 1.4 - Using Software for Sizing a Transmission Lines (www.compudraulic.com)
Fig. 1.5 - Use of Hydraulic Pipes in Industrial (Left) and Mobile (right) Applications
Fig. 1.6 - Hydraulic Pipes According to ANSI Standard
Fig. 1.7 - Pipe Connection by National Thread
Fig. 1.8 - Standard Spiral Thread versus Dryseal Thread
Fig. 1.9- Hydraulic Pipe Fittings
Fig. 1.10 - Use of Hydraulic Tubes in Industrial (Left) and Mobile (right) Applications
Fig. 1.11 - Material of Hydraulic Tubes
Fig. 1.12 - Manual Mandrills of Different Sizes for Tube Bending
Fig. 1.13 - 37O Single and Double Flaring According to SAE J533B
Fig. 1.14 - 45O Single and Double Flaring According to SAE J533B
Fig. 1.15 - Fitting Body Configurations (Courtesy of Bosch Rexroth)
Fig. 1.16 - Fitting Seal Configurations (Courtesy of Bosch)
Fig. 1.17 – Spiral Clearance on NPT Fittings (Courtesy of Brennan Industries)
Fig. 1.18 - National Pipe Thread (Courtesy of Gates)
Fig. 1.19 – Crossectional Views of NPTF and NPSM Fittings (Courtesy of Brennan Industries)
Fig. 1.20 - SAE Inverted Flare Thread (Courtesy of Gates)
Fig. 1.21 - British Standard Pipe Parallel Thread (Courtesy of Gates)
Fig. 1.22 - British Standard Pipe Tapered Thread (Courtesy of Gates)
Fig. 1.23 - Sectional Views for BSPT and PSPP Fittings (Courtesy of Brennan Industries)
Fig. 1.24 - GAZ 24° Metric French Thread (Courtesy of Gates)
Fig. 1.25 - DIN 24° Cone Metric German Thread (Courtesy of Gates)
Fig. 1.26 - EO-2 Fitting (Courtesy of Parker)
Fig. 1.27 - DIN 60° Cone Metric German Thread (Courtesy of Gates)
Fig. 1.28 - DIN 3852 Thread (Courtesy of Gates)
Fig. 1.29 - Japanese 30O Flare Thread (Courtesy of Gates)
Fig. 1.30 - Japanese Tapered Pipe Thread (Courtesy of Gates)
Fig. 1.31 - SAE JI926 Straight Thread O-Ring Boss (ORB)
Fig. 1.32 - Typical O-Ring Boss Fittings of Different Configurations (www.redl.com)
Fig. 1.33 – Sectional View of an ORB Fitting (Curtesy of Brennan Industries)
Fig. 1.34 - O-Ring Face Seal SAE J1453 (Courtesy of Gates)
Fig. 1.35 - Typical O-Ring Face Seal Fittings of Different Configurations (Courtesy of Parker)
Fig. 1.36 - O-Lok Fitting (Courtesy of Parker)
Fig. 1.37 - North American or Metric Flareless Assembly (Courtesy of Gates)
Fig. 1.38 - Typical North American or Metric Flareless Assembly (Courtesy of Parker)

Fig. 1.39 - How Cutting Rings Work (www.stauffusa.com)
Fig. 1.40 – Characteristics of 37O Flare Fitting (Courtesy of Gates)
Fig. 1.41 – Typical 37O Flared Fitting Configurations (Courtesy of Parker)
Fig. 1.42 – Sectional View of a 37O Flare Fitting Configurations
(Courtesy of Brennan Industries)
Fig. 1.43 - Triple-Lok Fitting (Courtesy of Parker)
Fig. 1.44 - SAE 45O Flare (Courtesy of Gates)
Fig. 1.45 - Typical JIC 45O Flare (Courtesy of Parker)
Fig. 1.46 - Japanese 30O Flare Thread (Courtesy of Gates)
Fig. 1.47 - Adaptors to Connect Fittings from Foreign Standards to SAE (Courtesy of Gates)
Fig. 1.48 - Typical Adapters (Courtesy of Parker)
Fig. 1.49 – Identifying Fittings by Measurements (Courtesy of Brennan Industries)
Fig. 1.50 – Identifying Fittings by Thread ID Kit (Courtesy of Brennan Industries)
Fig. 1.51 - Construction of Conventional Swivel Joints (Courtesy of Assufluid)
Fig. 1.52 - Typical Conventional Swivel Joints (Courtesy of Parker)
Fig. 1.53 - WEO Plug-In Swivel Joints (Courtesy of CEJN)
Fig. 1.54 - Construction of WEO Plug-In Swivel Joints (Courtesy of CEJN)
Fig. 1.55 - Operation of WEO Plug-In Swivel Joints (Courtesy of CEJN)
Fig. 1.56 - Examples of using Hydraulic Hoses in Mobile and Industrial Applications
Fig. 1.57 - Basic Construction of Hydraulic Hoes (Courtesy of Gates)
Fig. 1.58 - Spiraled versus Braded Reinforcement Layers in Hydraulic Hoses
Fig. 1.59 - Nominal Size of Hydraulic Hoses (Courtesy of Gates)
Fig. 1.60 - Minimum Straight Length for Bent Hoses
Fig. 1.61 - Straight Hose Assembly Length
Fig. 1.62 - Bent Hose Assembly Length
Fig. 1.63 - Hose Overall Assembly Length according to DIN 20066
Fig. 1.64 - Identification of Hose Couplings (Courtesy of Gates)
Fig. 1.65 - Permanent versus Reusable Hose Couplings (Courtesy of Gates)
Fig. 1.66 - Reusable Hose Couplings (Courtesy of Assufluid)
Fig. 1.67 - Permanent Hose Couplings (Courtesy of Assufluid)
Fig. 1.68 - Permanent Hose Couplings (Courtesy of Gates)
Fig. 1.69 - Hose Coupling Common Configurations (Courtesy of Gates)
Fig. 1.70 - Hose Coupling Configurations for Medium and High Pressure (Courtesy of Gates)
Fig. 1.71 - Typical Design of Preassembled Cimpable Hose Coupling
(Courtesy of Parker)
Fig. 1.72 - Hose Crimping Machine (Courtesy of Gates)
Fig. 1.73 - Prepare a Hose for Crimping
Fig. 1.74 – Measuring Crimp Diameter (Courtesy of Gates)
Fig. 1.75 - Flat Face Quick Connect Coupling (Courtesy of Assufluid)
Fig. 1.76 - Quick Connect Coupling with Single Non-Return Valve (Courtesy of Assufluid)
Fig. 1.77 - Quick Connect Coupling with Double Non-Return Valves (Courtesy of Assufluid)
Fig. 1.78 - Male and Female Parts of iLokTM Quick Connect Coupling (Courtesy of Gates)
Fig. 1.79 - Locking Mechanism of iLokTM Quick Connect Coupling (Courtesy of Gates)
Fig. 1.80 - Locking Mechanism of TLX Quick Connect Coupling (Courtesy of CEJN)

Fig. 1.81 - Universal Push-to-Connect (UPTC) (Courtesy of Parker)
Fig. 1.82 - Assembling of Universal Push-to-Connect (UPTC) (Courtesy of Parker)
Fig. 1.83- Sealing of Universal Push-to-Connect (UPTC) (Courtesy of Parker)
Fig. 1.84 - Disassembling of Universal Push-to-Connect (UPTC) (Courtesy of Parker)
Fig. 1.85 - Push-Lok Hose and Fitting Assembly (Courtesy of Parker)
Fig. 1.86 - Assembling and Disassembling Push-Lok Hose and Fitting Assembly (Courtesy of Parker)
Fig. 1.87 - Innovative Quick Connect Coupling (Courtesy of ArgoHyots)
Fig. 1.88 - Multi-X Quick Connect Coupling (Courtesy of CEJN)
Fig. 1.89 - Conceptual Operation of Multi-X Quick Connect Coupling (Courtesy of CEJN)
Fig. 1.90 - Metallic Spring and Armor Guards (Courtesy of Parker)
Fig. 1.91- Nonmetallic Spiral Wrap Guards (sapphirehydraulics.com)
Fig. 1.92- Hose Shields (www.epha.com)
Fig. 1.93- Hose Protection Sleeves (www.sealsaver.com)
Fig. 1.94- Hose Oil-Injection Protection Sleeve "Lifeguard" (Courtesy of Gates)
Fig. 1.95- Hose Wipe Restraint (Courtesy of Parker)
Fig. 1.96 - Real Time Hoses Condition Monitoring (www.hydrotechnik.com)
Fig. 1.97- Real Time Hoses Condition Monitoring Maximizes Hose Life (www.hydrotechnik.com)
Fig. 1.98 - Use of Adaptors (Courtesy of Gates)
Fig. 1.99 - Keyword for Selecting Hydraulic Hoses (Courtesy of Parker)
Fig. 1.100 - Hydraulic Hoses Pictogram (Courtesy of Parker)
Fig. 1.101 - Example of Hydraulic Hoses with Pictogram Marked (Courtesy of Parker)
Fig. 1.102 - Examples of Hoses for High Pressure (Courtesy of Parker)
Fig. 1.103 - Examples of Hoses for High Collapse External Pressure (Courtesy of Parker)
Fig. 1.104 - Example of Hoses for Cold Temperature (Courtesy of Parker)
Fig. 1.105 - Examples of Hoses for High Temperature (Courtesy of Parker)
Fig. 1.106 - Examples of Nonconductive Hydraulic Hoses (Courtesy of Parker)
Fig. 1.107 - Hydraulic Hoses for Extreme Conditions (Courtesy of Parker)
Fig. 1.108 - Use of fittings in Hoses (mac-hyd.com)
Fig. 1.109 - Standard Flange Dimensions
Fig. 1.110- Components of Flange Assembly
Fig. 1.111 - Hydraulic Pipe Flange Assembly
Fig. 1.112 - Hydraulic Tube and Hose Flange Assembly
Fig. 1.113 - Square Flanges (https://flanges-pipe.com)
Fig. 1.114 - Rubber Expansion Fittings (www.grainger.com)
Fig. 1.115 - Series 1620 Test Points (www.hydrotechnik.com)
Fig. 1.116 - Innovative Test Points (Courtesy of CEJN)
Fig. 1.117 - Construction of Measurement Hoses (www.hydrotechnik.com)
Fig. 1.118 - Measurement Hoses Supported by Anti-Buckling and Protection Covers (www.hydrotechnik.com)
Fig. 1.119 - Measurement Hoses Combined with Test Points (www.hydrotechnik.com)
Fig. 1.120 - Pressure Drop in Measurement Hoses (www.hydrotechnik.com)
Fig. 1.121 - Best Practices of Routing Measurement Hoses (www.hydrotechnik.com)
Fig. 1.122 – Hydraulic Manifolds (www.daman.com)

Chapter 2- Contamination Control in Hydraulic Transmission Lines
Fig. 2.1- Transmission Line Chop Saw
Fig. 2.2- Transmission Lines Projectile Cleaning (Courtesy of Ultra Clean Technologies)
Fig. 2.3- Compressed Air Requirements (Courtesy of Ultra Clean Technologies)
Fig. 2.4- Launchers (Courtesy of Gates)
Fig. 2.5- Projectiles (Courtesy of Ultra Clean Technologies)
Fig. 2.6- Cap Seals (Courtesy of Ultra Clean Technologies)
Fig. 2.7- Complete Kit (Courtesy of Ultra Clean Technologies)
Fig. 2.8- Hydraulic Hose Cleanliness (Courtesy of Ultra Clean Technologies)
Fig. 2.9- Hydraulic Hose Cutting Causes Contamination (Courtesy of Ultra Clean Technologies)
Fig. 2.10- Hydraulic Hose Cleaning Procedure (Courtesy of Ultra Clean Technologies)
Fig. 2.11- Hydraulic Hose Cleaning after Attaching Hose Ends (C. of Ultra Clean Technologies)
Fig. 2.12- Tube Bending and Flaring (Courtesy of Ultra Clean Technologies)
Fig. 2.13- Tube Projectile Cleaning (Courtesy of Ultra Clean Technologies)
Fig. 2.14- Pre-Stacked Clean Seal Capsules (Courtesy of Ultra Clean Technologies)
Fig. 2.15- Heat Source for Shrinking the Clean Seal Capsules (C. of Ultra Clean Technologies)
Fig. 2.16- Process of Applying Clean Seal Capsules (Courtesy of Ultra Clean Technologies)
Fig. 2.17- Process of Removing Clean Seal Capsules (Courtesy of Ultra Clean Technologies)
Fig. 2.18- Clean Seal Flanges (Courtesy of Ultra Clean Technologies)
Fig. 2.19- Examples of Hydraulic System Flushing Power Units
Fig. 2.20- Examples of Flushing Industrial Machine Function/Circuit

Chapter 3: Safety and Maintenance of Transmission Lines
Fig. 3.1- Bet Practices for Hydraulic Transmission Line Leakage Inspection (www.bondfluidaire.com)
Fig. 3.2- Keep Outer Surfaces including End Joints (mac-hyd.com)
Fig. 3.3- Hose Crimping Process (Courtesy from Parker)
Fig. 3.4- Hose Crimping Machines (Courtesy from Gates)
Fig. 3.5- Best Practices for Hydraulic Hose Routing
Fig. 3.6- Steps to Install Hose Assembly (Courtesy of Gates)
Fig. 3.7- Transmission Line Signs of Leak (Courtesy of American Technical Publisher)
Fig. 3.8- Determining the Thread Type of Hydraulic Fittings
Fig. 3.9- Best Practices for Assembling Flareless Fitting (Courtesy from Parker)
Fig. 3.10- Best Practices for Tube Flaring (Courtesy from Parker)
Fig. 3.11- Best Practices for Assembling 37O Flared Fitting (Courtesy from Parker)
Fig. 3.12- Flare Tight Seal (www.Flaretite.com)
Fig. 3.13- Perfect and Incorrect Tube Bends (www.aircraftsystemstech.com)
Fig. 3.14- Pipe Assembly (Courtesy from American Technical Publishers)
Fig. 3.15- Proper Hose Clamping
Fig. 3.16- Restrains for Hoses Clamping
Fig. 3.17- Clamping Collars for Tubes and Pipes (Courtesy of Assofluid)
Fig. 3.18- Guidelines for Tube Clamping Distances
Fig. 3.19- Guidelines to Avoid Mechanical Stresses (Courtesy from Parker Hannifin)
Fig. 3.20- Hose Burst Test

Fig. 3.21- Salt Spray Test (Courtesy from Gates)
Fig. 3.22- Hose Cleanliness Measurement (Courtesy from Gates)
Fig. 3.23- Hose Reliability Assessment Test
Fig. 3.24- Best Practices for Storage of Transmission Lines
Fig. 3.25 – Risks of Oil Injection (Courtesy of the International Hydraulic Safety Authority)
Fig. 3.26 – Safety Focus Card (Courtesy of the International Fluid Power Society)
Fig. 3.27 – Oil Injection Training Video (Courtesy of Fluid Power Training Institute)

Chapter 4: Troubleshooting and Failure Analysis

Fig. 4.1– Effective Method for Transmission Line leakage Inspection (Courtesy of Spectroline)
Fig. 4.2– Hose Leakage due to Line Not Cleaned Before Assembly (Courtesy of Parker)
Fig. 4.3– Hose Cover Abrasion (Courtesy of Gates)
Fig. 4.4– Hose Ends Detached due to Short Length (Courtesy of Gates)
Fig. 4.5– Hose Cover Cracked due to Poor Routing (Courtesy of Gates)
Fig. 4.6– Hose Twisting due to Improper Assembly (Courtesy of Gates)
Fig. 4.7– Hose Burned due to Direct Contact with Heat Sources (Courtesy of Gates)
Fig. 4.8– Hose Burst due to Exceeding Minimum Bend Radius (Courtesy of Gates)
Fig. 4.9– Hose Body Burst due to Exceeding lifetime
Fig. 4.10– Hose Failure due to Fluid Incompatibility (Courtesy of Parker)
Fig. 4.11– Hose Failure due to Short Length (Courtesy of Parker)
Fig. 4.12– Hose Failure due to Heat Aging (Courtesy of Parker)
Fig. 4.13– Hose Cracked due to Exposure to Paint Spray
Fig. 4.14– Tube and Pipe Burst due to Overpressure (Courtesy of Parker)
Fig. 4.15– Pipe and Tube Pin Holes due to Poor Material
Fig. 4.16– Pipe and Tube is Leaking due to Mechanical Stresses

Chapter 5: Energy Losses in Transmission Lines

Fig. 5.1- Flow Patterns
Fig. 5.2 - Laminar versus Turbulent Flow
Fig. 5.3- Reynold's Number
Fig. 5.4 - Fluid Line Design
Fig. 5.5- Recommended Fluid Speed inside a Conductor
Fig. 5.6- Examples where Fluid Compressibility are Considered in System Design
Fig. 5.7- Mass-Spring System
Fig. 5.8- Effect of Oil Compressibility
Fig. 5.9- Effect of Oil Compressibility on Pressure Change
Fig. 5.10- Continuity Equation
Fig. 5.11- Flow Rate and Flow Speed
Fig. 5.12- Ideal Fluid Flow
Fig. 5.13- Pressure Head
Fig. 5.14- Kinetic Head
Fig. 5.15- Practical Observation of Continuity Equation and Bernoulli's Equation
Fig. 5.16- Real Fluid Flow
Fig. 5.17- Line Losses

Fig. 5.18- Moody Diagram
Fig. 5.19- Frictional Pressure Losses Calculation in Metric System of Units
Fig. 5.20- Frictional Pressure Losses Calculation in English System of Units
Fig. 5.21- Local Losses in Fittings
Table 5.1 – Losses factor in Fittings
Fig. 5.22- Local Losses in Elbow, Empirical Data
Fig. 5.23- Local Losses in Orifices

Chapter 6: Modeling of Hydraulic Transmission Lines

Fig. 6.1- Modeling Pressure Drop in Hydraulic Transmission Lines
Fig. 6.2- Structure of a Generic Model for a Hydraulic Transmission Line
Fig. 6.3- Moody Diagram
Fig. 6.4- Modeling Pressure Losses due to Line Resistivity
Fig. 6.5- Modeling Pressure Losses due to Line Inductance
Fig. 6.6- Modeling Outlet Flow due to Line Capacitance
Fig. 6.7- Simulation Model for Hydraulic Transmission Line
Fig. 6.8- Model Structure under the Mask of Transmission Line Model
Fig. 6.9- Results of Case Study 1
Fig. 6.10- Results of Case Study 2
Fig. 6.11- Results of Case Study 3
Fig. 6.12- Structure of a Generic Model for a Hydraulic Fitting
Fig. 6.13- Losses Factor in Fittings
Fig. 6.14- Losses Factor in Elbows
Fig. 6.15- Simulation Model for Hydraulic Fitting
Fig. 6.16- Validation of Hydraulic Fitting Model
Fig. 6.17- Structure of a Generic Model for a Hydraulic Orifice
Fig. 6.18- Simulation Model for Hydraulic Orifice
Fig. 6.19- Validation of Hydraulic Fitting Model
Fig. 6.20- Example of Modeling a Transmission Line Assembly

Chapter 7: Water Hammer and Noise Control

Fig. 7.1 – How Water Hammer is Created in Transmission Lines
Fig. 7.2 – How Water Hammer is Created in a Transmission Lines
Fig. 7.3 – Shock Suppressors (Courtesy of Parker)
Fig. 7.4 –Suppressor Longitudinal Sectional View (Courtesy of Parker)
Fig. 7.5 –Suppressor Cross Sectional View (Courtesy of Parker)
Fig. 7.6 – Stainless Shock suppressor Cross Sectional View (Courtesy of Parker)
Fig. 7.7 – Examples of Minimizing Piston Pump Noise by using a Shock Suppressors
 (Courtesy of Parker)

APPENDIX B: LIST OF TABLES

Table 1.1 - Using Tables for Sizing a Transmission Lines
Table 1.2- ANSI Standard Schedules for Hydraulic Pipes
Table 1.3 - Examples of Sizing Hydraulic Pipes
Table 1.4 - ANSI B 31.3 Standard for Hydraulic Pipes Pressure Rating
Table 1.5 – Maximum Allowable Working Pressure (psi) for Carbon Steel Tubes
Table 1.6 - Minimum Bend Radius for Hydraulic Tubes of Different Sizes
Table 1.7 - 37O Single and Double Flaring Standard Dimensions According to SAE J533B
Table 1.8 - 45O Single and Double Flaring Standard Dimensions According to SAE J533B
Table 1.9- Size Range of Hydraulic Hoses (Courtesy of Gates)
Table 1.10 – Hoses Specifications in accordance with SAE Standard
Table 1.11 - SAE Standard for Hydraulic Hoses (Hydraulic & Pneumatic Magazine)
Table 1.12 - ISO 18752 Standard Pressure Classes for Hydraulic Hoses (Courtesy of Parker)
Table 1.13 - ISO 18752 Standard Pressure Classes for Hydraulic Hoses (Courtesy of Parker)
Table 1.14 - Example of Minimum Bend Radius Reported by the Hose Manufacturer (Courtesy of Parker)
Table 1.15 - Hose-Fluid Compatibility Chart (Courtesy of Gates)
Table 3.1- BP-Transmission Lines-02-Maintenance Scheduling
Table 3.2- Recommended Tightening Torque for 37° & 45° (Machined or Flared Fittings) (www.new-line.com)
Table 3.3- Recommended Tightening Torque for Flat-Face O-Ring Seal (Steel) (www.new-line.com)
Table 3.4- Recommended Tightening Torque for SAE Steel O-Ring Boss (www.new-line.com)
Table 4.1 – Hydraulic Transmission Lines Inspection Sheet
Table 4.2– Transmission Lines Troubleshooting Chart
Table 5.1 – Losses factor in Fittings

APPENDIX C: LIST OF REFERENCES

Hydraulic Systems Volume 1- Introduction to Hydraulics for Industry Professionals
Author: Dr. Medhat Kamel Bahr Khalil, 2016.

Hydraulic Systems Volume 2- Electro-Hydraulic Components and Systems
Author: Dr. Medhat Kamel Bahr Khalil, 2016.
Publisher: Compudraulic, USA. ISBN: 978-0-9977634-2-3

Hydraulic Systems Volume 3- Hydraulic Fluids and Contamination Control
Author: Dr. Medhat Kamel Bahr Khalil, 2016.
Publisher: Compudraulic, USA.
ISBN: 978-0-9977816-3-2

Hydraulic Systems Volume 4- Hydraulic Fluids Conditioning
Author: Dr. Medhat Kamel Bahr Khalil, 2022.
Publisher: Compudraulic, USA.
ISBN: 978-0-9977634-8-5

Hydraulic Systems Volume 5- Safety and Maintenance
Author: Dr. Medhat Kamel Bahr Khalil, 2022.
Publisher: Compudraulic, USA.
ISBN: 978-0-9977816-5-6

Hydraulic Systems Volume 6- Troubleshooting and Failure Analysis
Author: Dr. Medhat Kamel Bahr Khalil, 2022.
Publisher: Compudraulic, USA.
ISBN: 978-0-9977634-6-1

Hydraulic Systems Volume 7- Modeling and Simulation for Application Engineers
Author: Dr. Medhat Kamel Bahr Khalil, 2016.
Publisher: Compudraulic, USA.
ISBN: 978-0-9977816-3-2

R01- Basic Electronics for Hydraulic Motion Control
Author: Jack L. Johnson, PE 1992.
Publisher: Penton Publishing Inc. 1100 Superior Avenue. Cleveland, OH 44114.
ISBN No. 0-932905-07-2.

R02- Closed Loop Electro-hydraulics Systems Manual
Author: Vickers/Eaton.
Publisher: Vickers Inc. 1992.
Training Center, 2730 Research Drive, Rochester Hills, MI 48309-3570.
ISBN 0-9634162-1-9

R03- Bosch Automation Technology
Author: Werner Gotz, Steffen Haack, Ralph Mertlick.
Publisher: Bosch.
ISBN 3-933698-05-7.

R04- Electrohydraulic Proportional and Control Systems
Publisher: Bosch Automation 1999.
ISBN 0-7680-0538-8.

R05- Proportional and Servo Valve Technology – The Hydraulic Trainer Volume 2
Author: R. Edwards, J. Hunter, D. Kretz, F. Liedhegener, W. Schenkel, A. Schmitt.
Publisher: Mannesman Rexroth AG 1988. D-8770 Lohr a. Main.
ISBN   3-8023-0266-4.

R06- Proportional Hydraulics
Author: D. Scholz.
Publisher: Festo Didactic KG, Esslingen, Germany.

R07- Electricity, Fluid Power, and Mechanical Systems for Industrial Maintenance
Author: Thomas Kissell.
Publisher: Prentice Hall, Inc. 1999, Upper Saddle River, NJ 07458.
ISBN 0-13-896473-4.

R08- Fluid Power in Plant and Field – First Edition
Author: Charles S. Hedges, R.C. Womack.
Publisher: Womack Machine Supply Co. 1968.
Womack Educational Publication, 2010 Shea Road, Dallas, TX 75235.
ISBN   68-22573 (Library of Congress Card Catalog No.).

R09- Hydraulics, Fundamentals of Service
Author: Deere and Company.
Publisher: John Deere Publishing 1999.
Almon TIAC Bldg. Suite 104, 1300-19th Street, East Moline, IL 61244.
ISBN 0-86691-265-7.

R10- Industrial Hydraulics Troubleshooting
Author: James E. Anders, Sr.
Publisher: McGraw-Hill, Inc.
ISBN 0-07-001592-9.

R11- Power Hydraulics
Author: John Ashby.
Publisher: Prentice Hall 1989. Prentice Hall International, (UK) Ltd.
66 Wood Lane End, Hemel Hempstead, Hertfordshire, HP2 4RG.
ISBN 0-13-687443-6.

R12- Fluid Power with Application
Author: Anthony Esposito.
Publisher: Prentice Hall.
ISBN 0-13-060899-8.

R13- Hydraulic Component Design and Selection
Author: E.C. Fitch.
Publisher: BarDyne Inc. 5111 North Perkins Rd. Stillwater, OK 74075.
ISBN 0-9705922-3-X.

R14- Planning and Design of Hydraulic Power Systems – The Hydraulic Trainer, Vol. 3
Author: Mannesmann Rexroth GmbH.
Publisher: Mannesman Rexroth AG 1988.
D-97813 Lhr a. Main, Jahnsrtrabe 3-5 D-97816 Lohr a. Main.
ISBN   3-8023-0266-4.

R15- Logic Element Technology: Hydraulic Trainer, Volume 4
Author: Mannesmann Rexroth GmbH.
Publisher: Mannesmann Rexroth GmbH 1989.
.Postfach 340, D 8770 Lohr am Main, Telefon (09352) 180.
ISBN   3-8023-0291-5.

R16- Hydrostatic Drives with Control of the Secondary Unit. The Hydraulic Trainer, Volume 6
Author: Dr. Alfred Feuser, Rolf Kordak, Gerold Liebler.
Publisher: Mannesmann Rexroth GmbH 1989.
Postfach 340, D 8770 Lohr am Main.

R17- Control Strategies for Dynamic Systems: Design and Implementation
Author: John H. Lumkes, Jr.
Publisher: Marcel Dekker, Inc. 2002.
Marcel Dekker, Inc. 270 Madison Avenue, New York, NY 10016.
ISBN 0-8247-0661-7.

R18- Feedback Control Of Dynamic Systems
Author: Gene F. Franklin, J. David Powell, Abbas Emami-Naeini.
Publisher: Prentice-Hall, Inc.
Upper Saddle River, New Jersey.
ISBN 0-13-032393-4.

R19- Modeling and Analysis of Dynamic Systems
Author: Charles M. Close, Dean. Frederick
Rensselaer Polytechnic Institute
Publisher: John Wiley & Sons, Inc.
ISBN 0-471-12517-2.

R20- Design of Electrohydraulic Systems For Industrial Motion Control
Author: Jack L. Johnson, PE.
Milwaukee School of Engineering.
Publisher: Parker.
Copyright © Jack L. Johnson, PE 1991.

R21- Basic Pneumatics
Author: Kjell Evensen & Jul Ruud.
Publisher: AB Mecmann Stockholm 1991.
S-125 81 Stockholm, Sweden.
ISBN 91-85800*21-X.

R22- Basic Pneumatics: The Pneumatic Trainer, Volume 1
Author: Ing. –Buro J.P. Hasebrink.
D7761 Moos.
Editor: Mannesmann Rexroth Pneumatik GmbH.
Bartweg 13, W 3000 Hannover 91.

R23- Electro-Pneumatics: The Pneumatic Trainer, Volume 2
Author: Rolf Balla.
Publisher: Mannesmann Rexroth 1990, Pneumatik GmbH.
Publication No: RE 00 262/01.92.

R24- Pneumatics Theory and Applications
Author: Bosch Automation.
Publisher: Robert Bosch GmbH 1998.
Automation Technology Division, Training (AT/VSZ)
ISBN 1-85226-135-8.

R25- Fluid Power Engineering
Author: M. Galal Rabie.
Publisher: McGraw-Hill.
ISBN 978-0-07-162246-2.

R26- Air Motors Ideas with Air
Author: GAST Mfg. Co.
Publisher: GAST Mfg. Co. 1978.
P.O. Box 97, Benton Harbor, MI 49022.
Book No: Booklet #100.

R27- Air Motor Handbook
Author: GAST Mfg. Co.
Publisher: GAST Mfg. Co. 1978.
P.O. Box 117, Benton Harbor, MI 49022.

R28- Troubleshooting Hydraulic Components: Using Leakage Path Analysis Methods
Author: Rory S. McLaren.
Publisher: Rory McLaren Fluid Power Training 1993.
562 East 7200 South, Salt Lake City, UT 84171.
ISBN No. 0-9639619-1-8.

R29- Hydraulics Theory and Application From Bosch
Author: Werner Gotz.
Publisher: Robert Bosch GmbH.
Hydraulics Division K6, Postfach 30 02 40, D-7000 Stuttgart 30.
Federal Republic of Germany, Technical Publications Department, K6/VKD2.

R30- A Complete Guide to ISO and ANSI Fluid Power Symbols
Author: Fluid Power Training Institute.
Publisher: Fluid Power Training Institute 200.
562 East Fort Union Boulevard, Midvale, Utah 84047.

R31- How to Work Safely with Hydraulics
Author: Fluid Power Training Institute.
Publisher: Fluid Power Training Institute 2004.
562 East7200 South, Midvale, Utah 84047.

R32- How to Interpret Fluid Power Symbols
Author: Rory S. McLaren.
Publisher: Fluid Power Training Institute.
Rory S. McLaren 1995.
ISBN 0-9639619-2-6.

R33- Safe Hydraulics
Editor: Gates Rubber Company.
Copyright 1995.
Denver, CO 80217.

R34- Electronically Controlled Proportional Valves. Selection and Application
Author: Michael J. Tonyan.
Publisher: Marcel Dekker, Inc. 1985.
Marcel Dekker, Inc., 270 Madison Avenue, New York, NY 10016.
ISBN 0-8247-7431-0.

R35- Introduction to Closed-Loop Oil Systems
Author: Rory S. McLaren.
Publisher: Rory McLaren Fluid Power Training Institute.
7050 Cherry Tree Lane, P.O. Box 711201, Salt Lake City, UT 84171.

R36- Industrial Hydraulic Technology, Second Edition
Author: Parker Hannifin Corporation.
Publisher: Parker Hannifin Corporation 1997.
6035 Parkland Blvd, Cleveland, OH 44124-4141.
Publication No: Bulletin 0231-B1.

R37- Basic Principle and Components of Fluid Technology – The Hydraulic Trainer, Volume 1
Author: Mannesman Rexroth.
Publisher: Mannesman Rexroth AG 1988.
D-97813 Lhr a. Main, Jahnsrtrabe 3-5 D-97816 Lohr a. Main.
ISBN 3-8023-0266-4.

R38- Safe-T-Bleed Corporation Catalog
Publisher: Safe-T-Bleed Corporation 2001.
Catalog No. STB-PC-1201-1
R39- Industrial Hydraulics Manual – EATON
Publisher: Eaton Fluid Power Training.
ISBN: 0-9788022-0-9.

R40- Vickers-Mobile Hydraulic Manual – Fourth Edition 1998
Author: Vickers.
Publisher: Vickers Inc. 1999.
Training Center, 2730 Research Drive, Rochester Hills, MI 48309-3570.
ISBN No. 0-9634162-5-1.

R41- Industrial Fluid Power Text, Volume 2
Author: Charles S. Hedges, R.C. Womack.
Publisher: Womack Machine Supply Company 1972.
Womack Educational Publications, 2010 Shea Road, Dallas, TX 75235.
ISBN 66-28254 (Library of Congress Card Catalog No.).

R42- Fluid Power Hydraulics and Pneumatics
Author: R. Daines.
Publisher: The Good-heart Willcox Company, Inc.

R43- Hydraulics in Industrial and Mobile Applications
Publisher: ASSOFLUID, Italian Association of Manufacturing and Trading Companies in Fluid
Power Equipment and Components

R44- Fluid Power in Plant and Field – Second Edition
Author: Charles S. Hedges, R.C. Womack.
Publisher: Womack Machine Supply Co. 1968.
Womack Educational Publication, 2010 Shea Road, Dallas, TX 75235.
ISBN 68-22573.

R45- Mobile Hydraulics Manual
Author: Eaton.
Publisher: Eaton Corporation Training.
Eden Prairie, Minnesota.
ISBN 0-9634162-5-1.

R46- EH Control Systems
Author: F.D. Norvelle.

R47- Fluid Power Journal
Publisher: International Fluid Power Society.

R48- Fundamentals of Industrial Controls and Automation
Author: Lonnie L. Smith and Mike J. Rowlett.
Publisher: Womack Educational Publications.
Dallas, Texas.
ISBN: 0-943719-04-6.

R49- Lightning Reference Handbook
Publisher: Berendsen Fluid Power.

R50- Pneumatics Basic Level
Author: P. Croser, F. Ebel.
Publisher: Festo Didactic GmbH & Co.
R51- Electro-pneumatics Basic Level
Author: F. Ebel, G. Prede, D. Scholz.
Publisher: Festo Didactic GmbH & Co.

R52- Mechanical System Components
Author: James F. Thorpe.
Publisher: Allyn and Bacon.
Needham Heights, Massachusetts.
ISBN: 0-205-11713-9.

R53- Electrical Motor Controls for Integrated Systems, Third Edition
Author: Gary J. Rockis, Glen A. Mazur.
Publisher: American Technical Publishers, Inc.
ISBN: 0-8269-1207-9.

R54- Instrumentation, Fourth Edition
Author Franklyn W. Kirk, Thomas A. Weedon, Philip Kirk.
Publisher American Technical Publishers, Inc.
ISBN: 0-8269-3423-4.

R55- Introduction to Mechatronics and Measurement Systems, Second Edition
Author David G. Alciatore, Michael B. Histand.
Publisher McGraw-Hill, Inc.
ISBN: 0-07-240241-5.

R56- Study Guides for IFPS Certification

R57- Work Books from Coastal Training Technologies

R58- Industrial Hydraulic Manual – Fourth Edition 1999
Author: Vickers.
Publisher: Vickers Inc. 1999.
 Training center, 2730 Research Drive, Rochester hills, Michigan 48309-3570.
ISBN 0-9634162-0-0.

R59- Industrial Automation and Process Control
Author: John Stenerson.
Publisher: Prentice Hall.
ISBN 0-13-033030-2.

R60- Industrial Automated Systems
Author: Terry Bartelt.
Publisher: Delmar Cengage Learning.
ISBN: 10-1-4354-888-1.
R61- Introduction to Fluid Power
Author: James L. Johnson.
Publisher: Delmar Cengage Learning.
ISBN: 10-0-7668-2365-2.

R62- Summary for Engineers
Author: Dr. Abdel Nasser Zayed.
Publisher: Dr. Abdel Nasser Zayed .
ISBN: 977-03-0647-9.

R63- Mechanics of Materials
Author: Ferdinand P.Beer, E. Russell Johnston Jr., John T DeWolf.
Publisher: McGraw Hill Publishing .
ISBN: 0-07-365935-5.

R64- Oil Hydraulic System, Principles and Maintenance
Author: S. R. Majumdar.
Publisher: McGraw Hill.
ISBN 10: -0-07-140669-7.

R65- Contamination Control in Hydraulic and Lubricating Systems
Publisher: Pall

R66- Diagnosing Hydraulic Pump Failure
Publisher: Caterpillar.

R67- Oil Service Products Catalog
Publisher: Schroder Industries.

R68- Industrial Fluid Power Volume 1
Author: Charles S. Hedges.
Publisher: Womack Educational Publication.
ISBN: 0-9605644-5-4.

R69- Industrial Fluid Power Volume 2
Author: Charles S. Hedges.
Publisher: Womack Educational Publication.
ISBN: 0-943719-01-1.

R70- Industrial Fluid Power Volume 3
Author: Charles S. Hedges.
Publisher: Womack Educational Publication.
ISBN: 0-943719-00-3.

R71- Electrical Control of Fluid Power
Author: Charles S. Hedges.
Publisher: Womack Educational Publication.
ISBN 0-9605644-9-7.

R72- Hydraulic Cartridge Valve Technology
Author: John J. Pippenger, P.E.
Publisher: Amalgam Publishing Company.
Post Office Box 617, Jenks, OK 74037 USA.
ISBN: 0-929276-01-9.

R73- Noise Control of Hydraulic Machinery
Author: Stan Skaistis.
Publisher: Marcel Dekker, 270 Madison Avenue, New York, NY 10016.
ISBN: 0-8247-7934-7.

R74-Solenoid Valves
Author: Hydraforce

R75-HF Proportional Valve Manual
Author: Hydraforce

R76-Automatic Control for Mechanical Engineers
Author: M. Galal Rabie, Professor of Mechanical Engineering
ISBN: 977-17-9869-3,2010.

R77-Fluid Power System Dynamics
Author: W. Durfee, Z. Sun

Index

"

"Head", 183

A

Abrasive projectiles, 118
Adaptors, 49
Angle Gauge, 50
Assembly Overall Length, 65

B

Bar Manifolds, 111
Barlow's Formula, 24
Barlow's formula, 16
Boardman, 24
British, 32

Bulk Modulus, 219
Bulk Modulus., 178, 220
Burst Pressure, 16
Burst Test, 153

C

Caliper, 50
capacitance, 198
Carbon Steel, 17, 24
Celerity, 220
Central Tube, 57
Clean Cut, 115
Clean Seal, 122
Clean Seal Capsules, 122
Clean Seal Flange, 125
clearance, 21
Coefficient, 195
Coefficient of Fluid Friction, 199
Cold Flexibility Test, 153
Compact Spiral, 100
Compression Ring, 29
Continuity, 181

Coupling, 75
Cover Plates, 111
Crimp Diameter, 72
crimping equipment, 68
Cut Hose Length, 65
Cut Length, 138
Cutting Ring, 29

D

Density., 220
Depth, 138
Discharge Coefficient, 213
Discharge Factor, 194
Distribution Manifolds, 111
Dryseal, 21, 30

E

End Joints, 134
energy losses, 172
EO-2, 35
Equation, 181
Extremely High-Pressure, 92

F

factor, 24
field Attachments or Screw in
    Connectors, 68
Flared, 29

flared, 147
Flareless, 29
flareless, 145
Flaretite, 148
flaring., 26
Flat Face, 73
flexible hose, 133
flow rate, 181
flow speed, 181
fluorescent, 135
Flushing, 126, 136
Formula, 24
Frictional Losses, 186, 201

G

Global Variables, 198

H

High Pressure Collapse Resistant, 95
High-Pressure, 92
Hose Assembly, 134
Hose Couplings, 67
hose end, 67
Hose Guards, 85
Hose Insertion, 138
Hose Reliability Assessment Test, 155
Hose Whip Restraint, 87
Hose Whip Restraints, 87
Hydraulic Components Sizing
    Calculator, 15
hydraulic diameter, 175
Hydraulic Transmission Lines, 10

I

iLok, 75
Impulse Test, 153
In the O-Ring Face Seal, 39
inductance, 198
injection, 135
Inner Diameter, 10
inner diameters, 17
installation and maintenance, 136

J

Joint Industrial Council, 44
Joints, 52
Jumpers, 130

K

Kidney Wash, 126

L

Lame, 24
Laminar, 172
Launcher, 117
leakage, 135
LifeSense, 88
Local Losses, 186, 209

Low-Pressure, 92

M

Manifolds, 111
Maximum Allowable Working Pressure, 16
Measurements Hose Lines, 108
Medium-Pressure, 92
Mineral Oils, 127
minimum bend radius, 25
Module Plates, 111
Modulus of Elasticity, 178
Moody Diagram, 189, 200
Multi-X Quick Connect Couplings, 83

N

National Fire Protection Association, 39
National Pipe, 30
National Pipe Straight, 30
National Pipe Tapered, 30
Noise Control, 221
Nominal Size, 17
nonconductive, 93
North American Standard, 29
NPT, 21
NPTF, 21

■ Index

O

O-Lok, 40
O-Ring Boss, 38
O-Ring Face Seal, 29
Offline Filtration, 126
outer diameter, 17
Outer Protective Cover, 57
Overall Assembly Length, 138

P

Permanent Couplings, 68
Pickling, 126, 136
pipe, 133
pipes, 17
Polyglycols, 127
Pressure Proof Test, 153
Projectile Cleaning, 116
Projectile Cleaning., 136
Projectiles, 118
Push-Lok, 80

Q

Quick, 73
Quick-Connect, 82

R

Regular, 118
Reinforcement Layer, 57
Relative Roughness, 189, 200
resistance, 198
Reusable Couplings, 68
Reynolds Number, 128, 174
Reynolds's Number, 199

S

SAE Flanges, 101
safety, 24
Salt Fog Test, 154
Salt Spray Test, 154
Salt-Spray Tests, 90
Sandwich Plates, 111
schedule number, 17
scheduling, 134
Seat, 50
Shock Suppressor, 221
Spiral, 21
spiral, 21
Spiral tapered, 30
Square Flanges, 104
Stack Module, 111
standard, 153
Standard Pipe, 32
Standpipe, 42
Stiffness, 178
Straight, 30
Subplates, 111
Swivel, 28, 52

T

Tapered Fuel, 30
Tensile Strength, 17
tensile strength, 20
Test points, 106
thread end, 67
Thread Gauge, 143
Thread ID Kit, 51
Thread Pitch Gauge, 50
Throttling, 195
TLX Quick Connect Coupling, 76
TM, 75
Transmission Lines, 133
transmission lines, 160-161
Triple-Lok, 45
tube, 26, 133
Tubes, 23
Turbulent, 172

U

Ultimate Strength, 24
Universal Push-to-Connect, 77
UPTC, 77
UV light, 135

V

Vernier Caliper, 143

W

Water Hammer, 218
WEO Plug-In, 53
Whitworth, 32

www.ingramcontent.com/pod-product-compliance
Lightning Source LLC
Chambersburg PA
CBHW042029110726
48010CB00007B/283